McCance and Widdowson's

The Composition of Foods

Ministry of Agriculture, Fisheries and Food

Medical Research Council

McCance and Widdowson's

The Composition of Foods

Fourth revised and extended edition
of MRC Special Report No 297

By A. A. Paul and D. A. T. Southgate

London
HER MAJESTY'S STATIONERY OFFICE

Amsterdam New York Oxford
ELSEVIER/NORTH-HOLLAND BIOMEDICAL PRESS

First published 1940
Second edition 1946
Third edition 1960
Reprinted with amendments 1967
Fourth edition 1978

Sole distributors outside the UK and Eire
Elsevier/North-Holland Biomedical Press
335 Jan van Galenstraat, PO Box 211
Amsterdam, The Netherlands

Sole distributors for the USA and Canada
Elsevier/North-Holland Inc
52 Vanderbilt Avenue
New York, NY 10017, USA

Typographic design by HMSO

ISBN HMSO 0 11 450036 3*
ISBN Elsevier/North-Holland 0 444 80027 1

Preface to the fourth edition

This edition is the result of collaboration between the Ministry of Agriculture, Fisheries and Food and the Medical Research Council, with the analytical work and method development carried out by the Laboratory of the Government Chemist. This work was supported by a grant from the Department of Education and Science, the Department of Health and Social Security, the Ministry of Defence and the Ministry of Agriculture, Fisheries and Food from March 1971 to April 1975, and from the Ministry of Agriculture, Fisheries and Food from April 1975 onwards.

The work began in 1969 under the Ministry's Inter-departmental Committee on Food Composition acting through a Steering Panel. The members of this panel were as follows:

Dr D. A. T. Southgate	Medical Research Council (Convener)
Miss A. A. Paul Miss J. Robertson	Ministry of Agriculture, Fisheries and Food (Joint Secretaries)
Mr A. A. Christie	Laboratory of the Government Chemist
Mrs M. M. Disselduff	Department of Health and Social Security
Mr D. Kimber	Ministry of Agriculture, Fisheries and Food
Mrs G. M. Mann	Ministry of Defence (until December 1975)
Miss J. W. Marr	Medical Research Council
Dr E. M. Widdowson	Medical Research Council

The Steering Panel acted as an advisory group with a working group at the Dunn Nutritional Laboratory, Cambridge, consisting of Miss Paul and Dr Southgate, who were responsible for reviewing the literature, the preparation of the sampling and analytical protocols and the final compilation of the book, with advice from Dr Widdowson as the work progressed. The analytical work was carried out at the Laboratory of the Government Chemist under the supervision of Mr Christie. Mrs J. Thorn (Ministry of Agriculture, Fisheries and Food) joined the working group in Cambridge for short periods and helped in the revision of certain sections, notably the recipes for the cooked dishes. Miss J. Russell (Ministry of Agriculture, Fisheries and Food) has been responsible for the computerisation of the tables.

The sections dealing with cereals, milk and milk products, meat and fish have been extensively revised. Other sections have required a more limited revision, especially the sections on fruits, vegetables and nuts, where it has been possible to use a large amount of the information from earlier editions. In all, some 400 foods have been analysed either completely or partially for this edition. The recipes used in the calculations of the composition of cooked dishes have been converted to metric units and the composition of the dishes recalculated. At the same time a number of new dishes have been included. The tables giving the values for vitamins have been revised and extended. The amino acid tables have similarly been expanded. A new part giving the fatty acid composition of foods has been included and tables giving cholesterol values have also been added. The arrangement of the tables of vitamins, amino acids and fatty acids has been made similar to that used for the proximate constituents.

Committee on Food Composition
Ministry of Agriculture, Fisheries and Food
Horseferry Road, London SW1P 2AE

Medical Research Council
Park Crescent, London W1N 4AL

November 1976

Contents

Contents *continued*

Note from the compilers

Anyone who sets about the task of revising a standard work such as 'McCance and Widdowson' is bound to approach the work with a certain amount of awe and we are no exception. In our revision we have had the great advantage of being able to build on a very firm foundation. We have tried to retain all the features of the earlier editions which have proved themselves in over 35 years of use. Most of the changes in this edition have been necessary because of changes in methods of food production and food technology and because of an increased awareness of the importance of nutrients not included in earlier editions.

We have received advice and encouragement from many sources. This advice was often conflicting and we, as compilers, with the help of the Steering Panel have made some compromises which will not necessarily suit all the users of these tables. In the preparation of the final text and tables we have had to use a considerable amount of judgement and we alone must be held responsible for any errors in this respect. In this we have some comfort in Dr Widdowson's remark: 'The man who makes no mistakes does not usually make anything—he certainly does not make food tables.'

A. A. Paul
D. A. T. Southgate
MRC Dunn Nutrition Unit
Dunn Nutritional Laboratory
Cambridge

Acknowledgements

A revision such as this would not have been possible without the collaboration, assistance and advice received from a large number of people. Miss D. F. Hollingsworth and Dr J. P. Greaves played an important role in the initiation of the work in 1969.

The major part of the sampling of the meat and its cooking and preparation was carried out by domestic science and dietetic students under the direction of Miss J. C. Currie (Glasgow and West of Scotland College of Domestic Science), Miss M. M. Jamieson (The Polytechnic of North London), Miss B. M. Llewelyn (Llandaff College of Education, Cardiff), Mrs J. Powell (City of Bath Technical College), Miss A. P. Robotham (Northern Counties College, Newcastle upon Tyne) and Miss M. Waterworth (City of Liverpool F. L. Calder College of Education). Miss M. Cameron, Mrs E. D. Davies, Mrs B. H. Lake, Mrs M. L. McLeod, Mrs J. R. Salfield and Miss B. Waugh were also involved at the colleges in various parts of the meat study. Insulated boxes for transporting the samples were kindly loaned by the Army Medical Directorate. The preparation and cooking of fish and vegetable samples was carried out under the direction of Mrs N. Parker (Cambridgeshire College of Arts and Technology). Other samples were obtained by Mrs H. S. Butler, Miss E. O. Ellis, Miss S. Hunt, Miss P. Mumford, Mrs I. M. Leppington and Dr J. O'Hara May. The Indian dishes and recipes were prepared by Miss P. Gill. To each of these people we tender our sincere thanks for undertaking the large amount of work that was involved.

Many individuals have given us much advice and in addition have willingly supplied us with unpublished data. Special thanks are due to Dr J. W. G. Porter and Dr S. Y. Thompson (National Institute for Research in Dairying), Dr R. M. Love (Torry Research Station), Mr A. Cuthbertson (Meat and Livestock Commission), Dr R. W. Pomeroy and Mr J. M. Harries (Meat Research Institute), Dr R. A. Barton (Massey University, New Zealand), Mr P. J. Harkett and Miss M. B. Groom (Food Research Institute), Mr R. A. Knight (Flour Milling and Baking Research Association), Dr J. N. Davies (Glasshouse Crops Research Institute), Dr J. D. Henshall and Mr D. J. Cook (The Campden Food Preservation Research Assocation), Dr I. H. Burger (British Food Manufacturing Industries Research Association), Dr A. Sinclair (Zoological Society of London), the Milk Marketing Board, the Herring Industry Board and the Wine and Spirit Association of Great Britain. To all the others who are not mentioned by name thanks are no less due for help in many aspects of the work.

The collection of material on manufactured foods was greatly assisted by the cooperation of manufacturers in providing information on their products.

These included Batchelor's Foods Ltd, Beecham Products Ltd, Birds Eye Foods Ltd, Bovril Ltd, Cadbury Schweppes Ltd, The Cocoa, Chocolate and Confectionery Alliance, the Coca-Cola Export Corporation, Energen Foods Co. Ltd, Express Dairies Ltd, General Foods Ltd, H. J. Heinz Co. Ltd, Mars Ltd, Nabisco Ltd, R. Paterson & Sons Ltd, Quaker Oats Ltd, Rank Hovis McDougall Ltd, Unilever Research Laboratory, Van den Berghs and Jurgens Ltd, A. Wander Ltd and Weetabix Ltd. Useful correspondence and discussions have also been held with a number of other companies.

We would also like to thank the members of the staff of the Laboratory of the Government Chemist, the Dunn Nutritional Laboratory and the Ministry of Agriculture, Fisheries and Food for their part in the analytical work, calculations, provision of library facilities and typing.

Thanks are also due to the large number of dietitians who replied to the questionnaires sent out in the early stages of the revision.

All the values in the tables are expressed in metric units ; the imperial equivalents are as follows :

1 ounce (oz) =	28.35 g	100 g =	3.53 oz
1 pound (lb) =	453.6 g	1 kg =	2.2 lb (2lb 3oz)
1 pint (pt) =	568 ml	1 litre =	1.76 pt

General introduction

A knowledge of the composition of foods is essential in the dietary treatment and management of disease and in most quantitative studies of human nutrition.

The first edition *The Chemical Composition of Foods* in 1940 arose from a need to provide investigators, particularly those in Great Britain, with this information for a wide range of foods. The first edition was mostly compiled from a number of previous studies of the composition of foods made by Professor McCance and Dr Widdowson and their colleagues and especially from the study of meat and fish by McCance and Shipp (1933) and of fruits, vegetables and nuts by McCance, Widdowson and Shackleton (1936). The second edition, published in 1946, included some of the more important wartime and postwar foods but was otherwise little changed.

Some changes were made in the third edition, published in 1960 under the title *The Composition of Foods*. The range of nutrients was increased by the inclusion of sections giving values for vitamins and amino acids. The values for the vitamins were drawn mainly, but not exclusively, from the literature and were based on a very comprehensive search through and a critical appraisal of the literature by the late Dr W. I. M. Holman, compiled by Miss I. M. Barrett. The values for amino acids were compiled by Dr B. P. Hughes using a similar approach supplemented by analytical work.

The preparation of a fourth edition has followed the general principles used in the preparation of the third edition, that is using a combination of direct analysis and selection of values published in the literature. The approach adopted for this revision, however, has resulted in some changes in the way the book has been prepared. This is described in the following section because knowledge of the way in which the tables were prepared will help the reader to make the best use of the values in them.

The proximate composition, energy value, inorganic constituent and vitamin tables, the amino acid tables, the fatty acid tables and the cholesterol tables are also available in a form suitable for input to computers. The data from the tables are held on a paper tape file, which has been prepared using the International Reference Version of the ISO 7-bit coded character set, thereby making it internationally acceptable. While every effort has been made to include as much of the information from the tables as possible on the paper tape some parts, such as the footnotes, have had to be omitted. Where this has happened, indicators to refer users to the printed tables have been included. A booklet specifying the layout, contents, copyright position etc accompanies the paper tape file. The paper tape and booklet can be ordered from the Ministry of Agriculture, Fisheries and Food (Publications).

Method of preparing the fourth edition

The process of revision can be considered under a number of headings, although in practice the various aspects are interrelated.

Examination of the existing values

This involved a detailed consideration of the values published in the third edition. All the values were examined in relation to recent published data for comparable foods, wherever these were available. It is worth noting, however, that for many foods and constituents the values recorded in the third edition are the only ones available. This survey showed which of the published values should be checked and possibly updated by analyses of a new sample. A few little used items have not been reprinted in this edition. The examination also showed that a large amount of the material in the third edition was still valid and nearly half of the values in this fourth edition have been taken directly from the third edition, particularly those for the fruits and vegetables.

Selection of new foods for inclusion

Views were sought from dietitians (Paul and Southgate, 1970), the food manufacturers and many nutrition workers and food scientists about new foods that should be included. The number of items suggested by these major users of the tables was very large, amounting to over 500; however, many of these were closely related foods. The most frequently requested items from this list were selected. Other items from this list were selected because of their importance in the national diet according to statistics on food consumption collected by the Ministry of Agriculture, Fisheries and Food and the retail food trade, and by consultation with specialists (usually food scientists) working with particular groups of foodstuffs. This approach ensured that most items of importance would be covered by the tables and in addition that the tables would provide information on the composition of the range of foods within different food groups.

The selection of manufactured and processed foods for inclusion was difficult. Many products undergo frequent changes in formulation and it was considered undesirable that foods of this type should be included in food composition tables which might be printed several years after the sample was collected and analysed, and remain in print for a number of years. The manufactured food items have therefore been restricted to those with an established and stable composition. In general, proprietary names have only been used where the designation of the food without the use of these names would have been difficult. The inclusion of a proprietary name does not imply that the particular brand has any special nutritional value.

Some food items which form an important part of the diets of various groups of immigrants have also been included. As data on many of these foods can be found in the tables compiled by Platt (1962) only a few raw foods have been included. A number of cooked and canned items were however analysed.

Choice of nutrients to be covered

Guidance in the choice of which nutrients should be included in the fourth edition was obtained from dietitians and nutrition workers in industry and in academic and research fields. These views showed that only a very few of the constituents given in the third edition were not used. They also indicated that there was a need for tables giving the fatty acid composition of foods and for improving the coverage of vitamins and amino acids. Many requests were also received for cholesterol values and for trace elements, especially zinc.

Selection of values from the literature

A detailed examination of the published literature was first undertaken to establish whether or not reliable values were available for the food or the nutrients in question. Unpublished values for the composition of the food were also considered. Preference was given to values where the publication gave full details of the sample, its method of preparation and analysis, and where the results were presented in a detailed and acceptable form. The complete list of criteria which were used in assessing the published values are summarised in table 1 (Southgate, 1974). Only rarely, however, did the published sources meet all these criteria. It was also rare to find complete analyses, and in many cases it has been necessary to consider values for different nutrients obtained on different samples.

Table 1 *Details considered in the evaluation of compositional data*

Name of food	Common name, with local synonyms Systematic name with variety where known
Origin	Plants: Locality, with details of soil conditions and fertiliser treatments
	Animals: Locality and method of husbandry and slaughter (where applicable)
Sampling	Place and time of collection Number of samples How obtained Nature of sample (e.g. raw, prepared, deep frozen, prepacked etc)
Treatment of samples before analysis	Conditions and length of storage Preparative treatment, including details of material discarded as waste Method of cooking (where applicable)
Analysis	Details of material analysed Methods used, with appropriate references and details of any modifications
Method of expression of results	Statistical treatment of analytical values Whether expressed on an 'as purchased', 'edible matter' or 'dry matter' etc basis

All the reported values for each nutrient in each item were collated and considered in detail and, on the basis of this review, selected values for each nutrient were chosen. These selected values are not necessarily the mean of all published values and depend very greatly on the judgement of the compilers in their interpretation of the results obtained by different analysts, sometimes using different methods.

If the review of the literature showed that little or no information on the composition of a food was available, or that the values given in the third edition were not representative of present-day foods, arrangements were made for the direct analysis of the food, either completely or for a restricted number of nutrients.

Analysis

Detailed sampling and analytical protocols were devised for each item for which analyses were required. The scope of the sampling scheme varied with the importance of the food, for example meats and meat products were collected in six regional centres, whereas other foods were collected only in Cambridge and London as there was reasonable confidence that this would be representative of the country as a whole. In the case of flour use was made of samples which were collected on a national scale in the Voluntary Flour Sampling Scheme.

The samples have been analysed at the Laboratory of the Government Chemist for all the constituents except the unavailable carbohydrates (=dietary fibre) and a few other constituents where a confirmatory analysis for an item already in the tables was required. These additional analyses were carried out at the Dunn Nutritional Laboratory, Cambridge (Southgate, Bailey, Collinson and Walker, 1976).

The details of the methods used in the analysis of foods for this and previous editions are given in appendix 1 (p 313).

Arrangement of the tables

General features of the fourth edition

The tables are arranged in four sections. In the first three sections the same general arrangement has been used and throughout the tables the code number of a food is the same each time it appears. These code numbers are not the same as those used in previous editions. The first section contains the descriptions of the foods and gives values for the proximate constituents, energy value, inorganic constituents and vitamins in the foods. The values are given per 100 g. The second section gives the amino acid compositions and these are given as mg amino acid per g nitrogen. The third section gives values for fatty acid composition as g fatty acids per 100 g total fatty acids. The fourth section contains some subsidiary tables giving values for some constituents in a more limited range of foods. The section contains tables of values for the cholesterol and phytic acid in foods and notes on the iodine and organic acids in foods.

Within each section the foods are arranged in the following groups: cereals and cereal foods; milk and milk products; eggs; fats and oils; meat and meat products; fish and fish products; vegetables; fruits; nuts; sugars, preserves and confectionery; beverages and alcoholic beverages; sauces, soups and miscellaneous foods. The cooked dishes are included in the appropriate food groups and not given in separate sections as in previous editions.

This classification is practical rather than scientific and may appear rather arbitrary in some cases. This is inevitable, whatever system of grouping is used, and a full index is provided so that any food can be located easily.

In each food group related foods are listed together. In most cases an alphabetical order of listing within these subgroups has been used but in some cases some other arrangement has seemed more appropriate.

Conventions and symbols used in the tables

Throughout the tables a number of conventions have been adopted and a number of symbols are used which have a precise meaning.

Expression of values

All the values in the tables apply to the edible part of the food and are expressed per 100g. In the amino acid and fatty acid sections the values are expressed on nitrogen and total fatty acid bases respectively. Where a food is usually served with the inedible matter as an integral part of the food (for example, a chop with its bone) a separate series of values is given which apply to the composition of 100g of the food weighed with its inedible matter.

Edible matter

Where the food is purchased or served with inedible material a factor is given which shows the proportion of the edible matter in the food. In the case of cooked foods this proportion also takes into account any changes in weight in cooking and gives the proportion of cooked edible material derived from the corresponding original raw item. The proportional factor is simply the percentage value given in previous editions divided by 100. The use of this factor is described later (p 32). Where the factor is greater than one (for example, stewed dried fruit) there is a gain in weight in cooking.

Selected values

Where the compositional values have been derived from the literature, they are not usually the average of all published values but represent the result of a critical assessment of the literature. This assessment involved consideration of the nature and origins of the samples, the method of cooking or other preparation and finally the analytical methods used. The values used in the tables are therefore designated selected values.

Ranges

Wherever possible, ranges are given in parentheses for some nutrients, particularly vitamins. These are mainly for important foods and for those where an appreciable amount of the constituent is present. The values given for the range show the usual extent of variation and extreme values have been discounted. These ranges are given for guidance in order to help the user assess the confidence which should be given to values for intakes of these nutrients calculated from the tables. In many cases there is insufficient information for a range to be given, but this does not mean that the nutrient composition is constant or any less variable than that of a nutrient for which a range is given.

Symbols

0 In the tables 0 signifies that virtually none of the constituent is known to be present in the food.

Tr This indicates that a trace is known to be present. In some cases a measurement has been made, and in other cases the literature indicates that a small amount is probably present but usually below the limits of the method used for analysis. The use of Tr indicates that the amount of the nutrient in the food in most cases is not known to be of quantitative dietetic significance and the value zero may be used in data processing.

( ) Figures in parentheses are estimates taken from related foods or, more rarely, tentative values based on a limited number of published sources.

— A dash shows that no information is available, either from direct analysis or by inference from the literature, to enable a value for the nutrient to be given in the tables. *This symbol should not be assigned the value zero in data processing: rather a value from a related food should be used with qualification.*

Food code number

The items in section 1 have been numbered from 1, Arrowroot, through to 969, Yeast, bakers, dried. In the other sections the code numbers are prefixed 2—amino acid composition; 3—fatty acid composition and 4—cholesterol.

Where the sample analysed for amino acids, fatty acids or cholesterol corresponded exactly with the item given in section 1, this is indicated by the code number assigned to the values in these sections.

For example:
Turkey, raw, meat only, has the code number 340 in section 1; 2340 in section 2; 3340 in section 3 and 4340 in section 4. The amino acid composition 2340, however, applies to all turkey entries and similarly 3340 gives the fatty acid composition of all fat in turkey entries.

Many values in sections 2 and 3 were obtained on grouped samples which do not appear in section 1; in these cases the base number (that is, the last three digits) is not used in section 1 but appears with its appropriate prefix in the other sections. These grouped values may apply to many different items in section 1.

For example:
The base number 123 is not used in section 1 because the amino acid composition of the protein in all cows' milk products is virtually the same and the values given under 2123 apply to all cows' milk products; similarly 3123 gives the composition of the fat in all cows' milk products.

In section 1, cows' milk, fresh, whole; cows' milk, fresh, whole, Channel Island; and single, double and whipping creams have been given two numbers to allow for different vitamin values in summer and winter milks.

Description and number of samples

The information given under this heading describes the number and nature of the samples taken for analysis. The sources of values not based on direct analysis that have been derived either from the literature or by calculation are also indicated under this heading. The description of the sample also includes the method of cooking where appropriate. Where the calculated composition of a cooked dish is given, the corresponding recipe and details of the cooking method are given in appendix 4. For most foods a number of samples were purchased at different shops, supermarkets or other retail outlets. The samples were not analysed separately but, as in previous editions, pooled before analysis. When the composite sample was made up from a number of different brands of a food, the numbers of the individual brands purchased was related to the relative shares of the retail market held by those brands. The different brands may or may not have been purchased at the same shop.

Blank pages

In section 1 a set of blank pages has been inserted between each major food group so that additional foods may be entered by the user if required.

Modes of expression and determination of constituents

This section is mainly concerned with indicating the principles which are used in the determination of the constituents and in describing the modes of expression used in the tables. Details of the analytical procedures used for this and previous editions are given in appendix 1.

Proximate constituents

The classical nutritional terminology has been adopted to describe the major constituents of a food, that is the water, protein, fat and carbohydrates. In the third edition the term 'calorific constituents' was used but it did not seem appropriate in this edition to use the corresponding term derived from the approved SI unit.

Water

Water is usually measured as the loss after drying either in an oven or at a lower temperature under reduced pressure. The water content of food containing volatile constituents, such as essential oils or alcohol, cannot be measured in this way and in the case of the alcoholic beverages values for total solids are given instead.

The water content of many foods, especially those of plant origin, can vary over fairly wide limits and differences in the water content of different samples are frequently a cause of analytical discrepancies. Many of these are eliminated if compositions are given on a dry-weight basis, but it is inappropriate to use this method of expression in tables of food composition.

Total nitrogen

Total nitrogen is measured by a variant of the Kjeldahl method. In previous editions attempts were made for some foods to estimate the non-protein nitrogen. This has not been possible in the present edition except where significant amounts were known to be present in the form of urea, purine or pyrimidine derivatives in mushrooms, beverages and a few fishes.

Protein

Protein has been calculated from the total nitrogen values by the use of factors. In the case of foods containing urea, purine and pyrimidine derivatives the total nitrogen value has been corrected before applying the factor. The factors used in this edition are those suggested by the FAO/WHO Committee on Energy and Protein Requirements (FAO/WHO, 1973)—see table 2.

Table 2 *Factors for converting total nitrogen in foods to protein*

	Factor (per gN)		Factor (per gN)
Cereals		Nuts	
Wheat		Peanuts, Brazil nuts	5.41
Wholemeal	5.83	Almonds	5.18
Flours, except wholemeal	5.70	All other nuts	5.30
Macaroni	5.70		
Bran	6.31	Milk and milk products	6.38
Rice	5.95	Gelatin	5.55
Barley, oats, rye	5.83		
Soya	5.71	All other foods	6.25

The proportion of non-protein nitrogen is high in many foods, notably fish, fruits and vegetables. In most of these, however, this non-protein nitrogen is amino acid in nature and therefore little dietetic error is involved in the use of a factor applied to the total nitrogen, although protein in the strict sense is overestimated.

A few other foods (notably Bovril and Marmite) contain a high proportion of non-protein nitrogen, which is again mainly amino acid and peptide nitrogen. In these foods total nitrogen (minus purine-N) has also been multiplied by 6.25 to give a value for 'protein'. For nutritional purposes this seemed justifiable and means that the energy values for these foods are now, more correctly, considerably higher than in previous editions.

Fat

Fat in most foods is a mixture of triglycerides, phospholipids, sterols and related compounds. For this edition it was decided that, where possible, the standard procedure for estimating fat in each class of foodstuffs would be used and the values given refer to total lipid.

The value for the fat in a food is extremely dependent on the analytical method used in its measurement and even officially approved methods have given very conficting results in some cases. In general the observations recorded in earlier editions have been confirmed, that is that the Soxhlet procedure gives incomplete extraction of the fat from many foods and values obtained by this method are almost invariably lower than those obtained using more recent methods involving acid hydrolysis or mixed solvent extraction.

Carbohydrates

Carbohydrates (including sugars and starch) are in all cases expressed as *monosaccharides* and throughout the tables all values for carbohydrate are *available carbohydrate*: that is, the sum of the free sugars (glucose, fructose, sucrose, lactose, maltose and higher maltose homologues), dextrins, starch and glycogen expressed as monosaccharides. These are the carbohydrates which are digested and absorbed by man and which are glucogenic in man. In the tables the headings are as follows:

Sugars include free monosaccharides and disaccharides.

Lactose is also expressed as monosaccharide, which is equivalent to lactose measured as the monohydrate.

Starch includes starch and dextrins hydrolysed by amyloglucosidase enzyme preparations and excludes glucofructans and fructans that are not hydrolysed by this enzyme.

On hydrolysis a disaccharide such as sucrose gives 105 g monosaccharide per 100 g and a polysaccharide such as starch gives 110 g of the monosaccharide glucose per 100 g. This means that white sugar (item 843) for example contains 105 g of carbohydrate (expressed as monosaccharide) per 100 g of sugar. To convert the carbohydrate expressed as monosaccharide to the forms present in the food, the monosaccharides derived from disaccharides should be divided by 1.05 and those from polysaccharides by 1.10.

Dietary fibre

The term 'dietary fibre' has been used instead of the '*unavailable carbohydrates*' of earlier editions; for all practical purposes these two terms are synonymous. Dietary fibre is defined as the sum of the polysaccharides and lignin which are not digested by the endogenous secretions of the human gastrointestinal tract. This fraction has a variable composition as it is made up of several different types of polysaccharide (pectic substances, hemicelluloses and cellulose) and the non-carbohydrate lignin.

The values given in this edition for cereals and cereal foods and some vegetables and fruit are based on direct analyses; the remainder are based on the values taken from earlier editions, where an indirect but equally valid method of measurement was used.

Alcohol

The values for alcohol are given as g/100ml of alcoholic beverages. Pure ethyl alcohol has a specific gravity of 0.790 and these values can be converted to alcohol, by volume (that is ml/100ml), by dividing by 0.79 (see p28).

Energy value

The energy values of the foods are given as both kilocalories (kcal) and kilojoules (kJ). These energy values have been calculated from the amounts of protein, fat, carbohydrate and alcohol in the foods using energy conversion factors.

The third edition of these tables included a note by Dr E M. Widdowson, which has been reprinted in appendix 2, on the calculation of the calorific values of foods and of diets. This note gave an account of the historical background to the various systems of energy conversion factors and the reasons behind the choice of the factors for that edition, which were 4.1kcal/g protein, 9.3kcal/g fat and 3.75kcal/g carbohydrate (available, expressed as monosaccharides).

In this edition the conversion factors used are as follows: protein 4kcal/g; fat 9kcal/g; carbohydrate (available, expressed as monosaccharides) 3.75kcal/g and alcohol 7kcal/g.

These factors permit the calculation of the metabolisable energy of a mixed diet of the type eaten in the United Kingdom with an accuracy which is equal to, or greater than, the accuracy one could reasonably expect in dietary work using tables of food composition (Southgate and Durnin, 1970).

In the calculation of energy values in terms of kilojoules the factors recommended by the Committee on Metrication in the Nutritional Sciences of the Royal Society (1972) have been used: that is, protein 17kJ/g; fat 37kJ/g; carbohydrate (available, expressed as monosaccharides) 16kJ/g and alcohol 29kJ/g. The factors have been applied directly to the amounts of protein, fat, carbohydrate and alcohol in the foods. If the energy values in kJ are calculated from the kcal values directly using the conversion factor 4.184kJ/kcal, numerically different values are obtained. The differences are small, being at the most of the order of 1–2 per cent and frequently less; furthermore they are of no dietetic significance.

For consistency, however, it is preferable that the calculation should be carried out in the manner adopted in these tables. The kilojoule factors are in fact closer to the factors calculated from the studies of Southgate and Durnin (1970) and indeed of Atwater himself (Merrill and Watt, 1955) than are the conventional 4, 9 and 3.75kcal factors. Table 3 summarises the energy conversion factors used in the present edition; the table also includes factors for organic acids and glycerol based on the heats of combustion of these substances. The contribution of organic acids and glycerol to the energy has not, however, been included for the relevant foods in this edition because of lack of suitable analytical data.

Changes due to the use of new factors in this edition

The energy values in the present edition are slightly lower than values given in the previous editions for foods with identical compositions. The differences

Table 3 *Energy conversion factors*

	kcal/g	kJ/g
Protein	4	17
Fat	9	37
Carbohydrate[a]	3.75	16
Ethyl alcohol	7	29
Glycerol	4.31[b]	18.0[b]
Acetic acid	3.49[b]	14.6[b]
Citric acid	2.47[b]	10.3[b]
Lactic acid	3.62[b]	15.1[b]
Malic acid	2.39[b]	10.0[b]

[a] Value for available carbohydrate expressed as monosaccharide

[b] Values derived from heat of combustion

are due principally to the use of a lower factor for the energy value of fat. Again the differences are small (less than 1 per cent for most foods) and are of little true dietetic significance.

Contribution of unavailable carbohydrates

Any contribution to the metabolisable energy arising from the bacterial degradation of unavailable carbohydrates in the large intestine is discounted by the method of calculation of energy values. The products of degradation are mainly short-chain fatty acids and it is not clear how well these are absorbed. In any case their energy contribution would be small. Any effect of the unavailable carbohydrates or dietary fibre on the absorption of other energy-yielding nutrients has also been discounted.

Inorganic constituents

The values for the inorganic constituents are given in the tables as mg per 100g. They appear in the order sodium (Na); potassium (K); calcium (Ca); magnesium (Mg); phosphorus (P); iron (Fe); copper (Cu); zinc (Zn); sulphur (S) and chloride (Cl).

Sodium and *potassium* are now usually measured by flame photometry or atomic absorption spectrophotometry. Common salt is one of the most frequently used food additives and consequently the values for sodium are likely to be much more variable than for potassium. The calculation of sodium intakes from food tables is therefore prone to error, particularly if salt is used in cooking. Accurate values for sodium intakes can only be obtained by analysis of replicates of the diet. Nevertheless the values given should provide a reliable guide for planning low-sodium diets. If the values for these elements are required in millimoles (milliequivalents) then the values in mg should be divided by 23 and 39 for Na and K respectively.

Calcium can be measured chemically or by atomic absorption spectrophotometry. The calcium content of a particular food varies within fairly narrow limits. There are, however, several major uses of calcium salts as additives. These are first to wheat flour, where there is a statutory addition in the United Kingdom to all wheat flours except wholemeal. The raising agents in self-raising flour and in some baking powders also contain calcium. The latter are very rich sources of calcium and their use increases the calcium content of the food considerably.

Contamination of the edible portion of meat and fish by pieces of bone also causes variation in the calcium content; this is especially evident in the case of fish where the very fine bones are frequently eaten. Contamination from particles of bone is common in many foods where the carcase of the animal has been cut with a saw or scissors (for example brain).

The tap water in many areas contains appreciable amounts of calcium (and other inorganic salts) and foods which have been cooked in hard tap water may have their calcium contents enhanced from this source. However, vegetables, fruits and soups in the tables were cooked in distilled water, and and the beverages were also prepared with distilled water.

Magnesium is now usually measured by atomic absorption spectrophotometry. The concentration of magnesium in a particular food varies within narrow limits.

Phosphorus is usually measured colorimetrically as the phosphate. Polyphosphates are becoming increasingly used as food additives, particularly in meat products, and occasionally in evaporated canned milks. The amounts used are small and only minor increases in the total phosphate content result from the use of these compounds. Phosphates are used as raising agents, and some baking powders are rich sources of phosphorus. The values expressed as mg P can be converted to mg PO_4 by multiplying by 3.06.

Iron, copper and zinc may be measured colorimetrically or by atomic absorption spectrophotometry. The elements, as well as being derived from constituents occurring naturally in foods, may also be derived from metallic contamination either from processing machinery, kitchen knives, pots and pans or from particles of soil. The contamination of foods from one source or another is probably a major cause of variation but, even so, wide ranges of values are often reported in the literature under circumstances where precautions had been taken to avoid contamination. Iron compounds are added to a number of important foods; there is a statutory addition in the UK to all wheat flours except wholemeal, and many proprietary breakfast cereals and baby foods contain added iron. Copper and zinc compounds are rarely used as food additives *per se* and elevated values for copper and zinc are usually evidence of metallic contamination. Iron and copper salts are frequently used in animal feeding and these often result in elevated levels in some tissues used as foods. Marine organisms, specially molluscs and crustaceans, accumulate quite high concentrations of inorganic constituents.

Sulphur in foods is mostly derived from the sulphur-containing amino acids, cysteine and methionine, and the nitrogen:sulphur ratio in many groups of foods has been shown to give a reasonable prediction of the sulphur content from the total nitrogen. The sulphur in foods is usually estimated after oxidation of the sulphur to sulphate but a range of techniques have been studied in order to establish the best methods for the foods analysed for this edition (see p318). Sulphur dioxide is widely used as preservative and its use results in high observed values for total sulphur, but this is almost certainly unavailable.

Chloride is measured by titration. The chloride in foods is mainly associated with the alkali metals and generally varies with the sodium. The variations in the use of added common salt that are responsible for the variations in the sodium content of foods also produce great variability in the chloride contents. The values expressed as mg can be converted to millimoles (milliequivalents) by dividing by 35.5.

Vitamins

The vitamins for each food are given on two pages; the first gives figures for retinol, carotene, vitamin D, thiamin, riboflavin, nicotinic acid, vitamin C and vitamin E, and the second for vitamin B_6, vitamin B_{12}, folic acid, pantothenic acid and biotin. The figures given on the second page are based on more limited data than those given on the first and calculations using them should be interpreted accordingly. The second page also includes a panel containing notes which apply to the vitamins on both pages.

The comments below on methodology refer in the main to the criteria used in assessing values in the literature; the methods used for the analyses undertaken for this edition are given in appendix 1 (p 318).

Vitamin A: retinol and carotene

In the third edition values for vitamin A potency were calculated as the sum of the preformed vitamin and the contribution from any carotenoids present. In this edition these two constituents have been expressed separately as μg retinol and carotene respectively. The values for the latter in most foods are for β-carotene. Where no information is available on the proportions of the different carotenoids in the food, the value given is for total carotene. In these cases the total vitamin A activity will be slightly overestimated, but the error introduced in this way is only small. Calculation of the total vitamin A potency in terms of μg retinol equivalents can be made by applying a divisor to the carotene values. The numerical value of this divisor depends on the types of carotenoids present and is a measure of the relative efficiency with which the carotenoids are converted to retinol in the body. For the total carotenoids in the diet, experimental studies have shown that 6 is the best value to use for the divisor so that:

$$\text{Vitamin A potency } (\mu\text{g retinol equivalents}) = \mu\text{g retinol} + \frac{\mu\text{g } \beta\text{-carotene}}{6}$$

Very few studies of individual foods have been made which enable one to propose the correct divisor for each food. It is more correct to calculate the retinol equivalents contributed by carotenoids from the total dietary carotenoids, and values for retinol equivalents are not given in the tables for individual foods (see p 33). Values reported in international units have been recalculated according to the definition that one international unit of vitamin A potency is equal to 0.3μg retinol or 0.6μg β-carotene.

Retinol is estimated spectrophotometrically after saponification, extraction and chromatographic separation; carotenes are usually measured spectrophotometrically after extraction and chromatographic separation.

Vitamin D

In most foodstuffs the concentration of vitamin D is too low for measurement by chemical techniques and biological assay is the only method available for these foods at present. Values from the older literature using the rat have been used in the tables but very few recent reports of the amounts of vitamin D in foods are available. In foods where there are high concentrations naturally or where vitamin D has been added, some values obtained by gas–liquid chromatographic methods have been given. These are limited to concentrations of the order of that found in fatty fish. The values are expressed as μg cholecalciferol per 100g. Values reported in international units have been recalculated on the basis that 1 i u = 0.025μg cholecalciferol.

Thiamin

The values given are based on chemical measurements by the thiochrome method and microbiological assay using *Lactobacillus viridescens* (ATCC

12706) or *L. fermenti* (ATCC 9338). These procedures give similar results provided that the appropriate extraction technique has been used to release the thiamin from the coenzyme. The values are expressed as mg per 100 g.

Riboflavin

Preliminary acid or enzymatic hydrolyses using proteolytic and amylolytic enzymes are required to release riboflavin from the bound forms, riboflavin mononucleotide and flavin adenine dinucleotide, which occur in foods. Provided that the appropriate preliminary treatment has been used, fluorimetric methods and microbiological assay using *Lactobacillus casei* (ATCC 7469) or *Streptococcus zymogenes* (ATCC 10100) give similar results. The values are expressed as mg per 100 g.

Nicotinic acid

A preliminary acid or enzymatic hydrolysis of the food samples is required to release the nicotinic acid from its bound forms. The values given are based on measurements which include both nicotinic acid and nicotinamide determinations. The method of determination is either by the colorimetric reaction with cyanogen bromide or by microbiological assay using *Lactobacillus plantarum* (ATCC 8014). The values are expressed as mg nicotinic acid per 100 g and no attempt has been made to separate the bound and probably unavailable forms of nicotinic acid in cereals (see p 33).

The calculation of nicotinic acid equivalents, which includes the contribution from tryptophan, should be made on the total diet as follows:

$$\text{mg nicotinic acid equivalents} = \text{mg nicotinic acid} + \frac{\text{mg tryptophan}}{60}$$

The conversion of tryptophan to nicotinic acid takes place in the body with varying efficiency and the factor of 60 is an approximation based on a limited number of studies. There are no data on which to judge whether this factor applies equally to all foods; thus nicotinic acid equivalents are not given for individual foods, but the potential contribution from the tryptophan in the food, calculated as mg tryptophan divided by 60, is given for guidance (see p 33).

Vitamin C

This term has been preferred for the sum of ascorbic acid and dehydroascorbic acid, as both forms are active. In fresh foods the reduced form is the major one present and the proportions present as the dehydro-form are increased during cooking and processing. Values given refer to total ascorbic acid, that is reduced plus dehydroascorbic acid, wherever possible.

The values considered from the literature are based on a variety of methods. For foods where only the reduced form was judged to be present, values based on the titration with 2,6-dichlorophenolindophenol have been used together with those based on the more specific 2,4-dinitrophenylhydrazine method and the more recent fluorimetric method, which measure total ascorbic acid.

Vitamin E

The various forms of tocopherols found in foods have different biological activities. In most animal products the α-form is the only one present but in plant products, especially seeds and the oils from seeds, other forms are present. β-tocopherol has an activity of around 30 per cent, γ-tocopherol 15 per cent, and α-tocotrienol (ζ-tocopherol) 30 per cent of the α-form respectively. The other forms have activities of 5 per cent or less of α-tocopherol. In the column headed 'Vitamin E' the values are for α-tocopherol and information is given in the notes about the other major active forms where the data are available.

The figures given in the tables are based on methods involving chromatographic separation of the various tocopherols and colorimetric measurements and occasionally on more recent gas–liquid chromatography methods. The values are expressed as mg per 100 g.

Vitamin B_6

The various forms of pyridoxine, pyridoxal, pyridoxamine and their phosphates and some other conjugated forms all contribute to the vitamin B_6 activity in a food. A preliminary treatment with acid in an autoclave is usually necessary to achieve complete extraction. Values obtained by microbiological assay using *Saccharomyces carlsbergensis* (ATCC 9080), which measures all forms, have been used. The values are expressed as mg per 100 g.

Vitamin B_{12}

Values obtained by microbiological assay with *Lactobacillus leichmanni* (ATCC 7830) have been preferred. The values are given as μg per 100 g.

Folic acid

Values for folic acid are given as the amounts for free and total folic acid present. These are respectively the folic acid activity measured microbiologically with *L. casei* (ATCC 7649) before and after treatment with conjugase. Values reported in the literature before the use of ascorbic acid as an antioxidant in the extraction medium have not been used, as folic acid activity is underestimated if it is not protected during extraction.

The availability of the various forms and conjugates of folic acid is poorly understood at the present time and it is not clear whether the organisms used in its assay respond to the same forms that are available to man. Many foods contain conjugases which will degrade the higher conjugates quite rapidly during the normal course of the storage and preparation of foods for consumption or analysis. This means that values for free folate are very variable and of doubtful significance. It is therefore probable that the estimates of the folic acid in foods given in this edition will need revision in the future and they should therefore be used with this in mind when the amounts of folic acid in a diet are being assessed. The values are given as μg per 100 g.

Pantothenic acid

These data are based on values obtained by microbiological assay using *L. plantarum* or *S. carlsbergenis*, after release from the bound form in which it occurs by the double enzyme treatment, usually avian liver and pig intestinal phosphatase. The values are expressed as mg per 100 g.

Biotin

These values are also based on microbiological determinations using *L. plantarum* after acid hydrolysis to release the bound forms. Recent improvements in techniques have resulted in lower amounts of biotin found and these are believed to be more correct than the older ones. The values are given as μg per 100 g.

Amino acids

The amino acid compositions are given in section 2 of the tables, in which the foods are arranged in the same groups as in section 1. In this section the values are given as mg amino acid per g nitrogen. This method of expression means that a single series of figures can be given which applies to all the foods containing proteins with the same amino acid composition. For example a single entry suffices for all types of beef. Values for the amounts of amino acids in 100 g of food can be calculated using the total nitrogen values given in section 1 (see appendix 5); this calculation has been included in the computer tape version of these tables.

The amino acids are listed in the tables under their recognised three-letter abbreviations. These are as follows:

Essential		*Non-essential*	
Isoleucine	Ile	Arginine	Arg
Leucine	Leu	Histidine	His
Lysine	Lys	Alanine	Ala
Methionine	Met	Aspartic acid	Asp
Cystine	Cys	Glutamic acid	Glu
Phenylalanine	Phe	Glycine	Gly
Tyrosine	Tyr	Proline	Pro
Threonine	Thr	Serine	Ser
Tryptophan	Trp		
Valine	Val		

A wider space between columns separates the essential amino acids—that is, those required for the maintenance of nitrogen balance in adult man—from the non-essential. Cystine and tyrosine have been included in the former group because of their sparing effects on the requirements for methionine and phenylalanine respectively.

In assessing the values reported in the literature results obtained both by microbiological methods and by ion-exchange chromatography have been considered, although most recent analyses have been by chromatographic methods. It is interesting to note that despite the virtually complete automation in this field, and the fact that these analyses can be carried out very rapidly, the actual number of reliable analyses which have been reported for even major foodstuffs is still quite small.

In assessing the values in the literature the description of the conditions used for the hydrolyses of the proteins and the measurement of hydrolytic losses were frequently inadequate and preference has been given, where possible, to results reported by authors who have carried out serial hydrolyses to determine the optimum length of hydrolysis and who give estimates of the hydrolytic losses. Unfortunately many authors do not describe their hydrolytic conditions, the most critical part of amino acid analysis, in sufficient detail, and it is often not possible to decide whether the discrepancies between reported values are genuine or are analytical artefacts.

The values for the sulphur amino acids (cystine and methionine) have been taken from analyses where the intact protein has been oxidised with performic acid before hydrolysis.

The values for tryptophan are derived in the main from separate colorimetric analyses and, more rarely, from the results obtained by the ion-exchange chromatography of alkaline hydrolysates.

Fatty acids

The tables giving the fatty acid composition of foods are contained in section 3. The values are expressed as g of the individual fatty acids per 100 g total fatty acids. Not all the fatty acids present in trace amounts are given in the tables and therefore the sum of the values may be less than 100. The foods are arranged in the same groups as in section 1, although some condensation has been possible because foods with the same fatty acid composition can be covered by a single series of values.

The values given in the tables are virtually all based on gas–liquid chromatographic analysis, usually of the methyl esters of the fatty acids prepared from the total extracted lipid, and they include the fatty acids derived from both triglycerides and phospholipids.

Occasional difficulties were met in the interpretation of values in the literature where these were expressed in an unusual way and insufficient data were given to enable recalculation to the basis adopted. Values have therefore been preferred which were reported in terms of fatty acids per 100 g total fatty acids or as methyl esters per 100 g total methyl esters. These two methods of expression are very similar for most individual foods.

The fatty acids are listed under their carbon number and number of double bonds in the usual way (table 4). The common names for the most frequently

Table 4 *The most common fatty acids in foods*

	Carbon : Double bonds	Common name
Saturated		
	C4 : 0	Butyric
	C6 : 0	Caproic
	C8 : 0	Caprylic
	C10 : 0	Capric
	C12 : 0	Lauric
	C14 : 0	Myristic
	C16 : 0	Palmitic
	C18 : 0	Stearic
	C20 : 0	Arachidic
	C22 : 0	Behenic
	C24 : 0	Lignoceric
Mono-unsaturated		
	C16 : 1	Palmitoleic
	C18 : 1	Oleic
	C20 : 1	Eicosenoic
	C22 : 1	Erucic
Polyunsaturated		
	C18 : 2	Linoleic
	C18 : 3	Linolenic
	C20 : 4	Arachidonic

occurring fatty acids are also given in this table. In each food group the most appropriate arrangement of the column headings in the tables has been used. For example the tables for milk have more headings for the shorter-chain fatty acids and those for fish more headings to cover the long-chain polyunsaturated acids present. The fatty acids are grouped into saturated, monounsaturated and polyunsaturated groups, and at the end of each table the column headed 'Others' gives values for fatty acids, other than those listed, that are present in measurable amounts. A trace in these tables usually means that less than 0.1 g/100 g total fatty acids was present. The value 0 usually means that the fatty acid in question was not reported and it has been assumed that none was present.

When the fatty acids provided by a given weight of food are being calculated allowance must be made for the fact that the total fat in a food includes triglycerides, of which a proportion is glycerol (that is, not fatty acid), phospholipids and unsaponifiable components such as sterols.

In foods where the total fat is virtually all triglyceride a correction factor based on the average chain length of the fatty acids present is adequate. The factors for food containing appreciable amounts of phospholipids and unsaponifiable matter depend on the class of foodstuff. Some suggested values for these factors are given in table 5.

Table 5 *Conversion factors to be applied to total fat to give values for total fatty acids in the fat*

Wheat, barley and rye [a]		Beef [c] lean	0.916
whole grain	0.72	fat	0.953
flour	0.67	Lamb, take as beef	
bran	0.82	Pork [d] lean	0.910
Oats, whole [a]	0.94	fat	0.953
Rice, milled [a]	0.85	Poultry	0.945
Milk and milk products	0.945	Brain [d]	0.561
Eggs [b]	0.83	Heart [d]	0.789
Fats and oils		Kidney [d]	0.747
all except coconut	0.956	Liver [d]	0.741
coconut	0.942	Fish, fatty [e]	0.90
		white [e]	0.70
		Vegetables and fruit	0.80
		Avocado pears	0.956
		Nuts	0.956

[a] Weihrauch, Kinsella and Watt (1976)
[b] Posati, Kinsella and Watt (1975)
[c] Anderson, Kinsella and Watt (1975)
[d] Anderson (1976)
[e] Exler, Kinsella and Watt (1975)

These factors are used as in the following examples.

Palmitic acid in 100g goats milk containing 4.5g fat:

$4.5 \times 0.945 = 4.25$ g total fatty acids

$4.25 \times \frac{27}{100} = 1.15$ g palmitic acid

Stearic acid in 100g whole egg containing 10.9g fat:

$10.9 \times 0.83 = 9.05$ g total fatty acids

$9.05 \times \frac{9.3}{100} = 0.84$ g stearic acid

The fatty acids per 100g of food have been calculated on the computer tape version of these tables for all foods for which appropriate fatty acid data are available (see appendix 5).

Notes on food groups

Cereals and cereal products

The items in the cereals section are arranged in the following groups: grains, flours and starches; bread and rolls; breakfast cereals; biscuits; cakes, buns and pastries; and puddings.

The first group includes foods which are not strictly of cereal origin such as tapioca and soya. The wheat flour and breads are arranged in order of descending extraction rate but otherwise an alphabetical arrangement has been followed. Some groups include items whose composition has been calculated from recipes in addition to items which have been analysed as purchased. The 'pudding' group consists almost entirely of items calculated from recipes.

Moisture content

The composition of many cereals seems to be relatively constant and the major cause of variation is the moisture content. Most air-dry grains and flours have moisture contents of 10 to 14 per cent and, furthermore, above 15 per cent most cereals deteriorate on storage. Many proprietary breakfast cereals are packed at a moisture content of around 2 per cent but rapidly absorb some moisture once the packet has been opened. The values given refer to packets purchased in the usual way which were opened and mixed just before analysis.

Fortification

Wheat flours (other than wholemeal) are fortified with calcium, iron, thiamin and nicotinic acid in the UK at the present time. Self-raising flours are exempt from the requirement to add creta (calcium carbonate) if a calcium acid phosphate raising agent is used. The values given for flours are for those available in the UK; the composition of unfortified flours are given in the notes. Nicotinic acid is expressed as the total content and the question of availability of the nicotinic acid present naturally in these foods is discussed on p 33.

Carbohydrates

Many cereals contain glucofructans, at concentrations around 1 per cent in wheat for example. These polysaccharides are very readily hydrolysed and should probably be regarded as available carbohydrate. They are thus included in the 'sugars' in the tables.

Vitamin losses on cooking

Average losses of vitamins have been used to calculate the composition of cooked items. These are shown in table 6.

Table 6 *Percentage losses of vitamins in cereals during cooking*

	Boiling	Baking		Boiling	Baking
Thiamin	40	25*	Folic acid (free)	90	50
Riboflavin	40	15	(total)	50	50
Nicotinic acid	40	5	Biotin	40	0
Vitamin B_6	40	25	Pantothenic acid	40	25

* 15% in breadmaking
(For other vitamins the losses have been assumed to be zero)

Milk and milk products

This group includes milk, butter, cream, cheese and yogurt. The milks are arranged in a sequence which is not strictly alphabetical. The first item—milk, fresh, whole—provides the basic figures from which the composition of a number of milk products is derived by calculation.

No separate values are given for infant milk formulations. This is because of the present state of flux concerning these foods (Department of Health and Social Security, 1974) and it was considered unwise to include products which were likely to be superseded before or shortly after the tables were published.

In the cheese group the items are grouped according to their types, with a few individual cheeses which do not fall strictly within these classifications given separately. This was done because the variations observed for a named cheese were of the same order as those within the group type.

Modes of expression

The values are expressed per 100g edible portion. A very small error is involved if these values are applied to liquid unconcentrated milks measured by volume, that is per 100ml. Milk has a specific gravity of about 1.03 so that 100ml = 103g.

Non-protein nitrogen

Milks contain some non-protein nitrogen but no allowance for this has been made in the calculated protein values in the tables.

Values for fat-soluble vitamins

The vitamin content of milk fat shows a seasonal variation with regard to the fat-soluble vitamins (retinol, carotene, vitamin D and vitamin E). Two sets of values are given for these vitamins in whole milk and cream to allow for the seasonal differences. In the heading 'summer' refers to milks between May and October and 'winter' between November and April. In the case of processed milk and milk products, where the values for the fat-soluble vitamins have been calculated from the fat content of these items, an average value for the vitamins in milkfat has been used. The one exception is cheese, for which the summer values have been used.

The values used were as follows:

Fat-soluble vitamins μg/g fat

Non-Channel Island breeds	Retinol	Carotene	Vitamin D	Vitamin E
Summer	9.3	5.8	0.0078	25
Winter	6.8	3.3	0.0038	19
Average	8.1	4.6	0.0058	22
Channel Island breeds				
Summer	7.9	12.8	0.0078	25
Winter	5.8	7.3	0.0038	19

Vitamin losses on heat treatment

The values given for the vitamins in milks which have been heat treated have been calculated from the values for raw milk using measured losses reported in the literature. The values for the percentage losses used were as shown in table 7. The amounts of the vitamins in other forms of processed milk were measured directly. The other vitamins show little or no loss under these conditions.

Table 7 *Percentage losses of vitamins on processing milk*

	Thiamin	Riboflavin	Vitamin B_6	Vitamin B_{12}	Folic acid	Vitamin C	Vitamin E
Pasteurisation	10	0	0	0	5	25	0
Sterilization	20	0	20	20	30	60	0
Ultra-high temperature (UHT)	10	0	10	5	20	30	0
UHT stored 3 months	10	0	35	20	>50	100	0
Boiled (from pasteurised milk)	0	10	10	5	20	30–70	20

For other vitamins the losses have been assumed to be zero.

B vitamins in cheese

The values for the B vitamins in cheeses show great variability. This is due to the synthesis of these vitamins by the microorganism involved in cheese production. The values therefore vary according to the stage of maturity of the cheese and, particularly for the soft cheeses, with the proportion of rind incorporated in the sample, because the concentration of many B vitamins is very much higher in the rind than in the body of the cheese.

Eggs

The values for whole egg are taken in the main from the large study described by Tolan *et al.* (1974). The average values for the composition of eggs obtained in this study were very similar to those reported in previous editions. The values for cooking losses are shown in table 8.

Table 8 *Percentage losses of vitamins in eggs during cooking*

	Boiled	Fried	Poached	Omelette	Scrambled
Thiamin	10	20	20	5	5
Riboflavin	5	10	20	20	20
Vitamin B_6	10	20	20	15	15
Folic acid	10	30	35	30	30
Pantothenic acid	10	20	20	15	15

For other vitamins the losses have been assumed to be zero.

Fats and oils

This group includes the major fats and oils; the values for butter have been repeated in this group as a comparison with other fats is frequently made. The proximate section for this group is rather simple as most of these foods are virtually pure triglyceride and the amounts of nitrogen and inorganic constituents are very low. This, however, is not the case in the fatty acid section, where the compositions of the different oils are given separately. But it is important to bear in mind that most oils show a very wide range of fatty acid composition depending on variety, growing conditions and maturity of the oil seed.

Composition of compounded fats including margarines

These fats are mixtures prepared by the manufacturer to have the desired physical and other properties. In several cases these properties can be achieved by several different blends and it is normal commercial practice to adjust the blending according to the availability of the different oil ingredients, which will alter the fatty acid composition of the product. If accurate fatty acid data are required for specific products it is usually better to consult the manufacturer if analytical facilities are not at one's disposal.

Meat and meat products

The items are arranged in the following groups: bacon; beef; lamb; pork; veal; poultry and game; offal; and meat products and dishes. This last group includes the items whose compositions are derived from recipes and which were given in a separate section in previous editions. Within these groupings the foods are arranged alphabetically.

Sources of values

The values for the composition of an average dressed carcase which are given in the bacon, beef, lamb and pork sections are derived by calculation from dissection data from the Meat and Livestock Commission (see appendix 6) and other information, including the new analytical values for raw meats described below.

For the most part the values given in this section are based on the results of new analyses of a large number of samples which were collected from different parts of the country (Paul and Southgate, 1977). Samples were purchased at six centres (London, Bath, Cardiff, Liverpool, Newcastle and Glasgow). Each centre purchased six samples of each item, three at each centre were cooked and three were used to provide the raw samples for analysis. Where two methods of cooking were used the number of cuts or products purchased was increased accordingly. The bacon and carcase meats were separated into lean, separable fat and inedible matter. The lean meat from each particular item was pooled for analysis and the separable fat was combined in a similar way to provide the fat sample for analysis. Some meat products were not available at all the regional centres and in these cases the number of samples purchased was usually increased at the centres where they were available to provide a sample of sufficient size. In the case of liver additional samples were purchased in London to assess the importance of seasonal factors for vitamin A content and for the losses of vitamin C during cooking.

The lean-to-fat ratio in meat

The major variable affecting the composition of meat is the proportion of lean to fat and it is extremely difficult to define the average degree of the fatness for a particular joint, or to be sure that this would be very helpful to the user of the tables who is concerned with the food and nutrient intake of an individual. For this reason values for 'lean only' and 'lean and fat' are given for virtually all the items in the bacon, beef, lamb and pork groups. The percentage of separable lean in the edible portion of the different cuts analysed is given in the description of the sample. If the meat consumed has a different percentage of lean then its composition should be computed from the 'lean only' values and the values for 'fat, average, raw or cooked'.

In the sampling scheme, care was taken to select typical cuts with regard to degree of fatness and the lean and fat values should provide a reasonable guide to the average composition of an item. However, for accurate use of the tables it is essential to measure the lean and fat separately in any bacon or carcase meat consumed.

Composition of processed meats

Meat technology in this area is in a particularly rapid state of development at the present time. In the case of bacon it has been possible to cover some of the very recent changes in a small subsidiary study. When the values for processed meats are being used it is essential to bear in mind that the composition of some of these products may have changed considerably over the interval between analysis and publication of these tables.

Effect of cooking on vitamins

For many of the items purchased and analysed for this edition, it has been possible to derive values for the percentage losses of vitamins on cooking. These observed values have been used to calculate losses in foods for which direct values were not available. The observed losses are summarised in table 9.

Table 9 *Percentage losses of vitamins in meat during cooking (average loss with range)*

	Roasting, frying and grilling	Stewing and boiling*
Thiamin	20 (0–40)	60 (40–70)
Riboflavin	20 (0–30)	30 (0–40)
Nicotinic acid	20 (10–30)	50 (30–70)
Vitamin B_6	20 (0–40)	50 (30–60)
Pantothenic acid	20	40 (30–50)
Folic acid (free)	—	30 } liver and kidney
(total)	—	30 } only†
	All methods	
Vitamin B_{12}	20 (10–50)	
Biotin	10 (0–30)	
Vitamin C	20 (0–30) liver only	
Vitamin A	0	
Vitamin E	20 (0–40)	

* These losses refer to the meat alone: the water-soluble vitamins are leached into the gravies and liquors, which means that if these are used the overall losses with these methods of cooking are smaller.

† The content of folic acid in other meats is too low to make meaningful calculations of losses.

B vitamins in fat

In most cases the separable fat samples were not analysed for the B vitamins. The technical problems associated with these measurements in the presence of large amounts of fat are quite considerable and it is often very difficult to obtain good replicate analyses. The values for the B vitamins in the lean and fat items may therefore be slight underestimates of the true amounts present. In most cases the contribution from the B vitamins in the fat would be less than 10 per cent.

Cooking losses in sausages

The composition of cooked sausages can be affected by the cooking procedure. Losses of water and to a certain extent fat are reduced if the sausage is not pricked during cooking and is cooked slowly to prevent bursting or extrusion at the ends. The figures in these tables refer to sausages that were cooked after pricking.

Fish and fish products

In considering the values for fish it is important to bear in mind that at present fish are drawn from a wild population. They are in fact one of the few remaining foods which are obtained by what is in effect hunting. This means that their composition is probably more variable than that of foods drawn from domesticated inbred stock whose nutrition has been closely controlled.

The fish section in the previous editions was very extensive and included a large number of species which are rarely eaten. There is, however, considerable variation in composition within one species and this variation is often considerably greater than that between species. In the fourth edition only the fish species which constitute the major part of fish landing in the UK have been included and in many cases combined values for groups of species are given. The systematic names for fish are given in appendix 3.

The items are arranged in several groups: white fish, fatty fish, cartilagenous fish, crustacea, molluscs and fish products and dishes. Many items have been retained from the previous edition, and particular attention was given to those fish which were cooked by being dipped in batter and breadcrumbs and fried. A limited number of comparative analyses have shown that the shape and size of the portion of fish fried has a much greater effect on its composition than the composition of the batter (with or without the crumbs) and thus these older items were considered to be reasonable ones to retain.

Fat content

The fat contents of many fish show considerable seasonal changes and it is difficult to assign definite values. The actual fat content of fish normally landed and consumed shows less variation because the fish tend to be caught during a limited part of the annual cycle; the values used are therefore based on the fat content of the fish during the period when the major landings of the species are made.

Calcium and phosphorus values

In fish with fine bones it is often difficult to remove the bones completely, whether before analysis or before consumption. The calcium and phosphorus content of these fish is thus more variable than in a fish which can be boned easily. The values in the tables are based on samples which have been prepared for consumption in the normal way.

Accumulation of metals

The crustaceans and molluscs tend to accumulate many cations from their environment, and the concentration of iron, copper and zinc reported in these fish shows very wide variation, depending on the source of the samples and the levels of metallic contamination to which they have been exposed.

Vitamin content

There is a general scarcity in the literature of values for the vitamin content of fish. Very often the sources of the samples are not well documented and many of the values cited in the literature are based on a very small number of analyses. The compilations prepared by other authors have been considered when the values given in these tables were being selected, but where the source of the data used in the compilation could not be examined directly the values have been used in only a few instances. There is a special need for work in this area, particularly on vitamin levels in crustacea and molluscs, about which there seems to be little information.

Effects of cooking on vitamins

Table 10 gives the cooking losses which were used to calculate the vitamin content of cooked fish. They are based mainly on the losses found in the samples of cod analysed for this edition. The values are thus more tentative than those given for other food groups and should be used only as a guide. No information on vitamin losses during frying (apart from those analysed for this edition) are available.

Table 10 *Percentage losses of vitamins during cooking fish*

	Poaching	Baking	Frying/grilling
Thiamin	10	30	20
Riboflavin	0	20	20
Nicotinic acid	10	20	20
Vitamin B_6	0	10	20
Pantothenic acid	20	20	20
Folic acid (free)	50	30	0
(total)	0	20	0
Vitamin B_{12}	0	10	0
Biotin	10	10	10
Vitamin C	—	—	20*
Vitamin E	(0)	0	0

* Used for roe

Vegetables

The vegetable section is based for the most part on the values published in earlier editions. All the figures have been compared with those in more recent published work; this comparison has shown that the values given in the previous edition are still representative.

New analyses have been carried out to provide values for a few additional foods and for a selected number of nutrients in foods already included. These selected analyses have been designed to provide additional values for dietary fibre ('unavailable carbohydrates' in previous editions) and for some vitamins and the effects of cooking on those vitamins. The new items include frozen peas, instant potato powder, sweet corn, peppers and some vegetables especially important for immigrants. Some cooked pulse dishes are given (calculated from recipes) and for these the general term 'dahl' has been used.

The vegetables are arranged in alphabetical order (as in previous editions) as it was found that subgrouping them was impracticable. The systematic names for vegetables are given in appendix 3.

Edible matter as a proportion of the weight purchased

The amount of inedible material removed from a vegetable before it is cooked depends on a number of factors: the condition of the vegetable, damage during harvesting, contamination of leaves with soil and not least the personal idiosyncracies of the person preparing the vegetable. The values given are those measured in the samples described, which were usually purchased at a variety of retail outlets. Where a vegetable is purchased 'prepacked' much of the waste has already been removed and the edible matter will usually represent a higher proportion of the weight purchased than that given in the tables.

Moisture

The water content is the major variable affecting the proximate composition of vegetables. This will depend on the conditions under which the vegetable has been stored. It is difficult to give figures which will allow for possible variations in water content due to storage, and the user of the tables should bear in mind that a wilted vegetable will have a lower moisture content than that shown.

Protein

Part of the total nitrogen in vegetables is non-protein nitrogen and is largely made up of free amino acids and their amides. For nutritional purposes it is therefore reasonable to consider all the total nitrogen as if it were in the form of protein.

Vitamins

The full range of vitamins was not measured in the new items and in most cases the vitamin values in these items were taken from the literature. The vitamins measured were thiamin, riboflavin, nicotinic acid, vitamin C and folic acid (that is, carotenes, biotin, pantothenic acid and vitamin B_{12} were not measured). Losses of vitamins under different cooking methods are given in table 11.

Table 11 *Percentage losses of vitamins in vegetables during cooking*

	Root vegetables	Leafy vegetables	Seeds
Carotene	0	0	0
Thiamin	25	40	30
Riboflavin	30	40	30
Nicotinic acid	30	40	30
Vitamin C	40	70	50
Vitamin E	0	0	0
Vitamin B_6	40	40	40
Folic acid (free)	90	90	90
(total)	50	20–40	50
Pantothenic acid	30	30	30
Biotin	30	30	30

These are representative values and actual losses depend on:

(*a*) volume of water used to cook the vegetables
(*b*) time of cooking
(*c*) state of division of the food

Vitamin C

Where sufficient information was available ranges are given to provide guidance to the user. The values for vitamin C in potatoes are dependent on the time that the tuber has been stored and, in the cooked potato, on the method of cooking. The vitamin C values for potatoes are given as a range, the lowest representing the value for a potato stored for 8–9 months and cooked with the greatest loss; the higher value represents the value in a freshly harvested tuber cooked with the minimum of loss.

Folic acid

In many vegetables the value for free folate (that is, measured before conjugase treatment) is higher in the raw than in the cooked state. This is probably due to the fact that conjugases present in the vegetable are inactivated during cooking whereas in the raw vegetable they are active during the extraction of the plant tissues.

Inorganic constituents

The concentration of inorganic constituents in vegetables is affected by soil and fertiliser treatment during growth and in the cooked vegetable by leaching into the cooking water. The variations due to cultural conditions affect the trace elements more than the major inorganic constituents. The data in the literature, however, were insufficient for values representing the effects of this variable to be given. Losses during cooking are increased if the vegetable is cut into small pieces or cooked in a large volume of water for a long time. The values given apply to the samples cooked as described. Vegetables may also acquire constituents from tap water, for example Ca, Mg, Fe and trace metals Fe, Cu, Zn and Al from pots and pans. These effects are extremely variable and if accurate intakes of inorganic constituents are required all cooking should be carried out in distilled water in glass or stainless steel vessels and if possible the calculations should be supplemented by analyses.

Amino acids and fatty acids

Values for the amino acids and fatty acids in vegetables have been drawn almost completely from the literature. The amino acid data in particular are very limited indeed. This is particularly unfortunate because these foods are widely used in low protein diets and a more extensive knowledge of their amino acid composition would be useful for those involved in this type of diet therapy.

Fruit

The values for fruits are based on the values published in previous editions; as was the case with the vegetables, recently published values showed that the values given in earlier editions were still representative. Some additional canned fruits have been included together with values for some 'tropical' fruits which are becoming more common. The systematic names of the fruits are given in appendix 3.

Stewed fruits

The values given for the stewed fruits have been recalculated using additional experimental values for the proportion of water added to the fruit and making a correction for the evaporative losses in stewing (10 per cent). All fruits were stewed in the minimum of water and the composition of the cooked fruit was calculated, as in previous editions, on the basis of measured values for the ratio of cooked to raw weights. The ratio of cooked weight (including inedible stones) to raw weight was 1.05:1 for loganberries and raspberries; 2:1 for dried figs and prunes; 3:1 for dried apricots and peaches; and 1.3:1 for all other fruits.

The earlier editions included fruit stewed without sugar. In this edition values are also given for fruit stewed with a given amount of sugar (12g per 100g fruit). The amount of sugar added will vary with personal preference and the fruit being cooked but it was felt that these new inclusions would be useful for the 'average' user, although where control of carbohydrate intake is essential the 'stewed without sugar' values are still required by the user.

Vitamin C

Where information was available ranges for this vitamin are given in addition to the selected value. The vitamin C in fruits is extremely variable and is dependent on the level of illumination that the individual fruit has received. This means that appreciable variation can occur in fruit from the same tree or bush and within the same fruit.

Amino acids and fatty acids

The comments in the vegetable section apply to the fruits with even more force.

Nuts

The values for nuts are with one exception derived from the previous edition. The new item is peanut butter, although an estimate of the salt in salted peanuts has also been made. The systematic names for the nuts are given in appendix 3.

Sugars, preserves and confectionery

Here the items are arranged according to the three subgroups. Many values have been taken from the third edition as most of the sweets were analysed for that edition.

The items analysed for this edition include lemon curd (starch-based) as purchased, mincemeat and chocolate. The composition of the other items appears to be still representative and falls within the ranges reported in the literature or obtained from evidence provided by the manufacturers.

Carbohydrate

The values are given in terms of monosaccharides and this means that white sugar, which is virtually pure sucrose, has a value of 105 g/100 g carbohydrate because one molecule of water is taken up on hydrolysis to give glucose and fructose. Many products in this group contain 'glucose syrup', which contains glucose, maltose and higher maltose homologues. These higher oligosaccharides will analyse as sugar in most methods but in others they appear in the starch fraction. Nutritionally there are only minor differences between these higher oligosaccharides and starch but the user should be aware that these products may yield analytical anomalies.

Water

It is particularly difficult to obtain values for the water content of products very rich in sugar and some, such as boiled sweets, contain water of crystallisation, which may or may not be considered as part of the water content of the food.

Beverages

These have been arranged in two subgroups. The first includes the concentrated powders, which are usually diluted with milk or water before consumption; the second comprises the soft drinks, and fruit and vegetable juices that are either diluted with water or are purchased ready for consumption. The values are expressed on a 100 g basis, as accurate measurement by volume is difficult with both effervescent and syrupy beverages.

The first subgroup includes coffee and tea and infusions made from them. Values for the diluted concentrated powders have not been given because of the wide range of methods recommended and because many nutritional workers are concerned with total milk intake, which is often considered separately.

Dietary fibre

Coffee and cocoa products seem to contain high concentrations of dietary fibre and in particular a 'high' apparent lignin content. This is due to the presence of the ill-defined component humic acid which analyses as lignin. It seems improbable that these products will prove to be useful sources of dietary fibre.

Fruit drinks

Only one example is given as the type of fruit used has little or no influence on its composition.

Vitamin C

Many beverages which are ostensibly fruit-based contain little or no vitamin C unless it is added during their manufacture. The user should check the label of any beverage of this sort to establish whether or not vitamin C has been added.

Alcoholic beverages

The items in this group are arranged into a number of subgroups: beers, ciders, wines, liqueur wines (in other words fortified wines), vermouths, liqueurs and spirits.

All values (with the exception of specific gravity) are given as g per 100 ml (w/v). The specific gravities are given so that calculations can be made if the beverages are measured by weight.

It has not been possible to cover all the different vintages or types of wine available and the values for the wines were obtained on typical examples. They should provide a reasonable guide for nutritional purposes but should not be regarded as a definitive one for the composition of wines.

The alcoholic strengths of beverages can be expressed in several different ways. The common method on the European continent is to use degrees Gay-Lussac, which are the percentage alcohol by volume (v/v). In the United Kingdom the proof system is used; in this, proof spirit is defined as having a specific gravity (SG^{20}_{20}) of 0.91702 and contains 49.276 per cent alcohol (w/w) or 57.155 per cent (v/v) (that is, about 45.2 g per 100 ml).

The approximate alcohol contents of various proof strengths are as follows (Customs and Excise, 1954):

°Proof	Alcohol (g/100 ml)
10	4.6
20	9.1
30	13.6
40	18.2
50	22.7
60	27.2
70	31.7

Sauces, pickles, soups, condiments and miscellaneous foods

This group includes in addition to the major named items a few miscellaneous items which are difficult to assign to other food groups. The first subgroup, sauces and pickles, includes items whose composition has been calculated from recipes and others which have been analysed directly. Some values have been retained from the third edition and others have been analysed for this edition.

The second subgroup, soups, includes many new items and a few whose composition has been calculated. The concentrated (condensed) and dried soups are given as purchased and as diluted ready for consumption. In the calculation of the composition of the dilutions of dried soups, a correction for evaporative losses has been made.

The third subgroup includes condiments, ingredients and a number of items such as yeast, Marmite, Bovril and Oxo. These last three items contain a large amount of non-protein nitrogen, which is largely in the form of amino acids and peptides. Nutritionally these behave as protein and it is therefore appropriate to multiply the total amino nitrogen by 6.25 to get some measure of their weight. This is not 'protein' in the strict sense but no nutritional errors are involved in treating it as such.

The composition of cooked dishes

The compositions of a range of cooked dishes are included in the tables. In contrast to the previous editions these are now given with the major food groups rather than in separate sections. For the fourth edition a complete revision of all the items based on calculations from recipes has been made. Some items have been deleted and a number of items with very similar compositions have been combined, for example milk puddings. About twenty new dishes have been added including some foreign dishes such as pizza, moussaka and cheesecake.

The composition of these cooked dishes has been calculated, as in previous editions, from the recipes, the composition of the ingredients and the change in weight on cooking. The change in weight on cooking has been assumed to be due either to the evaporation of water or to a gain by absorption. The composition of dishes where the method of preparation involves a change in fat content in addition to water content cannot be calculated directly in this way and in these cases the cooked dishes were analysed for fat and water before the calculations were made.

All the recipes are given in metric quantities and the calculations have been made using the compositions of the ingredients given in this edition. The composition of the new items has been based on experimentally determined changes in weight on cooking. The dishes have been prepared on at least two occasions and the average of these two sets of data has been used in the calculations.

The composition of the items given in the third edition have been recalculated from the metricated recipes on the assumption that the final consistency, and therefore water content, would be the same as that found previously.

Where flour was used as an ingredient, plain flour was used and baking powder was added for cakes and some puddings. The baking powder used was a proprietary preparation whose composition is given in the tables (item 956). This preparation contains calcium acid phosphate, sodium bicarbonate, sodium pyrophosphate and flour; the use of another raising agent (for example sodium bicarbonate) will result in a different composition in the cooked dish with respect to Na, Ca and P. Margarine fortified with vitamins A and D was used in the preparation of the dishes. The use of an unfortified margarine or other fat would affect the values for these two vitamins.

In the calculations an egg has been assumed to weigh 50g; a level teaspoon refers to the standard 5ml spoon and has been taken to hold 5g salt and 3.5g baking powder.

The recipes are given in appendix 4 and their numbering corresponds to the numbering of the item in the tables.

Calculation of the composition of dishes prepared from other recipes
The method of calculation is as follows. The weights of the raw ingredients are used to calculate the total amounts of nutrients in the dish. A correction for wastage due to ingredients left on utensils and in the vessels used in preparation is made at this stage. The weight of the raw dish is then measured, using a scale weighing to about 1 g (a less accurate scale may be used if the total weight of ingredients is over 500g). The dish is then cooked and the dish reweighed. (A minor correction to allow for the difference between weighing the dish hot and at room temperature is not usually necessary.) The difference in weight is taken as being accounted for by water and the composition of the cooked dish is calculated as follows. Divide the total nutrients in the dish calculated from the raw ingredients by the weight of the cooked dish and multiply by 100. The water content of the raw ingredients *less* the loss in weight on cooking divided by the weight of the cooked dish gives the water content of the cooked dish if it is required. An example is:

Egg Custard

Ingredient	Amount in recipe g	Amounts contributed Protein g	Fat g	Carbohydrate etc. g
Milk	500	16.5	19.0	23.5
Egg	100	12.3	10.9	Tr
Sugar	30	0	0	31.5
Vanilla essense	to taste	Ignored for calculation		
Total in recipe (a)	630	28.8	29.9	55.0
Cooked weight	500			
Composition of cooked dish (per 100g) (b)	—	5.8	6.0	11.0

(a) = sum of nutrients in ingredients
(b) = (a) divided by cooked weight × 100

If some of the ingredients are left in the mixing bowl or not used then the (a) values should be multiplied by weight used divided by weight of total ingredients.

Using the tables

Tables of food composition are used for many different purposes and this section is intended to guide the reader in the use of the tables and to give some idea of the accuracy of calculations based on the values in them.

The first of the main uses of tables of food composition is the calculation of nutrient intakes from records of food consumption. These calculations may be for large groups of people such as the population of a country, where the records of food consumption are derived from food supplies measured in raw commodities at the wholesale level, or for the individual, where food intake has been measured as consumed in the prepared state. In the United Kingdom at least there are several different levels of calculation using food consumption data between these two extremes.

The second major use is the formulation of diets or food supplies that will provide a specified intake of nutrients. Again, calculations are made at several different levels ranging from an international agency's estimates of the desirable food supplies for a country or region to the dietitian's formulation of a diet for an individual according to the prescription of a physician.

Another use is the calculation of the nutrient composition of a manufactured food from its ingredients.

Each of these types of usage has its own requirements for food items and range of nutrients in the food tables and, furthermore, requires a different level of accuracy from the calculations. It is, however, extremely difficult to cover all these requirements in a single compilation and the tables in this book are designed to provide information from which the users must make their own selection.

Variation in composition of foods

The first point that the user must bear in mind is that very few foods have a constant composition. Anyone who uses a table of food composition as an oracle is misleading himself. Foods are biological materials and as such show a considerable variation in composition. Manufactured foods are usually subjected to quality control, which may apply to some aspects of their composition, and they might therefore be expected to have a more constant composition. All manufacturers, however, permit some tolerances and in practice manufactured foods may show as much variation in composition as some unprocessed foods. These variations in composition depend on the type of foodstuff and the nutrient in question. The main features of the variations in the concentrations of different nutrients have been described in the section on modes of expression.

The values in the tables are in the main analytical results obtained from the analysis of representative samples of the food items. One should therefore expect them to reflect the average composition of the food; isolated samples on the other hand may well have a different composition. This means that calculations from food composition data should intrinsically be more accurate when they are concerned with large groups of people (although not necessarily where they are drawn from one institution for example). Where studies involving individuals are concerned one should achieve an improvement in accuracy by extending the period of study and thus in effect increasing the size of the food sample consumed. This is borne out by comparisons of calculated and analysed values for the composition of mixed diets, where agreement between these two is better for a collection taken over seven days than for a single meal.

Probable levels of accuracy using food tables

The variations in the composition of foods thus have a considerable influence on the level of accuracy that one can expect from food tables. Accuracy will also depend to a certain extent on the appropriateness of the items in the tables. Guidance in this area can only be given in a general fashion because the number of experimental studies where this accuracy has been tested experimentally is very limited.

In general, when comparisons are made over a period of several days for an individual's diet under metabolic balance conditions the agreement between calculated and analysed values for protein, carbohydrate, energy, K, Ca, Mg and P is usually within 5–10 per cent. Calculated fat intakes often show a great discrepancy but this is sometimes attributable to the method of analysis used for fat (see p8). Calculated intakes of sodium and iron may differ greatly from analysed values owing to variations in added salt and contamination respectively. Little information is available concerning the vitamins except for vitamin C, where intakes calculated from tables can be rather inaccurate. The amino acid composition of a diet, however, is quite accurately estimated by calculation (Hughes, 1959). No information is available for comparisons of fatty acid compositions.

The calculation of nutrient intakes

The first stage in the calculation of nutrient intakes is deciding which item in the tables corresponds with the item consumed, a process often known as 'coding'. This is a procedure that needs to be approached with thought and is one reason why these tables include a fairly detailed description of the samples used to obtain the values. Difficulties arise where there is no corresponding item in the tables or where the description of either the food consumed or the item in the tables is inadequate and the coder cannot be sure that these two do in fact correspond.

In the usual dietary calculation (where a food item is part of a diet) the best procedure is to use a related food for items that are not listed. The user should, however, first check in the index (p407 *et seq.*) because the food may be listed under a synonym or it may be included in another food group. The choice of the related food can be guided by texture or consistency if no other information is available. Where there is uncertainty about whether items correspond, in most calculations little error is involved by assuming that they do (except for foods which may or may not contain added constituents such as salt).

The number of manufactured foods included in this edition is limited for reasons discussed earlier (p2), and the user is advised to consult the manufacturer concerned if a food forms an important part of the diet or accurate control of sugar or sodium intakes, for example, is essential.

Cooked dishes

Many users have felt in the past that because they used different recipes the composition of their cooked dishes would be very different from those given in the tables. This may be true but often these differences are quite small. However, if different recipes are used regularly it may be useful to calculate the composition of dishes from these recipes. The procedure is relatively simple and is described at the end of the section on the recipes (p334).

Calculation from 'as purchased' weights

All the values in the tables refer to the composition of the edible matter and where consumption data are available on the basis of food 'as purchased' it is necessary to correct these weights before making calculations (see p5).

For all foods where there is a loss through wastage or a change in weight during cooking, the column headed 'Edible matter, proportion of weight purchased' gives a factor by which the 'as purchased' weight should be multiplied to give the weight of edible matter.
For example:

100g streaky bacon rashers as purchased will give
100 × 0.85 = 85g edible raw bacon and
100 × 0.51 = 51g edible cooked (fried) bacon

Vitamin equivalents

Over recent years it has become customary to express most vitamin values in terms of the weight of the pure vitamin rather than in 'international units'.

In the case of two vitamins, vitamin A and nicotinic acid, where activity can come from more than one substance having different activities, a system of weight equivalents has been suggested.

In the present UK recommendations for nutrient intakes (Department of Health and Social Security, 1969) the recommended intakes for these two vitamins are expressed in equivalents and the user may need to make calculations on these terms in order to compare nutrient intakes with the recommendations.

The method of calculating retinol equivalents is given on p12. The divisor of 6, which is used to calculate the retinol equivalents from the β-carotene, is an average value for mixed diets. Many foods should have different divisors, depending on the carotenes present in the food. The DHSS report (Department of Health and Social Security, 1969) specifically recommends that the β-carotene in milk and milk products should be divided by 2, while the FAO/WHO recommendation (FAO/WHO, 1965) is to use the divisor of 6 for all foods. Until sufficient information is available on individual foods, it is better to use the divisor of 6 for the sum of the β-carotenes in the diet as this has some experimental justification.

If one needs to calculate retinol equivalents for direct comparison with the DHSS recommendations the calculation is as follows:

$$\mu\text{g retinol equivalents} = \mu\text{g retinol} + \frac{\mu\text{g } \beta\text{-carotene from milk and milk products}}{2} + \frac{\mu\text{g } \beta\text{-carotene from other foods}}{6}$$

In practice the use of a divisor of 2 rather than 6 for milk and milk products gives a difference of about 100 μg (7 per cent of the total retinol equivalents) in the average British diet.

The method for calculating the potential contribution of tryptophan to the nicotinic acid equivalents in the diet has general agreement and is described on p13. In the DHSS recommendations, however, it is suggested that the nicotinic acid naturally occurring in cereals should be discounted because it is probably unavailable.

Direct comparison with the DHSS recommended intakes of nicotinic acid should therefore be made using the following calculations.

$$\text{Nicotinic acid equivalents (mg)} = \left[\text{Total nicotinic acid (mg)} - \text{Nicotinic acid naturally present in cereals (mg)}\right] + \frac{\text{mg tryptophan}}{60}$$

The naturally occurring nicotinic acid in cereals contributes about 2 mg nicotinic acid to the average British diet (about 7 per cent of the total intake of nicotinic acid equivalents: Paul, 1969).

Availability of nutrients

The values given in the tables have been obtained by chemical or microbiological analyses and give the total amount of the constituent in the food. With the exception of carbohydrate and the calculated energy content no attempt has been made to give any estimate of availability.

In the case of carbohydrate, the 'available carbohydrates' consist of carbohydrates which are chemically distinct from the 'unavailable carbohydrates'. The energy values are for metabolisable energy (p 9) and make allowance for losses of energy-yielding constituents in urine and faeces. Many foods contain unavailable forms of some nutrients, of which the best documented are iron, nicotinic acid and lysine. As nutritional studies progress, however, it is becoming apparent that the question of availability should be extended to many other nutrients, especially amino acids and the trace elements. The availability of many inorganic constituents is also influenced by other constituents of the diet. It is difficult to devise analytical procedures that can be used to measure availability because it is essentially a physiological concept. Furthermore, the availability of iron, for example, depends on the iron status of the individual and it would therefore not be possible to give a meaningful value for available iron in the tables which would apply to any individual.

Each nutrient poses special problems and it was not considered that there were enough experimentally based data to give values for 'available nutrients' in this edition of the tables. But the user must be aware that differences in availability do exist and that the total chemical (or microbiological) values given in these tables may, for many nutrients, represent the maximum that could be available to the body, and that the actual amount may be much smaller.

Conclusion: the need for the combined use of tables and text

It is extremely difficult to present in the tables themselves all the various factors that should be considered when the tabulated values are used. This could be done only by the use of an inordinate number of qualifying footnotes, which would make the tables difficult to use. This means that it is important to be thoroughly familiar with the text and to use tables and text together. It is especially important for the user to be familiar with the symbols and conventions. This approach to the tables should lead to a more accurate use of the data and resolve many of the problems which appear to have arisen with the previous editions. Certainly many of the inquiries that have been received relating to the earlier editions would not have arisen had the text been referred to first.

References to text

Anderson, B. A. (1976) Comprehensive evaluation of fatty acids in foods. VII. Pork products. *J. Amer. diet. Ass.*, **69**, 44–49

Anderson, B. A., Kinsella, J. A., and Watt, B. K. (1975) Comprehensive evaluation of fatty acids in foods. II. Beef products. *J. Amer. diet. Ass.*, **67**, 35–41

Customs and Excise (1954) *Specific gravity of spirits*. HMSO, London

Department of Health and Social Security (1969) *Recommended intakes of nutrients for the United Kingdom*. Reports on Public Health and Medical Subjects, No. 120. HMSO, London

Department of Health and Social Security (1974) *Present day practice in infant feeding*. Reports on Health and Social Subjects, No. 9. HMSO, London

Exler, J., Kinsella, J. E., and Watt, B. K. (1975) Lipids and fatty acids of important finfish. New data for nutrient tables. *J. Amer. Oil Chem. Soc.*, **52**, 154–159

FAO/WHO (1967) *Requirements of vitamin A, thiamine, riboflavine and niacin*. Report of a Joint FAO/WHO Expert Group. FAO Nutrition Meeting Report Series, No. 41; WHO Technical Report Series, No. 362

FAO/WHO (1973) *Energy and protein requirements*. Report of a Joint FAO/WHO *Ad Hoc* Expert Committee. FAO Nutrition Meetings Report Series, No. 52; WHO Technical Report Series, No. 522

Hughes, B. P. (1959) The amino-acid composition of three mixed diets. *Brit. J. Nutr.*, **13**, 330–337

McCance, R. A., and Shipp, H. L. (1933) *The chemistry of flesh foods and their losses on cooking*. Medical Research Council Special Report Series, No. 187. HMSO, London

McCance, R. A., Widdowson, E. M., and Shackleton, L. R. B. (1936) *The nutritive value of fruits, vegetables and nuts*. Medical Research Council Special Report Series, No. 213. HMSO, London

Merrill, A. L., and Watt, B. K. (1955) *Energy value of foods—basis and derivation*. US Department of Agriculture, Agriculture Handbook No. 74. Washington, DC

Paul, A. A. (1969) The calculation of nicotinic acid equivalents and retinol equivalents in the British diet. *Nutrition, Lond.*, **23**, 131–136

Paul, A. A., and Southgate, D. A. T. (1970) Revision of *The composition of foods*. Some views of dietitians. *Nutrition, Lond.*, **24**, 21–24

Paul, A. A., and Southgate, D. A. T. (1977) A study of the composition of retail meat: dissection into lean, separable fat and inedible portion. *J. hum. Nutr.* **31**, 259–272

Platt, B. S. (1962) *Tables of representative values of foods commonly used in tropical countries*. Medical Research Council Special Report Series, No. 302. HMSO, London

Posati, L. P., Kinsella, J. E., and Watt, B. K. (1975) Comprehensive evaluation of fatty acids in foods. III. Eggs and egg products. *J. Amer. diet. Ass.*, **67**, 111–115

Royal Society (1972) *Metric units, conversion factors and nomenclature in nutritional and food sciences*. Report of the subcommittee on metrication of the British National Committee for Nutritional Sciences

Southgate, D. A. T. (1974) *Guide lines for the preparation of tables of food composition*. S. Karger, Basel

Southgate, D. A. T., and Durnin, J. V. G. A. (1970) Calorie conversion factors: an experimental reassessment of the factors used in the calculations of the energy value of human diets. *Brit. J. Nutr.*, **24**, 517–535

Southgate, D. A. T., Bailey, B., Collinson, E., and Walker, A. F. (1976) A guide to calculating of intakes of dietary fibre. *J. hum. Nutr.*, **30**, 303–313

Tolan, A., Robertson, J., Orton, C. R., Head, M. J., Christie, A. A., and Millburn, B. A. (1974) Studies on the composition of food. 5. The chemical composition of eggs produced under battery, deep litter and free range conditions. *Brit. J. Nutr.*, **31**, 185–200

Weihrauch, J. L., Kinsella, J. E., and Watt, B. K. (1976) Comprehensive evaluation of fatty acids in foods. VI. Cereal products. *J. Amer. diet. Ass.*, **68**, 335–340

The tables

Symbols (see p 5)
Tr Trace
( ) Estimated value
— No information available

Section I
Description of foods, proximate composition, energy value, inorganic constituents and vitamins (per 100 g)

All carbohydrate values are for available carbohydrate as monosaccharides

Note: A set of blank pages has been inserted between each major food group so that additional foods may be entered by the user if required.

No	Food	Description and number of samples	Water g	Sugars g	Starch and dextrins g	Dietary fibre g	Total nitrogen g
	Grains, flours and starches						
1	**Arrowroot**	2 samples from different shops	12.2	Tr	94.0	—	0.07
2	**Barley** pearl, raw	2 samples from different shops	10.6	Tr	83.6	6.5	1.35
3	boiled	2 samples from different shops, boiled in water	69.6	Tr	27.6	2.2	0.46
4	**Bemax**	Stabilized wheat germ; mixed sample	6.0	16.0	28.7	—	4.54
5	**Bran** wheat	Analytical and literature sources	8.3	3.8	23.0	44.0	2.24
6	**Cornflour**	3 samples from different shops	12.5	Tr	92.0	—	0.09
7	**Custard powder**	Taken as cornflour, except for Na and Cl	12.5	Tr	92.0	—	0.09
9	**Flour** wholemeal (100%)	Data from Voluntary Flour Sampling	14.0	2.3[a]	63.5	9.6	2.26
10	brown (85%)	Scheme, 1974–5, Ministry of	14.0	1.9[a]	66.9	7.5	2.25
11	white (72%) breadmaking	Agriculture, Fisheries and Food; cake	14.5	1.5[a]	73.3	3.0	1.98
12	household, plain	and biscuit flour are similar in composi-	13.0	1.7[a]	78.4	3.4	1.72
13	self-raising	tion to household plain flour	13.0	1.4[a]	76.1	3.7	1.63
14	patent (40%)	Mixed sample	14.1	1.4[a]	76.6	—	1.89
15	**Macaroni** raw	2 samples from different shops	10.4	Tr	79.2	—	2.41
16	boiled	2 samples from different shops, boiled in water	71.5	Tr	25.2	—	0.75
17	**Oatmeal** raw	Coarse, medium and fine; 2 samples of each	8.9	Tr	72.8	7.0	2.12
18	**Porridge**	60g oatmeal and 2 level teaspoons salt per 500ml water	89.1	Tr	8.2	0.8	0.24

[a] Includes the glucofructan levosin

No	Food	Energy value kcal	Energy value kJ	Protein (see p7) g	Fat g	Carbo-hydrate g	Na mg	K mg	Ca mg	Mg mg	P mg	Fe mg	Cu mg	Zn mg	S mg	Cl mg
	Grains, flours and starches															
1	**Arrowroot**	355	1515	0.4	0.1	94.0	5	18	7	8	27	2.0	0.22	—	2	7
2	**Barley** pearl, raw	360	1535	7.9	1.7	83.6	3	120	10	20	210	0.7	0.12	(2.0)	110	110
3	boiled	120	510	2.7	0.6	27.6	1	40	3	7	70	0.2	0.04	(0.7)	36	36
4	**Bemax**	347	1465	26.5	8.1	44.7	4	1000	17	300	930	10.0	1.20	—	—	80
5	**Bran** wheat	206	872	14.1	5.5	26.8	28	1160	110	520	1200	12.9	1.34	16.2	65	150
6	**Cornflour**	354	1508	0.6	0.7	92.0	52	61	15	7	39	1.4	0.13	Tr	1	71
7	**Custard powder**	354	1508	0.6	0.7	92.0	320	61	15	7	39	1.4	0.13	Tr	1	480
9	**Flour** wholemeal (100%)	318	1351	13.2	2.0	65.8	3	360	35	140	340	4.0	0.40	3.0	—	38
10	brown (85%)	327	1392	12.8	2.0	68.8	4	280	150[a]	110	270	3.6[a]	0.35	2.4	—	45
11	white (72%) breadmaking	337	1433	11.3	1.2	74.8	3	130	140[b]	36	130	2.2[b]	0.22	0.9	110	62
12	household, plain	350	1493	9.8	1.2	80.1	2	140	150[b]	20	110	2.4[b]	0.17	0.7	—	45
13	self-raising	339	1443	9.3	1.2	77.5	350[c]	170	350[c]	42	510[c]	2.6[b]	0.15	0.6	—	45
14	patent (40%)	347	1480	10.8	1.3	78.0	3	100	110	19	89	1.7	0.11	—	110	60
15	**Macaroni** raw	370	1574	13.7	2.0	79.2	26	220	26	57	150	1.4	0.07	(1.0)	95	31
16	boiled	117	499	4.3	0.6	25.2	8	67	8	18	47	0.5	0.02	(0.3)	29	10
17	**Oatmeal** raw	401	1698	12.4	8.7	72.8	33	370	55	110	380	4.1	0.23	(3.0)	160	73
18	**Porridge**	44	188	1.4	0.9	8.2	580	42	6	13	43	0.5	0.03	(0.3)	18	890

[a] These are values for fortified flour. Unfortified brown flour would contain about 20 mg Ca and 2.5 mg Fe per 100 g

[b] These are values for fortified flour. Unfortified white flour would contain about 15 mg Ca and 1.5 mg Fe per 100 g

[c] The amount present will depend on the nature and level of the raising agent used

No	Food	Retinol µg	Carotene µg	Vitamin D µg	Thiamin mg	Riboflavin mg	Nicotinic acid mg	Potential nicotinic acid from tryptophan mg Trp ÷ 60	Vitamin C mg	Vitamin E mg
	Grains, flours and starches									
1	**Arrowroot**	0	0	0	(Tr)	(Tr)	(Tr)	0.1	0	(Tr)
2	**Barley** pearl, raw	0	0	0	0.12	0.05	2.5	2.3	0	0.2[a]
3	boiled	0	0	0	Tr	Tr	(0.9)	0.8	0	Tr
4	**Bemax**	0	0	0	1.45	0.61	5.8	5.3	0	11.0[b]
5	**Bran** wheat	0	0	0	0.89	0.36	29.6	3.0	0	1.6[c]
6	**Cornflour**	0	0	0	(Tr)	(Tr)	(Tr)	0.1	0	0
7	**Custard powder**	0	0	0	(Tr)	(Tr)	(Tr)	0.1	0	0
9	**Flour** wholemeal (100%)	0	0	0	0.46	0.08	5.6	2.5	0	1.0[d]
10	brown (85%)	0	0	0	0.42[e]	0.06	4.2[e]	2.6	0	(Tr)
11	white (72%) breadmaking	0	0	0	0.31[f]	0.03	2.0[f]	2.3	0	Tr
12	household, plain	0	0	0	0.33[f]	0.02	2.0[f]	2.0	0	Tr
13	self-raising	0	0	0	0.28[f]	0.02	1.5[f]	1.9	0	Tr
14	patent (40%)	0	0	0	(0.32[f])	(0.02)	(2.0[f])	2.2	0	Tr
15	**Macaroni** raw	0	0	0	0.14	0.06	2.0	2.8	0	(Tr)
16	boiled	0	0	0	0.01	0.01	0.3	0.9	0	(Tr)
17	**Oatmeal** raw	0	0	0	0.50	0.10	1.0	2.8	0	0.8[g]
18	**Porridge**	0	0	0	0.05	0.01	0.1	0.3	0	(0.1)

No	Food	Vitamin B_6 mg	Vitamin B_{12} µg	Folic acid Free µg	Folic acid Total µg	Pantothenic acid mg	Biotin µg
	Grains, flours and starches						
1	**Arrowroot**	(Tr)	0	Tr	Tr	(Tr)	(Tr)
2	**Barley** pearl, raw	0.22	0	9	20	0.5	—
3	boiled	Tr	0	Tr	(3)	(0.2)	(Tr)
4	**Bemax**	0.95	0	(260)	(330)	(1.7)	—
5	**Bran** wheat	1.38	0	130	260	2.4	14
6	**Cornflour**	(Tr)	0	Tr	Tr	(Tr)	(Tr)
7	**Custard powder**	(Tr)	0	Tr	Tr	(Tr)	(Tr)
9	**Flour** wholemeal (100%)	0.50	0	25	57	0.8	7
10	brown (85%)	(0.30)	0	23	51	(0.4)	(3)
11	white (72%) breadmaking	0.15	0	14	31	0.3	1
12	houshold, plain	0.15	0	14	22	0.3	1
13	self-raising	0.15	0	11	19	0.3	1
14	patent (40%)	(0.10)	0	(5)	(10)	(0.3)	(1)
15	**Macaroni** raw	0.06	0	4	11	(0.3)	(1)
16	boiled	(0.01)	0	Tr	(2)	(Tr)	(Tr)
17	**Oatmeal** raw	0.12	0	11	60	1.0	20
18	**Porridge**	0.01	0	1	6	0.1	(2)

Notes

[a] Also contains 0.3 mg β-tocopherol and 1.2 mg α-tocotrienol per 100 g.

[b] α-tocopherol constitutes about half the total tocopherols present.

[c] Also contains 1.0 mg β-tocopherol and 1.1 mg α-tocotrienol per 100 g.

[d] Also contains 0.7 mg β-tocopherol and 0.4 mg α-tocotrienol per 100 g. There is a loss of tocopherols on storage of about 60% over 13 weeks.

[e] These are levels for fortified flour. Unfortified flour would contain 0.30 mg thiamin and 1.7 mg nicotinic acid per 100 g.

[f] These are levels for fortified flour. Unfortified flour would contain 0.10 mg thiamin and 0.7 mg nicotinic acid per 100 g.

[g] Also contains 0.1 mg γ-tocopherol and 1.0 mg α-tocotrienol per 100 g.

No	Food	Description and number of samples	Water g	Sugars g	Starch and dextrins g	Dietary fibre g	Total nitrogen g
	***Grains, flours and starches** contd*						
19	**Rice** polished, raw	5 samples from different shops	11.7	Tr	86.8	2.4	1.09
20	boiled	5 samples from different shops, boiled in water	69.9	Tr	29.6	0.8	0.37
21	**Rye** flour (100%)	Commercial grist of all-English rye	15.0	Tr	75.9	—	1.40
22	**Sago** raw	2 samples from different shops	12.6	Tr	94.0	—	0.04
23	**Semolina** raw	2 samples from different shops, coarse and fine	14.0	Tr	77.5	—	1.87
24	**Soya** flour, full fat	Mixed sample	7.0	11.2	12.3	11.9	6.45
25	low fat	Mixed sample	7.0	13.4	14.8	14.3	7.94
26	**Spaghetti** raw	6 samples from different shops	10.5	2.7	81.3	—	2.39
27	boiled	6 samples from different shops, boiled in water	71.7	0.8	25.2	—	0.74
28	canned in tomato sauce	6 large cans from different shops	83.1	3.4	8.8	—	0.30
29	**Tapioca** raw	4 varieties, medium pearl, seed pearl, coarse and flake	12.2	Tr	95.0	—	0.07
	Bread and Rolls						
	Bread						
30	wholemeal	Analytical and calculated values, mixed samples	40.0	2.1	39.7	8.5	1.51
31	brown	Calculated from brown flour	39.0	1.8	42.9	5.1	1.56
32	Hovis	Bulked sample	40.0	2.4	42.7	4.6	1.70
33	white	Analytical and calculated values, mixed samples	39.0	1.8	47.9	2.7	1.40

No	Food	Energy value kcal	Energy value kJ	Protein (see p7) g	Fat g	Carbo-hydrate g	Na (mg)	K (mg)	Ca (mg)	Mg (mg)	P (mg)	Fe (mg)	Cu (mg)	Zn (mg)	S (mg)	Cl (mg)
	Grains, flours and starches contd															
19	**Rice** polished, raw	361	1536	6.5	1.0	86.8	6	110	4	13	100	0.5	0.06	1.3	78	27
20	boiled	123	522	2.2	0.3	29.6	2	38	1	4	34	0.2	0.02	0.4	27	9
21	**Rye** flour (100%)	335	1428	8.2	2.0	75.9	(1)	410	32	92	360	2.7	0.42	(2.8)	—	—
22	**Sago** raw	355	1515	0.2	0.2	94.0	3	5	10	3	29	1.2	0.03	—	1	13
23	**Semolina** raw	350	1489	10.7	1.8	77.5	12	170	18	32	110	1.0	0.15	—	92	71
24	**Soya** flour, full fat	447	1871	36.8	23.5	23.5	1	1660	210	240	600	6.9	—	—	—	—
25	low fat	352	1488	45.3	7.2	28.2	1	2030	240	290	640	9.1	—	—	—	—
26	**Spaghetti** raw	378	1612	13.6	1.0	84.0	5	160	23	35	120	1.2	0.27	(1.0)	97	63
27	boiled	117	499	4.2	0.3	26.0	2	50	7	11	37	0.4	0.08	(0.3)	30	20
28	canned in tomato sauce	59	250	1.7	0.7	12.2	500	130	21	11	30	0.4	0.13	—	—	800
29	**Tapicoa** raw	359	1531	0.4	0.1	95.0	4	20	8	2	30	0.3	0.07	—	4	13
	Bread and rolls															
	Bread															
30	wholemeal	216	918	8.8	2.7	41.8	540	220	23	93	230	2.5	0.27	2.0	81	860
31	brown	223	948	8.9	2.2	44.7	550	210	100	75	190	2.5	0.23	1.6	85	880
32	Hovis	228	968	9.7	2.2	45.1	580	210	150	60	190	4.5	0.18	—	88	790
33	white	233	991	7.8	1.7	49.7	540[a]	100	100	26	97	1.7	0.15	0.8	79	890[a]

[a] Batch loaves contain 610mg Na and 1000mg Cl per 100g

No	Food	Retinol µg	Carotene µg	Vitamin D µg	Thiamin mg	Riboflavin mg	Nicotinic acid mg	Potential nicotinic acid from tryptophan mg Trp ÷ 60	Vitamin C mg	Vitamin E mg
	Grains, flours and starches *contd*									
19	**Rice** polished, raw	0	0	0	0.08	0.03	1.5	1.5	0	0.3 [a]
20	boiled	0	0	0	0.01	0.01	0.3	0.5	0	(0.1)
21	**Rye** flour (100%)	0	0	0	0.40	0.22	1.0	1.6	0	0.8 [b]
22	**Sago** raw	0	0	0	(Tr)	(Tr)	(Tr)	Tr	0	(Tr)
23	**Semolina** raw	0	0	0	(0.10)	(0.02)	(0.7)	2.2	0	(Tr)
24	**Soya** flour, full fat	0	—	0	0.75	0.31	2.0	8.6	0	—
25	low fat	0	—	0	0.90	0.36	2.4	10.6	0	—
26	**Spaghetti** raw	0	0	0	0.14	0.06	2.0	2.8	0	—
27	boiled	0	0	0	0.01	0.01	0.3	0.9	0	—
28	canned in tomato sauce	0	Tr	0	(0.01)	(0.01)	(0.3)	0.4	Tr	—
29	**Tapioca** raw	0	0	0	(Tr)	(Tr)	(Tr)	0.1	0	(Tr)
	Bread and Rolls									
	Bread									
30	wholemeal	0	0	0	0.26	0.06	3.9	1.7	0	(0.2)
31	brown	0	0	0	0.24	0.06	2.9	1.8	0	(Tr)
32	Hovis	0	0	0	0.52	0.10	3.9	2.0	0	—
33	white	0	0	0	0.18	0.03	1.4	1.6	0	Tr

No	Food	Vitamin B_6 mg	Vitamin B_{12} µg	Folic acid Free µg	Folic acid Total µg	Pantothenic acid mg	Biotin µg	Notes
	***Grains, flours and starches** contd*							
19	**Rice** polished, raw	0.30	0	15	29	0.6	3	[a] Also contains 0.3 mg γ-tocopherol per 100 g.
20	boiled	(0.05)	0	3	(6)	(0.2)	(1)	[b] Also contains 0.4 mg β-tocopherol, 0.3 mg γ-tocopherol and 1.5 mg α-tocotrienol per 100 g.
21	**Rye** flour (100%)	0.35	0	31	78	1.0	6	
22	**Sago** raw	(Tr)	0	(Tr)	(Tr)	(Tr)	(Tr)	
23	**Semolina** raw	(0.15)	0	(20)	(25)	(0.3)	(1)	
24	**Soya** flour, full fat	0.57	0	—	—	1.8	—	
25	low fat	0.68	0	—	—	2.1	—	
26	**Spaghetti** raw	0.06	0	4	13	(0.3)	(1)	
27	boiled	(0.01)	0	Tr	(2)	(Tr)	(Tr)	
28	canned in tomato sauce	(0.01)	0	Tr	(2)	(Tr)	(Tr)	
29	**Tapioca** raw	(Tr)	0	(Tr)	(Tr)	(Tr)	(Tr)	
	Bread and Rolls							
	Bread							
30	wholemeal	0.14	0	22	39	0.6	6	
31	brown	0.08	0	21	36	0.3	3	
32	Hovis	0.09	0	—	20	0.3	2	
33	white	0.04	0	6	27	0.3	1	

No	Food	Description and number of samples	Water g	Sugars g	Starch and dextrins g	Dietary fibre g	Total nitrogen g
	***Bread** contd*						
34	white, fried	Mean of 3 pooled samples	4.0	1.7	49.6	(2.2)	1.33
35	toasted	Mean of 3 pooled samples	24.0	2.1	62.8	(2.8)	1.69
36	dried crumbs	Mean of several samples	9.7	2.6	74.9	(3.4)	2.03
37	currant	4 samples from different shops	37.7	13.0	38.8	(1.7)	1.12
38	malt	3 varieties	39.0	18.6	30.8	—	1.46
39	soda	Recipe p334	34.2	3.0	53.3	2.3	1.37
40	**Rolls** brown, crusty	Mean of 3 pooled samples	28.6	2.1	55.1	(5.9)	2.02
41	soft	1 pooled sample	31.0	1.9	46.0	(5.4)	2.05
42	white, crusty	Mean of 5 pooled samples	28.8	2.1	55.1	(3.1)	2.04
43	soft	Mean of 6 pooled samples	28.8	1.9	51.7	(2.9)	1.72
44	starch reduced	5 samples from different shops (Energen)	8.5	1.6	44.1	(2.0)	7.72
45	**Chapatis** made with fat	6 samples	28.5	1.8	46.5	3.7	1.42
46	made without fat	Analysed and calculated values	45.8	1.6	42.1	(3.4)	1.28
	Breakfast cereals						
47	**All-Bran**	6 packets of the same brand (Kellogg's)	2.3	15.4	27.6	26.7	2.40
48	**Cornflakes**	6 packets of the same brand (Kellogg's)	3.0	7.4	77.7	11.0	1.38
49	**Grapenuts**	6 packets of the same brand (General Foods)	3.9	9.5	66.4	7.0	1.85
50	**Muesli**	10 packets, 4 brands	5.8	26.2	40.0	7.4	2.21
51	**Puffed Wheat**	6 packets of the same brand (Quaker)	2.5	1.5	67.0	15.4	2.44
52	**Ready Brek**	6 packets of the same brand (Lyons)	6.3	2.2	67.7	7.6	2.12
53	**Rice Krispies**	6 packets of the same brand (Kellogg's)	3.8	9.0	79.1	4.5	0.99

No	Food	Energy value kcal	Energy value kJ	Protein (see p7) g	Fat g	Carbo-hydrate g	Na mg	K mg	Ca mg	Mg mg	P mg	Fe mg	Cu mg	Zn mg	S mg	Cl mg
	Bread *contd*															
34	white, fried	558	2326	7.6	37.2[a]	51.3	500	100	90	22	79	1.8	0.13	(0.8)	75	820
35	toasted	297	1265	9.6	1.7	64.9	640	130	110	28	100	2.2	0.16	(0.8)	95	1040
36	dried crumbs	354	1508	11.6	1.9	77.5	760	150	130	34	120	2.8	0.20	(0.9)	110	1140
37	currant	250	1063	6.4	3.4	51.8	160	250	90	25	120	2.7	0.09	(0.8)	59	280
38	malt	248	1054	8.3	3.3	49.4	280	380	94	78	250	3.6	0.06	(0.8)	110	530
39	soda	264	1122	8.0	2.3	56.3	410	270	150	20	110	1.7	0.13	0.6	—	480
40	**Rolls** brown, crusty	289	1229	11.5	3.2	57.2	(640)	240	120	86	220	3.2	0.26	1.8	170	(1030)
41	soft	282	1194	11.7	6.4	47.9	(620)	220	120	80	200	2.5	0.24	1.7	190	(990)
42	white, crusty	290	1231	11.6	3.2	57.2	(630)	120	120	30	110	2.1	0.17	0.9	150	(1040)
43	soft	305	1291	9.8	7.3	53.6	(630)	110	120	28	100	1.8	0.16	0.9	130	(1040)
44	starch reduced	384	1631	44.0	4.1	45.7	650	130	47	63	190	4.0	0.52	—	—	(980)
45	**Chapatis** made with fat	336	1415	8.1	12.8	50.2	130	160	66	41	130	2.3	0.20	1.1	—	250
46	made without fat	202	860	7.3	1.0	43.7	120	150	60	37	120	2.1	0.20	1.0	—	230
	Breakfast cereals															
47	**All-bran**	273	1156	15.1	5.7	43.0	1670	1070	74	370	900	12.0	1.2	8.4	—	2440
48	**Cornflakes**	368	1567	8.6	1.6	85.1	1160	99	3	14	47	0.6	0.03	0.3	—	1780
49	**Grapenuts**	355	1510	10.8	3.0	75.9	660	270	37	78	240	5.2	0.25	2.1	—	1080
50	**Muesli**	368	1556	12.9	7.5	66.2	180	600	200	100	380	4.6	0.41	2.2	—	330
51	**Puffed Wheat**	325	1386	14.2	1.3	68.5	4	390	26	140	350	4.6	0.56	2.8	—	50
52	**Ready Brek**	390	1651	12.4	8.7	69.9	23	390	64	120	420	4.9	0.41	2.7	—	66
53	**Rice Krispies**	372	1584	5.9	2.0	88.1	1110	160	7	50	150	0.7	0.12	1.1	—	1700

[a] The fat content depends on the conditions of frying; thin slices pick up more fat than thick ones

No	Food	Retinol µg	Carotene µg	Vitamin D µg	Thiamin mg	Riboflavin mg	Nicotinic acid mg	Potential nicotinic acid from tryptophan mg Trp ÷ 60	Vitamin C mg	Vitamin E mg
	***Bread** contd*									
34	white, fried	0	0	0	—	—	—	1.6	0	—[a]
35	toasted	0	0	0	(0.17)	(0.04)	(1.8)	2.0	0	Tr
36	dried crumbs	0	0	0	(0.20)	(0.04)	(2.1)	2.4	0	Tr
37	currant	0	0	0	(0.18)	(0.03)	(1.4)	1.3	0	Tr
38	malt	0	0	0	—	—	—	1.7	0	—
39	soda	10	10	0.01	(Tr)	0.06	1.3	1.7	0	Tr
40	**Rolls** brown, crusty	0	0	0	0.23	0.12	2.6	2.4	0	(Tr)
41	soft	0	0	0	0.23	0.15	2.7	2.4	0	(Tr)
42	white, crusty	0	0	0	0.23	0.07	1.5	2.4	0	Tr
43	soft	0	0	0	0.25	0.08	1.4	2.0	0	Tr
44	starch reduced	0	0	0	—	—	—	9.0	0	—
45	**Chapatis** made with fat	—	—	—	0.26	0.04	1.7	1.7	0	—
46	made without fat	0	0	0	0.23	0.04	1.5	1.5	0	Tr
	Breakfast cereals									
47	**All-bran**	0	0	0	0.75	2.8	49.0	3.2	0	2.0[c]
48	**Cornflakes**	0	0	0	1.8[b]	1.6[b]	21.0[b]	0.9	0	0.4
49	**Grapenuts**	1330[f]	0	3.5[f]	1.2	1.6	17.0	2.2	0	1.6[d]
50	**Muesli**	0	0	0	0.33	0.27	2.7	3.0	Tr	3.2
51	**Puffed Wheat**	0	0	0	Tr	0.06	5.2	2.9	0	1.7[e]
52	**Ready Brek**	0	0	0	1.5	0.09	9.4	2.8	0	1.2
53	**Rice Krispies**	0	0	0	2.3	1.7	24.0	1.3	0	0.6

No	Food	Vitamin B_6 mg	Vitamin B_{12} µg	Folic acid Free µg	Folic acid Total µg	Pantothenic acid µg	Biotin µg
	***Bread** contd*						
34	white, fried	—	0	—	—	—	—
35	toasted	—	0	—	—	(0.4)	(2)
36	dried crumbs	—	0	—	—	—	—
37	currant	(0.04)	0	(6)	(30)	(0.3)	(1)
38	malt	—	0	—	—	—	—
39	soda	0.09	0	6	9	0.3	1
40	**Rolls** brown, crusty	0.15	0	(21)	(36)	(0.3)	(3)
41	soft	0.14	0	(21)	(36)	(0.3)	(3)
42	white, crusty	0.06	0	(7)	(27)	(0.3)	(1)
43	soft	0.06	0	(7)	(27)	(0.3)	(1)
44	starch reduced	—	0	—	—	—	—
45	**Chapatis** made with fat	(0.21)	0	9	15	(0.3)	(2)
46	made without fat	(0.19)	0	8	14	(0.3)	(2)
	Breakfast cereals						
47	**All-bran**	0.83	0	27	100	—	—
48	**Cornflakes**	0.03	0	6	7	—	—
49	**Grapenuts**	2.81	5[f]	30	41	—	—
50	**Muesli**	0.14	0	11	48	—	—
51	**Puffed Wheat**	0.14	0	16	19	—	—
52	**Ready Brek**	1.5	0	10	53	—	—
53	**Rice Krispies**	0.19	0	6	14	—	—

Notes

[a] The vitamin E content will depend on the fat used for frying.

[b] These cornflakes are fortified. Unfortified cornflakes contain a trace of thiamin, 0.03 mg riboflavin and 0.6 mg nicotinic acid per 100 g.

[c] Also contains 1.6 mg γ-tocopherol per 100 g.

[d] Also contains 0.6 mg γ-tocopherol per 100 g.

[e] Also contains 2.5 mg γ-tocopherol per 100 g.

[f] Data supplied by the manufacturer.

No	Food	Description and number of samples	Water g	Sugars g	Starch and dextrins g	Dietary fibre g	Total nitrogen g
	***Breakfast cereals** contd*						
54	**Shredded Wheat**	6 packets of the same brand (Nabisco)	7.6	0.4	67.5	12.3	1.81
55	**Special K**	6 packets of the same brand (Kellogg's)	2.7	9.6	68.6	5.5	3.15
56	**Sugar Puffs**	6 packets of the same brand (Quaker)	1.8	56.5	28.0	6.1	1.01
57	**Weetabix**	6 packets of the same brand (Weetabix)	3.8	6.1	66.5	12.7	1.96
	Biscuits						
58	**Chocolate** full coated	7 different kinds	2.2	43.4	24.0	3.1	1.00
59	**Cream crackers**	6 packets	4.3	Tr	68.3	(3.0)	1.66
60	**Crispbread** rye	8 packets of the same brand (Ryvita)	6.4	3.2	67.4	11.7	1.61
61	wheat, starch reduced	8 packets of the same brand (Energen)	4.9	7.4	29.5	4.9	7.94
62	**Digestive** plain	3 brands	4.5	16.4	49.6	(5.5)	1.68
63	chocolate	10 packets, 5 plain chocolate, 5 milk chocolate	2.5	28.5	38.0	3.5	1.17
64	**Ginger nuts**	10 packets, 6 brands	3.4	35.8	43.3	2.0	0.98
65	**Home made**	eg Easter, Imperial; basic recipe p 334	8.4	26.8	38.7	1.7	1.10
66	**Matzo**	6 packets, Rakusens, Superfine, tea	6.7	4.2	82.4	3.9	1.85
67	**Oatcakes**	6 packets, 4 brands	5.5	3.1	59.9	4.0	1.71
68	**Sandwich**	10 packets, custard creams and similar types	2.6	30.2	39.0	1.2	0.87
69	**Semi-sweet**	10 packets, Osborne, Rich Tea, Marie	2.5	22.3	52.5	2.3	1.18
70	**Short-sweet**	10 packets, shortcake, Lincoln	2.6	24.1	38.1	1.7	1.08
71	**Shortbread**	Recipe p 334	5.0	17.2	48.3	2.1	1.08
72	**Wafers** filled	9 packets, assorted	2.3	44.7	21.3	1.6	0.82
73	**Water biscuits**	3 brands	4.5	2.3	73.5	(3.2)	1.89

No	Food	Energy value kcal	Energy value kJ	Protein (see p7) g	Fat g	Carbo-hydrate g	Na mg	K mg	Ca mg	Mg mg	P mg	Fe mg	Cu mg	Zn mg	S mg	Cl mg
	***Breakfast cereals** contd*															
54	**Shredded Wheat**	324	1378	10.6	3.0	67.9	8	330	38	130	340	4.2	0.40	2.3	—	53
55	**Special K**	388	1650	18.0[a]	2.5	78.2	880	190	42	52	190	20.0	0.26	1.9	—	1460
56	**Sugar Puffs**	348	1482	5.9	0.8	84.5	9	160	14	55	140	2.1	0.23	1.5	—	41
57	**Weetabix**	340	1444	11.4	3.4	70.3	360	420	33	120	300	7.6	0.54	2.1	—	570
	Biscuits															
58	**Chocolate** full-coated	524	2197	5.7	27.6	67.4	160	230	110	42	130	1.7	0.25	0.8	—	250
59	**Cream crackers**	440	1857	9.5	16.3	68.3	610	120	110	25	110	1.7	—	(0.6)	87	830
60	**Crispbread** rye	321	1367	9.4	2.1	70.6	220	500	50	100	310	3.7	0.38	3.1	—	370
61	wheat, starch reduced	388	1642	45.3	7.6	36.9	610	210	60	61	220	5.4	0.47	2.8	—	980
62	**Digestive** plain	471	1981	9.8	20.5	66.0	440	160	110	32	130	2.0	0.23	(0.6)	72	430
63	chocolate	493	2071	6.8	24.1	66.5	450	210	84	41	130	2.1	0.24	1.0	—	410
64	**Ginger nuts**	456	1923	5.6	15.2	79.1	330	220	130	25	87	4.0	0.16	0.5	—	320
65	**Home made**	469	1971	6.4	22.0	65.5	220	88	82	12	84	1.5	0.11	0.5	—	340
66	**Matzo**	384	1634	10.5	1.9	86.6	17	150	32	20	100	1.5	0.16	0.7	—	80
67	**Oatcakes**	441	1855	10.0	18.3	63.0	1230	340	54	100	420	4.5	0.37	2.3	—	1290
68	**Sandwich**	513	2151	5.0	25.9	69.2	220	120	100	13	82	1.6	0.07	0.5	—	290
69	**Semi-sweet**	457	1925	6.7	16.6	74.8	410	140	120	17	84	2.1	0.08	0.6	—	520
70	**Short-sweet**	469	1966	6.2	23.4	62.2	360	110	87	15	85	1.8	0.11	0.6	—	490
71	**Shortbread**	504	2115	6.2	26.0	65.5	270	91	97	13	75	1.5	0.12	0.5	—	440
72	**Wafers** filled	535	2242	4.7	29.9	66.0	70	160	73	22	83	1.6	0.16	0.6	—	150
73	**Water biscuits**	440	1859	10.8	12.5	75.8	470	140	120	19	87	1.6	0.08	(0.6)	100	680

[a] N × 5.7

No	Food	Retinol µg	Carotene µg	Vitamin D µg	Thiamin mg	Riboflavin mg	Nicotinic acid mg	Potential nicotinic acid from tryptophan mg Trp ÷ 60	Vitamin C mg	Vitamin E mg
	***Breakfast cereals** contd*									
54	**Shredded Wheat**	0	0	0	0.27	0.05	4.5	2.1	0	1.0[a]
55	**Special K**	0	0	0	1.7	1.9	21.0	3.7	0	0.5[b]
56	**Sugar Puffs**	0	0	0	Tr	0.03	2.5	1.2	0	0.2[c]
57	**Weetabix**	0	0	0	1.0	1.5	12.0	2.3	0	1.8[d]
	Biscuits									
58	**Chocolate** full coated	0	Tr	0	0.03	0.13	0.5	1.2	0	1.3[e]
59	**Cream crackers**	0	0	0	(0.13)	(0.08)	(1.5)	1.9	0	—
60	**Crispbread** rye	0	0	0	0.28	0.14	1.1	1.8	0	0.5
61	wheat, starch reduced	0	0	0	0.15	0.10	3.9	9.3	0	0.5
62	**Digestive** plain	0	0	0	(0.13)	(0.08)	(1.5)	2.0	0	—
63	chocolate	0	Tr	0	0.08	0.11	1.3	1.4	0	1.0[f]
64	**Ginger nuts**	0	0	0	0.10	0.03	0.9	1.1	0	1.5
65	**Home made**	240	0	2.18	0.14	0.06	1.0	1.0	0	2.2
66	**Matzo**	0	0	0	0.11	0.03	0.9	2.2	0	Tr
67	**Oatcakes**	0	0	0	0.32	0.09	0.7	2.3	0	2.0[g]
68	**Sandwich**	0	0	0	0.14	0.13	1.1	1.0	0	3.4
69	**Semi-sweet**	0	0	0	0.13	0.08	1.5	1.4	0	1.4
70	**Short-sweet**	0	0	0	0.16	0.04	0.9	1.3	0	1.3
71	**Shortbread**	230	140	0.23	0.15	0.01	1.1	1.3	0	0.6
72	**Wafers** filled	0	0	0	0.09	0.08	0.5	1.0	0	1.9
73	**Water biscuits**	0	0	0	(0.11)	(0.03)	(0.9)	2.2	0	—

No	Food	Vitamin B_6 mg	Vitamin B_{12} µg	Folic acid Free µg	Folic acid Total µg	Panto-thenic acid mg	Biotin µg
	***Breakfast cereals** contd*						
54	**Shredded Wheat**	0.24	0	10	29	—	—
55	**Special K**	0.16	0	11	37	—	—
56	**Sugar puffs**	0.05	0	8	12	—	—
57	**Weetabix**	0.24	0	14	50	—	—
	Biscuits						
58	**Chocolate** full coated	0.04	0	—	—	—	—
59	**Cream crackers**	(0.06)	0	(13)	—	—	—
60	**Crispbread** rye	0.29	0	(17)	(40)	(1.1)	(7)
61	wheat, starch reduced	0.22	0	—	—	—	—
62	**Digestive**, plain	(0.06)	0	(8)	—	—	—
63	chocolate	0.08	0	—	—	—	—
64	**Ginger nuts**	0.07	0	(8)	—	—	—
65	**Home made**	0.07	0	5	7	0.3	4
66	**Matzo**	0.06	0	—	—	—	—
67	**Oatcakes**	0.10	0	8	—	(1.0)	(20)
68	**Sandwich**	0.04	0	—	—	—	—
69	**Semi-sweet**	0.06	0	8	—	—	—
70	**Short-sweet**	0.05	0	8	—	—	—
71	**Shortbread**	0.07	0	5	7	—	—
72	**Wafers** filled	0.03	0	—	—	—	—
73	**Water biscuits**	(0.06)	0	(8)	—	—	—

Notes

[a] Also contains 2.0 mg γ-tocopherol per 100 g.

[b] Also contains 0.5 mg γ-tocopherol per 100 g.

[c] Also contains 1.4 mg γ-tocopherol per 100 g.

[d] Also contains 2.2 mg γ-tocopherol per 100 g.

[e] Also contains 1.3 mg γ-tocopherol per 100 g.

[f] Also contains 1.0 mg γ-tocopherol per 100 g.

[g] Also contains 1.4 mg γ-tocopherol per 100 g.

No	Food	Description and number of samples	Water g	Sugars g	Starch and dextrins g	Dietary fibre g	Total nitrogen g
	Cakes						
74	**Fancy iced cakes**	10 different types	12.7	54.0	14.8	2.4	0.66
75	**Fruit cake** rich	eg Christmas, Dundee; recipe p 334	20.6	46.7	11.6	3.5	0.62
76	rich, iced	Coated with marzipan and royal icing; recipe p 334	17.9	54.2	7.8	3.4	0.71
77	plain	10 cakes, 4 brands	19.5	43.1	14.8	2.8	0.89
78	**Gingerbread**	Recipe p 335	19.0	31.8	30.9	1.3	1.03
79	**Madeira cake**	10 cakes, 4 brands	20.2	36.5	21.9	1.4	0.94
80	**Rock cakes**	Basic recipe, rubbing in method; recipe p 335	15.6	31.3	28.9	2.4	0.92
81	**Sponge cake** with fat	eg Victoria, chocolate, orange; basic recipe, creaming method; recipe p 335	14.9	30.5	22.7	1.0	1.06
82	without fat	Basic recipe, whisking method; recipe p 335	30.0	30.9	22.7	1.0	1.64
83	jam filled	10 cakes, 3 brands; sandwich and Swiss roll	24.5	47.7	16.5	1.2	0.74
	Buns and pastries						
84	**Currant buns**	20 samples from 4 different shops	28.6	14.0	40.5	—	1.30
85	**Doughnuts**	16 samples from 4 different shops	26.4	15.0	33.8	—	1.03
86	**Eclairs**	With chocolate icing and dairy cream filling; recipe p 335	34.8	26.3	11.9	—	0.73
87	**Jam tarts**	Recipe p 335	19.2	37.5	25.3	1.7	0.61
88	**Mince pies**	Recipe p 335	11.5	30.0	31.7	2.9	0.74

No	Food	Energy value		Protein	Fat	Carbo-hydrate	mg									
		kcal	kJ	g	g	g	Na	K	Ca	Mg	P	Fe	Cu	Zn	S	Cl
	Cakes															
74	**Fancy iced cakes**	407	1717	3.8	14.9	68.8	250	170	44	30	120	1.4	0.25	0.7	—	230
75	**Fruit cake** rich	332	1403	3.7	11.0	58.3	170	430	75	26	73	1.8	0.33	—	—	240
76	rich, iced	352	1487	4.1	11.5	62.0	120	360	71	37	86	1.6	0.24	—	—	170
77	plain	354	1490	5.1	12.9	57.9	250	390	60	25	110	1.7	0.25	0.5	—	320
78	**Gingerbread**	373	1573	6.1	12.6	62.7	210	470	210	48	91	3.8	0.20	—	—	410
79	**Madeira cake**	393	1652	5.4	16.9	58.4	380	120	42	12	120	1.1	0.10	0.5	—	500
80	**Rock cakes**	394	1658	5.4	16.3	60.2	480	210	390	16	300	1.4	0.17	0.4	—	260
81	**Sponge cake** with fat	464	1941	6.4	26.5	53.2	350	82	140	10	150	1.4	0.10	0.6	—	400
82	without fat	301	1276	10.0	6.7	53.6	82	120	74	13	160	1.9	0.11	1.1	—	110
83	jam filled	302	1280	4.2	4.9	64.2	420	140	44	14	220	1.6	0.20	0.5	—	260
	Buns and pastries															
84	**Currant buns**	302	1279	7.4	7.6	54.5	100	180	90	22	65	2.5	0.03	—	73	200
85	**Doughnuts**	349	1467	6.0	15.8	48.8	60	110	70	16	55	1.9	0.11	—	56	89
86	**Eclairs**	376	1569	4.1	24.0	38.2	160	92	48	16	68	1.0	0.15	0.4	—	250
87	**Jam tarts**	384	1616	3.5	14.9	62.8	230	110	62	13	47	1.6	0.18	—	—	360
88	**Mince pies**	435	1826	4.3	20.7	61.7	340	150	76	14	55	1.7	0.17	0.4	—	540

No	Food	Retinol µg	Carotene µg	Vitamin D µg	Thiamin mg	Riboflavin mg	Nicotinic acid mg	Potential nicotinic acid from tryptophan mg Trp ÷ 60	Vitamin C mg	Vitamin E mg
	Cakes									
74	**Fancy iced cakes**	0	—	0	0.01	0.04	0.2	0.8	0	—
75	**Fruit cake** rich	120	10	1.14	0.08	0.08	0.5	0.8	0	1.4
76	rich, iced	80	7	0.77	0.12	0.13	0.5	0.4	Tr	2.4
77	plain	—	—	—	0.08	0.07	0.6	1.0	0	—
78	**Gingerbread**	130	2	1.24	(Tr)	0.07	0.7	1.3	Tr	1.2
79	**Madeira cake**	—	—	—	0.06	0.11	0.5	1.1	0	—
80	**Rock cakes**	180	7	1.59	(Tr)	0.07	0.8	1.1	Tr	1.6
81	**Sponge cake** with fat	300	0	2.76	(Tr)	0.12	0.6	1.6	0	2.7
82	without fat	80	0	1.02	0.11	0.24	0.6	2.7	0	0.9
83	jam filled	—	—	—	0.04	0.07	0.4	0.9	0	—
	Buns and pastries									
84	**Currant buns**	0	—	0	(0.18)	(0.03)	(1.4)	1.5	0	—
85	**Doughnuts**	0	—	0	—	—	—	1.2	0	—
86	**Eclairs**	200	70	0.91	0.05	0.09	0.3	1.0	Tr	1.2
87	**Jam tarts**	70	Tr	0.64	0.08	0.01	0.6	0.7	4	0.6
88	**Mince pies**	90	5	0.80	0.11	0.02	0.9	0.8	0	0.8

No	Food	Vitamin B_6 mg	Vitamin B_{12} µg	Folic acid Free µg	Folic acid Total µg	Pantothenic acid mg	Biotin µg	Notes
	Cakes							
74	**Fancy iced cakes**	—	0	—	—	—	—	
75	**Fruit cake** rich	0.13	Tr	4	4	0.2	4	
76	rich, iced	0.10	Tr	5	10	0.2	3	
77	plain	—	0	—	—	—	—	
78	**Gingerbread**	0.06	Tr	5	6	0.3	4	
79	**Madeira cake**	—	0	—	—	—	—	
80	**Rock cakes**	0.09	Tr	4	7	0.3	3	
81	**Sponge cake** with fat	0.06	Tr	6	7	0.5	8	
82	without fat	0.08	1	10	10	0.8	15	
83	jam filled	—	0	—	—	—	—	
	Buns and pastries							
84	**Currant buns**	—	0	—	—	—	—	
85	**Doughnuts**	—	0	—	—	—	—	
86	**Eclairs**	0.04	Tr	5	5	0.3	4	
87	**Jam tarts**	0.04	0	2	4	0.1	1	
88	**Mince pies**	0.08	0	3	5	0.1	1	

No	Food	Description and number of samples	Water g	Sugars g	Starch and dextrins g	Dietary fibre g	Total nitrogen g
	***Buns and pastries** contd*						
89	**Pastry, choux** raw	Recipe p 336	61.2	0.4	19.5	0.8	0.92
90	cooked	Recipe p 336	40.4	0.7	30.0	1.3	1.42
91	**Pastry, flaky** raw	Recipe p 336	29.6	0.8	35.0	1.5	0.77
92	cooked	Recipe p 336	7.3	1.1	46.3	2.0	1.02
93	**Pastry, shortcrust** raw	Recipe p 336	19.5	1.0	47.2	2.0	1.03
94	cooked	Recipe p 336	6.9	1.2	54.6	2.4	1.20
95	**Scones**	Recipe p 336	21.5	6.1	49.8	2.1	1.30
96	**Scotch pancakes**	Drop scones; recipe p 336	40.7	8.1	32.5	1.4	1.13
	Puddings						
97	**Apple crumble**	Recipe p 336	52.3	23.8	13.2	2.5	0.32
98	**Bread and butter pudding**	Recipe p 336	67.2	12.0	5.2	0.6	0.99
99	**Cheesecake**	Recipe p 337	34.7	13.9	10.1	0.9	0.70
100	**Christmas pudding**	Recipe p 337	39.4	39.1	8.5	2.0	0.88
101	**Custard,** egg	Baked or sauce; recipe p 337	76.8	11.0	0	0	0.91
102	made with powder	Recipe p 337	74.7	11.5	5.3	—	0.60
103	**Custard tart**	Recipe p 337	48.0	6.0	23.6	1.0	0.97
104	**Dumpling**	Recipe p 337	60.2	0.5	24.7	1.0	0.51

No	Food	Energy value kcal	Energy value kJ	Protein g	Fat g	Carbohydrate g	Na mg	K mg	Ca mg	Mg mg	P mg	Fe mg	Cu mg	Zn mg	S mg	Cl mg
	***Buns and pastries** contd*															
89	**Pastry, choux** raw	214	893	5.5	13.0	19.9	260	70	51	9	84	1.1	0.07	0.5	—	390
90	cooked	330	1379	8.5	20.1	30.7	390	110	78	14	130	1.8	0.11	0.8	—	600
91	**Pastry, flaky** raw	427	1780	4.4	30.6	35.8	350	67	68	11	52	1.1	0.09	0.3	—	560
92	cooked	565	2356	5.8	40.5	47.4	470	88	90	15	68	1.5	0.12	0.4	—	740
93	**Pastry, shortcrust** raw	455	1900	5.9	27.8	48.2	410	85	92	15	69	1.5	0.11	0.4	—	660
94	cooked	527	2202	6.9	32.2	55.8	480	99	110	17	79	1.8	0.13	0.5	—	760
95	**Scones**	371	1562	7.5	14.6	55.9	800	140	620	19	470	1.5	0.12	0.6	—	480
96	**Scotch pancakes**	283	1193	6.7	11.6	40.6	400	320	120	16	110	1.3	0.10	0.6	—	510
	Puddings															
97	**Apple crumble**	208	878	1.8	6.9	37.0	68	100	28	5	30	0.6	0.09	0.2	—	110
98	**Bread and butter pudding**	159	668	6.1	7.8	17.2	150	200	130	18	140	0.7	0.08	0.7	69	250
99	**Cheesecake**	421	1747	4.2	34.9	24.0	260	120	67	11	87	0.7	0.10	0.5	—	360
100	**Christmas pudding**	304	1279	5.2	11.6	47.6	240	390	87	34	93	1.9	0.26	0.5	58	380
101	**Custard** egg	118	497	5.8	6.0	11.0	78	170	130	14	140	0.5	0.04	0.7	66	130
102	made with powder	118	496	3.8	4.4	16.8	76	170	140	14	110	0.1	0.03	0.4	35	140
103	**Custard tart**	287	1199	5.9	16.9	29.6	250	130	110	15	100	1.0	0.08	0.5	—	390
104	**Dumpling**	211	885	2.9	11.7	25.2	400	43	160	8	120	0.8	0.06	0.2	—	450

No	Food	Retinol µg	Carotene µg	Vitamin D µg	Thiamin mg	Riboflavin mg	Nicotinic acid mg	Potential nicotinic acid from trytophan mg Trp ÷ 60	Vitamin C mg	Vitamin E mg
	***Buns and pastries** contd*									
89	**Pastry, choux** raw	150	0	1.42	0.10	0.12	0.5	1.4	0	1.4
90	cooked	230	0	2.19	0.12	0.16	0.8	2.1	0	2.1
91	**Pastry, flaky** raw	150	0	1.33	0.15	0.01	0.9	0.9	Tr	1.4
92	cooked	200	0	1.76	0.14	0.01	1.1	1.2	Tr	1.8
93	**Pastry, shortcrust** raw	130	0	1.19	0.20	0.01	1.2	1.2	0	1.2
94	cooked	160	0	1.38	0.17	0.01	1.3	1.4	0	1.4
95	**Scones**	150	5	1.23	(Tr)	0.08	1.2	1.5	Tr	1.3
96	**Scotch pancakes**	150	7	1.26	(Tr)	0.12	0.9	1.5	Tr	1.3
	Puddings									
97	**Apple crumble**	70	25	0.66	0.07	0.02	0.4	0.3	6	0.8
98	**Bread and butter pudding**	70	30	0.53	0.05	0.20	0.3	1.5	Tr	0.4
99	**Cheesecake**	280	110	0.94	0.05	0.10	0.3	1.0	2	—
100	**Christmas pudding**	20	20	0.21	0.08	0.10	0.7	1.1	0	1.1
101	**Custard** egg	60	20	0.37	0.05	0.25	0.1	1.5	Tr	0.3
102	made with powder	40	20	0.03	0.05	0.21	0.1	0.9	Tr	0.1
103	**Custard tart**	100	10	0.78	0.10	0.11	0.6	1.4	Tr	0.8
104	**Dumpling**	10	10	0	(Tr)	0.01	0.6	0.6	0	0.2

No	Food	Vitamin B_6 mg	Vitamin B_{12} µg	Folic acid Free µg	Folic acid Total µg	Pantothenic acid mg	Biotin µg	Notes
	***Buns and pastries** contd*							
89	**Pastry, choux** raw	0.06	Tr	10	12	0.5	6	
90	cooked	0.08	Tr	8	9	0.6	10	
91	**Pastry, flaky** raw	0.07	0	6	10	0.1	Tr	
92	cooked	0.07	0	5	7	0.1	Tr	
93	**Pastry, shortcrust** raw	0.09	0	8	13	0.2	1	
94	cooked	0.08	0	5	8	0.2	1	
95	**Scones**	0.08	Tr	5	8	0.2	2	
96	**Scotch pancakes**	0.07	Tr	5	7	0.5	4	
	Puddings							
97	**Apple crumble**	0.04	0	2	4	0.1	Tr	
98	**Bread and butter pudding**	0.05	Tr	5	6	0.5	7	
99	**Cheesecake**	0.02	Tr	3	3	—	—	
100	**Christmas pudding**	0.11	Tr	4	8	0.3	4	
101	**Custard** egg	0.05	Tr	7	8	0.6	7	
102	made with powder	0.05	Tr	4	5	0.4	2	
103	**Custard tart**	0.06	Tr	5	6	0.4	4	
104	**Dumpling**	0.03	0	2	3	0.1	Tr	

No	Food	Description and number of samples	Water g	Sugars g	Starch and dextrins g	Dietary fibre g	Total nitrogen g
	Puddings *contd*						
105	**Fruit pie** individual, with pastry top and bottom	10 pies, as purchased, 3 brands; apple, blackcurrant, blackberry, apricot	22.9	30.9	25.8	2.6	0.75
106	**Fruit pie** with pastry top	eg Apple, gooseberry, plum, rhubarb; recipe p 338	60.4	14.7	12.9	2.2	0.34
107	**Ice cream** dairy	10 family sized packets, 2 brands	64.4	22.6	2.2	—	0.58
108	non-dairy	10 family sized packets, 2 brands	65.7	19.7	1.0	—	0.52
109	**Jelly** packet, cubes	8 samples, assorted flavours	29.9	62.5	0	—	1.10
110	made with water	Recipe p 338	84.0	14.2	0	—	0.25
111	made with milk	Recipe p 338	78.7	16.0	0	—	0.47
112	**Lemon meringue pie**	Recipe p 338	35.0	24.8	21.6	0.7	0.75
113	**Meringues**	Without cream; recipe p 338	2.0	95.6	0	0	0.85
114	**Milk pudding**	eg Rice, sago, semolina, tapioca; recipe p 338	71.8	10.9	9.5	—	0.65
115	canned, rice	10 cans, 4 brands	77.6	8.9	5.8	—	0.53
116	**Pancakes**	Recipe p 338	43.4	16.6	19.6	0.9	1.00
117	**Queen of puddings**	Recipe p 339	54.8	28.9	4.6	0.3	0.77
118	**Sponge pudding** steamed	Recipe p 339	28.5	18.9	27.1	1.2	0.99
119	**Suet pudding** steamed	Recipe p 339	36.6	14.0	26.6	1.1	0.77
120	**Treacle tart**	Recipe p 339	21.0	33.6	27.7	1.2	0.65
121	**Trifle**	Recipe p 339	64.5	18.7	5.6	—	0.57
122	**Yorkshire pudding**	Recipe p 339	56.4	3.8	22.0	1.0	1.12

No	Food	Energy value		Protein	Fat	Carbo-hydrate	mg									
		kcal	kJ	g	g	g	Na	K	Ca	Mg	P	Fe	Cu	Zn	S	Cl
	***Puddings** contd*															
105	**Fruit pie** individual, with pastry top and bottom	369	1554	4.3	15.5	56.7	210	120	51	12	64	1.2	0.10	0.5	—	260
106	**Fruit pie** with pastry top	180	756	2.0	7.6	27.6	110	170	48	9	32	0.6	0.10	0.1	—	190
107	**Ice cream** dairy	167	704	3.7	6.6	24.8	80	180	140	13	100	0.2	(0.03)	0.4	—	140
108	non-dairy	165	691	3.3	8.2	20.7	70	150	120	11	90	0.3	(0.03)	0.4	—	140
109	**Jelly** packet, cubes	259	1104	6.1	0	62.5	25	25	32	4	7	1.7	0.16	—	37	30
110	made with water	59	251	1.4	0	14.2	6	6	7	1	2	0.4	0.04	—	8	7
111	made with milk	86	363	2.8	1.6	16.0	27	66	59	6	42	0.4	0.04	—	21	48
112	**Lemon meringue pie**	323	1359	4.5	14.6	46.4	200	81	46	9	70	1.0	0.09	0.4	—	310
113	**Meringues**	380	1620	5.3	0	95.6	110	91	4	6	20	0.1	0.05	Tr	110	100
114	**Milk pudding**	131	552	4.1	4.2	20.4	55	160	130	14	110	0.1	0.03	—	37	110
115	canned, rice	91	386	3.4	2.5	14.7	50	140	93	11	80	0.2	0.03	0.4	—	95
116	**Pancakes**	307	1286	6.1	16.3	36.2	50	140	120	14	120	0.9	0.07	0.6	—	94
117	**Queen of puddings**	216	910	4.8	7.9	33.5	150	110	80	12	100	0.7	0.07	0.5	57	230
118	**Sponge pudding** steamed	344	1443	5.9	16.4	46.0	310	88	210	11	190	1.2	0.09	0.5	—	260
119	**Suet pudding** steamed	333	1394	4.4	18.1	40.6	470	91	240	14	180	0.9	0.08	0.4	—	510
120	**Treacle tart**	371	1563	3.8	14.0	61.3	360	150	65	14	51	1.5	0.11	—	—	420
121	**Trifle**	160	674	3.5	6.1	24.3	50	150	82	14	87	0.7	0.09	0.4	—	88
122	**Yorkshire pudding**	215	902	6.8	10.1	25.8	600	160	130	20	130	1.0	0.08	0.7	—	940

No	Food	Retinol µg	Carotene µg	Vitamin D µg	Thiamin mg	Riboflavin mg	Nicotinic acid mg	Potential nicotinic acid from tryptophan mg Trp ÷ 60	Vitamin C mg	Vitamin E mg
	***Puddings** contd*									
105	**Fruit pie**, individual, with pastry top and bottom	0	(Tr)	0	0.05	0.02	0.4	0.9	(Tr)	—
106	**Fruit pie** with pastry top	40	75	0.32	0.05	0.02	0.5	0.4	9 *(2–20)*	0.5
107	**Ice cream** dairy	14	0	Tr	0.04	0.18	0.1	0.9	0	0.4
108	non-dairy	—	0	0	0.04	0.15	0.1	0.8	0	1.2
109	**Jelly** packet, cubes	0	0	0	0	0	0	0	0	0
110	made with water	0	0	0	0	0	0	0	0	0
111	made with milk	13	8	0.01	0.02	0.07	Tr	0.3	Tr	Tr
112	**Lemon meringue pie**	100	0	0.98	0.07	0.08	0.5	1.1	5	1.0
113	**Meringues**	0	0	0	0	0.25	0.1	1.6	0	0
114	**Milk pudding**	30	20	0.02	0.04	0.14	0.1	1.0	(Tr)	0.1
115	canned, rice	(30)	(20)	(0.02)	0.03	0.14	0.2	0.7	0	(0.1)
116	**Pancakes**	40	10	0.23	0.13	0.19	0.6	1.4	Tr	0.3
117	**Queen of puddings**	80	30	0.38	0.04	0.15	0.2	1.2	Tr	0.4
118	**Sponge pudding** steamed	180	Tr	1.65	(Tr)	0.09	0.7	1.4	Tr	1.6
119	**Suet pudding** steamed	20	20	0.01	(Tr)	0.06	0.6	0.9	Tr	Tr
120	**Treacle tart**	70	0	0.60	0.09	0.01	0.7	0.6	0	0.6
121	**Trifle**	50	60	0.17	0.05	0.14	0.2	0.8	1	0.3
122	**Yorkshire pudding**	40	10	0.26	0.11	0.18	0.6	1.6	Tr	0.3

No	Food	Vitamin B_6 mg	Vitamin B_{12} µg	Folic acid Free µg	Folic acid Total µg	Panto-thenic acid mg	Biotin µg	Notes
	***Puddings** contd*							
105	**Fruit pie** individual, with pastry top and bottom	—	0	—	—	—	—	
106	**Fruit pie** with pastry top	0.03	0	3	4	0.1	Tr	
107	**Ice cream** dairy	0.02	Tr	2	2	—	—	
108	non-dairy	0.02	Tr	2	2	—	—	
109	**Jelly** packet, cubes	0	0	0	0	0	0	
110	made with water	0	0	0	0	0	0	
111	made with milk	0.02	Tr	2	2	0.2	1	
112	**Lemon meringue pie**	0.05	Tr	5	5	0.3	5	
113	**Meringues**	Tr	Tr	Tr	Tr	0.2	Tr	
114	**Milk puddings**	0.05	Tr	3	4	0.3	2	
115	canned, rice	0.02	Tr	—	—	—	—	
116	**Pancakes**	0.08	Tr	5	6	0.5	5	
117	**Queen of puddings**	0.03	Tr	3	5	0.4	6	
118	**Sponge pudding** steamed	0.05	Tr	5	7	0.3	5	
119	**Suet pudding** steamed	0.04	Tr	3	6	0.2	2	
120	**Treacle tart**	0.04	0	3	5	0.1	1	
121	**Trifle**	0.06	Tr	5	6	0.4	3	
122	**Yorkshire pudding**	0.07	Tr	5	6	0.4	5	

No	Food	Description and number of samples	Water g	Sugars g	Starch and dextrins g	Dietary fibre g	Total nitrogen g

No	Food	Energy value		Protein	Fat	Carbo-hydrate	mg									
		kcal	kJ	g	g	g	Na	K	Ca	Mg	P	Fe	Cu	Zn	S	Cl

No	Food	Retinol µg	Carotene µg	Vitamin D µg	Thiamin mg	Riboflavin mg	Nicotinic acid mg	Potential nicotinic acid from tryptophan mg Trp ÷ 60	Vitamin C mg	Vitamin E mg

No	Food	Vitamin B_6 mg	Vitamin B_{12} µg	Folic acid Free µg	Folic acid Total µg	Panto-thenic acid mg	Biotin µg	Notes

No	Food	Description and number of samples	Water g	Lactose g	Other sugars g	Total nitrogen g
	Milk, cows'					
124 125	fresh, whole	Data from Milk Marketing Board and literature sources	87.6	4.7	0	0.52
127 128	fresh, whole, Channel Islands	Data from Milk Marketing Board and literature sources	86.3	4.7	0	0.56
129	sterilised	As raw milk with calculated vitamin losses	87.6	4.7	0	0.52
130	longlife (UHT treated)	As raw milk with calculated vitamin losses	87.6	4.7	0	0.52
131	fresh, skimmed	Calculated on the basis of 0.1% fat content	90.9	5.0	0	0.53
132	condensed, whole, sweetened	8 cans of the same brand	25.8	10.2	45.3	1.30
133	condensed, skimmed, sweetened	2 samples; vitamins calculated from condensed, whole, sweetened milk	27.0	(12.0)	(48.0)	1.55
134	evaporated, whole, unsweetened	10 cans, 4 different brands	68.6	11.3	0	1.35
135	dried, whole	Mixed sample, 3 different types	2.9	39.4	0	4.12
136	dried, skimmed	10 tins, 4 different brands	4.1	52.8	0	5.70
137	**Milk, goats'**	Literature sources	87.0	4.6	0	0.52
138	**Milk, human** mature	Mixed sample from 96 mothers about 1 month *post partum*	87.1	7.2	0	0.20
139	transitional	Mixed sample from 15 mothers on 10th day *post partum*	90.2	6.9	0	0.31
140	**Butter** salted	Analytical (6 samples) and literature sources	15.4	Tr	0	0.07

No	Food	Energy value		Protein (N × 6.38)	Fat	Carbo-hydrate	mg									
		kcal	kJ	g	g	g	Na	K	Ca	Mg	P	Fe	Cu	Zn	S	Cl
	Milk, cows'															
124 125	fresh, whole	65	272	3.3	3.8	4.7	50 *(35–90)*	150 *(110–170)*	120 *(110–130)*	12 *(9–14)*	95 *(90–100)*	0.05 *(0.03–0.06)*	0.02 *(0.01–0.03)*	0.35 *(0.2–0.6)*	30 —	95 *(90–110)*
127 128	fresh, whole, Channel Islands	76	316	3.6	4.8	4.7	50	140	120	12	95	0.05	0.02	0.35	30	100
129	sterilised	65	274	3.3	3.8	4.7	50	140	120	12	95	0.05	0.02	0.35	30	100
130	longlife (UHT treated)	65	274	3.3	3.8	4.7	50	140	120	12	95	0.05	0.02	0.35	30	100
131	fresh, skimmed	33	142	3.4	0.1	5.0	52	150	130	12	100	0.05	0.02	0.36	31	100
132	condensed, whole, sweetened	322	1362	8.3	9.0	55.5	130	390	280	27	220	0.20	(0.04)	1.0	81	(260)
133	condensed, skimmed, sweetened	267	1139	9.9	0.3	60.0	180	500	380	38	270	0.29	0.03	1.2	94	310
134	evaporated, whole, unsweetened	158	660	8.6	9.0	11.3	180	390	280	28	250	0.20	(0.04)	1.1	84	(350)
135	dried, whole	490	2051	26.3	26.3	39.4	440	1270	1020	84	740	0.40[a]	0.14	3.2	240	810
136	dried, skimmed	355	1512	36.4	1.3	52.8	550	1650	1190	117	950	0.40	0.20	4.1	320	(1100)
137	**Milk, goats'**	71	296	3.3	4.5	4.6	40	180	130	20	110	0.04	0.05	0.30	—	130
138	**Milk, human** mature	69	289	1.3	4.1	7.2	14	58	34	3	14	0.07	0.04	0.28	—	42
139	transitional	67	281	2.0	3.7	6.9	48	68	25	2	16	0.07	0.04	—	—	86
140	**Butter** salted	740	3041	0.4	82.0	Tr	870[b]	15	15	2	24	0.16	0.03	0.15	9	1340[b]

[a] Proprietary brands for infant feeding are usually fortified to higher levels

[b] The added salt content of butter may vary between 0 and 3 g per 100 g; this is an average figure for salted butter. Unsalted butter contains 7 mg Na and 10 mg Cl per 100 g

No	Food	Retinol µg	Carotene µg	Vitamin D µg	Thiamin mg	Riboflavin mg	Nicotinic acid mg	Potential nicotinic acid from tryptophan mg Trp ÷ 60	Vitamin C mg	Vitamin E mg
	Milk, cows'									
124	fresh, whole [a], summer	35	22	0.030	0.04	0.19 [b]	0.08	0.78	1.5 [c]	0.10
125	winter	26	13	0.013						0.07
					(0.03–0.06)	*(0.15–0.23)*	*(0.06–0.13)*			
127	fresh, whole, Channel Islands [a], summer	38	61	0.038	0.04	0.19 [b]	0.08	0.84	1.5 [c]	0.12
128	winter	28	35	0.018						0.09
129	sterilised	31	18	0.022	0.03	0.19 [b]	0.08	0.78	0.8 [d]	0.09
130	longlife (UHT treated)	31	18	0.022	0.04	0.19	0.08	0.78	1.5 [d]	0.09
131	fresh, skimmed [a]	Tr	Tr	Tr	0.04	0.20	0.08	0.80	1.6 [c]	Tr
132	condensed, whole, sweetened	99	49	0.088	0.08	0.48	0.22	1.95	2.0	0.42
133	condensed, skimmed, sweetened	Tr	Tr	Tr	0.10	0.58	0.26	2.33	2.4	Tr
134	evaporated, whole, unsweetened	84	48	0.088 [e]	0.06	0.51	0.28	2.03	1.0	0.56
135	dried, whole	290 [f]	170	0.24 [f]	0.33	1.1	0.60	6.18	10.0 [f]	0.61
136	dried, skimmed	Tr	Tr	0	0.42	1.6	1.2	8.55	6.0	Tr
137	**Milk, goats'**	40	0	0.060	0.04	0.15	0.19	0.78	1.5	—
138	**Milk, human** mature	60	0	0.025	0.02	0.03	0.22	0.47	3.7	0.34
139	transitional	—	—	—	—	—	—	—	—	—
140	**Butter** salted	750	470	0.76	Tr	Tr	Tr	0.11	Tr	2.0
		(520–970)	*(350–650)*	*(0.63–1.0)*						

No	Food	Vitamin B_6 mg	Vitamin B_{12} µg	Folic acid Free µg	Folic acid Total µg	Panto thenic acid mg	Biotin µg
	Milk, cows'						
124 125	fresh, whole[a]	0.04	0.3	4	5	0.35	2.0
127 128	fresh, whole, Channel Islands[a]	0.04	0.3	4	5	0.35	2.0
129	sterilised	0.03	0.2	3[d]	4[d]	0.35	2.0
130	longlife (UHT treated)[b]	0.04[g]	0.2[h]	4[d]	5[d]	0.35	2.0
131	fresh, skimmed[a]	0.04	0.3	4	5	0.36	2.0
132	condensed, whole, sweetened	0.02	0.5	4	8	0.85	3.0
133	condensed, skimmed, sweetened	0.02	0.5	5	10	1.0	3.6
134	evaporated, whole, unsweetened	0.04	Tr	4	7	0.85	3.0
135	dried, whole	0.23	2.0	32	40	2.7	10
136	dried, skimmed	0.25	3.0	14	21	3.5	16
137	**Milk, goats'**	0.04	Tr	1	(1)	0.34	2.0
138	**Milk, human** mature	0.01	Tr	3	5	0.25	0.7
139	transitional	—	—	—	—	—	—
140	**Butter** salted	Tr	Tr	Tr	Tr	Tr	Tr

Notes

[a] The small losses due to pasteurisation (see p 20) make no difference to the average composition of milk, except for vitamin C (see note c). These values are applicable to raw, pasteurised and homogenised milk.

[b] Milk that has not been exposed to light. There is a loss on exposure to sunlight of 10% per hour.

[c] As delivered to the home. This falls to 1.0 mg after 12 hours and 0.5 mg after 24 hours. Raw milk contains 2.0 mg.

[d] There may be a total loss on storage.

[e] This is an unfortified value. Most brands are fortified to a level of 2.8 µg.

[f] Proprietary brands for infant feeding are usually fortified to higher levels.

[g] There is a 35% loss on storage.

[h] There is a 20% loss on storage.

Boiled milk

For losses on boiling milk see p 20.

No	Food	Description and number of samples	Water g	Lactose g	Other sugars g	Total nitrogen g
	Cream					
142 143	single	3 samples; vitamins calculated	71.9	3.2	0	0.38
145 146	double	3 samples; vitamins calculated	48.6	2.0	0	0.24
148 149	whipping	Calculated on the basis of 35% fat content	61.5	2.5	0	0.30
150	sterilised, canned	10 cans, 6 brands; vitamins calculated	69.8	2.7	0	0.40
	Cheese					
151	Camembert type	Soft ripe cheese, eg Camembert, Brie	47.5	Tr	0	3.58
152	Cheddar type	Hard cheese, eg Cheddar, Cheshire, Gruyère, Emmental	37.0	Tr	0	4.08
153	Danish Blue type	Blue veined cheese, eg Danish Blue, Roquefort	40.5	Tr	0	3.61
154	Edam type	Semi-hard cheese, eg Edam, Gouda, St Paulin	43.7	Tr	0	3.82
155	Parmesan	3 samples; vitamins from literature	28.0	Tr	0	5.50
156	Stilton	3 samples; vitamins from literature	28.2	Tr	0	4.02
157	cottage cheese	12 samples, 3 brands; contains added cream	78.8	1.4	0	2.14
158	cream cheese	3 samples	45.5	Tr	0	0.49
159	processed cheese	10 samples, 4 brands	43.8	Tr	0	3.37
160	cheese spread	6 samples; vitamins calculated	51.0	0.9	0	2.87
	Yogurt low fat					
161	natural	10 samples, 2 brands	85.7	4.6	1.6[a]	0.78
162	flavoured	10 samples; mixed banana, raspberry, strawberry	79.0	4.8	9.2[b]	0.79
163	fruit	30 samples, 2 brands; mixed strawberry, raspberry, orange, blackcurrant, pineapple	74.9	3.3	14.6[b]	0.76
164	hazelnut	6 samples, 2 brands	73.4	3.2	13.3[b]	0.82

[a] Galactose [b] Mainly sucrose

No	Food	Energy value kcal	Energy value kJ	Protein (N × 6.38) g	Fat g	Carbo-hydrate g	Na mg	K mg	Ca mg	Mg mg	P mg	Fe mg	Cu mg	Zn mg	S mg	Cl mg
	Cream															
142 143	single	212	876	2.4	21.2	3.2	42	120	79	6	44	0.31	0.20	0.26)	—	72
145 146	double	447	1841	1.5	48.2	2.0	27	79	50	4	21	0.20	0.13	(0.17)	—	46
148 149	whipping	332	1367	1.9	35.0	2.5	34	100	63	5	27	0.25	0.16	(0.21)	—	58
150	sterilised, canned	230	950	2.6	23.3	2.7	56	(120)	(80)	(6)	(44)	(0.30)	(0.20)	(0.26)	—	140
	Cheese															
151	Camembert type	300	1246	22.8	23.2	Tr	1410	110	380	17	290	0.76	0.08	3.0	—	2320
152	Cheddar type	406	1682	26.0	33.5	Tr	610	120	800	25	520	0.40	0.03	4.0	230	1060
153	Danish Blue type	355	1471	23.0	29.2	Tr	1420	190	580	20	430	0.17	0.09	—	—	2390
154	Edam type	304	1262	24.4	22.9	Tr	980	160	740	28	520	0.21	0.03	4.0	—	1640
155	Parmesan	408	1696	35.1	29.7	Tr	760	150	1220	50	770	0.37	—	4.0	250	1110
156	Stilton	462	1915	25.6	40.0	Tr	1150	160	360	27	300	0.46	0.03	—	230	1720
157	cottage cheese	96	402	13.6	4.0[a]	1.4	450	54	60	6	140	0.10	0.02	0.47	—	670
158	cream cheese	439	1807	3.1	47.4	Tr	300	160	98	10	100	0.12	(0.04)	0.48	—	480
159	processed cheese	311	1291	21.5	25.0	Tr	1360	82	700	24	490	0.50	0.50	3.2	—	1020
160	cheese spread	283	1173	18.3	22.9	0.9	1170	150	510	25	440	0.69	0.09	—	—	760
	Yogurt low fat															
161	natural	52	216	5.0	1.0	6.2	76	240	180	17	140	0.09	0.04	0.60	—	180
162	flavoured	81	342	5.0	0.9	14.0	64	220	170	17	140	0.16	0.10	0.64	—	160
163	fruit	95	405	4.8	1.0	17.9	64	220	160	17	140	0.24	0.07	0.63	—	150
164	hazelnut	106	449	5.2	2.6	16.5	70	240	180	20	140	0.23	0.09	0.69	—	160

[a] Cottage cheese made without added cream contains about 0.4 g fat per 100 g

No	Food	Retinol µg	Carotene µg	Vitamin D µg	Thiamin mg	Riboflavin mg	Nicotinic acid mg	Potential nicotinic acid from tryptophan mg Trp ÷ 60	Vitamin C mg	Vitamin E mg
	Cream									
142	single, summer	200	125	0.165	0.03	0.12	0.07	0.57	1.2	0.5
143	winter	145	70	0.081						0.4
145	double, summer	450	280	0.376	0.02	0.08	0.04	0.36	0.8	1.2
146	winter	330	160	0.183						0.9
148	whipping, summer	325	205	0.273	0.02	0.09	0.05	0.45	0.9	0.8
149	winter	240	115	0.133						0.7
150	sterilised, canned	190	110	0.135	0.01	0.10	0.06	0.60	Tr	0.5
	Cheese									
151	Camembert type	215	135	0.181	0.05 [a]	0.60	0.80	5.37	0	(0.6)
						(0.30–0.90)	*(0.05–2.0)*			
152	Cheddar type	310	205	0.261	0.04	0.50	0.10	6.12	0	0.8
					(0.02–0.08)	*(0.30–0.80)*	*(0.01–0.20)*			
153	Danish Blue type	270	170	0.228	0.03	0.60	0.90	5.42	0	(0.7)
						(0.40–0.80)	*(0.1–2.3)*			
154	Edam type	215	135	0.179	0.04	0.40	0.06	5.73	0	(0.8)
							(0.02–0.19)			
155	Parmesan	325	195	0.274	0.02	0.50	0.30	8.25	0	0.9
156	Stilton	370	230	0.312	0.07	0.30	—	6.03	0	(1.0)
157	cottage cheese	32	18	0.023	0.02	0.19	0.08	3.21	0	—
158	cream cheese	385	220	0.275	(0.02)	(0.14)	(0.08)	0.74	0	1.0
159	processed cheese	240	120	0.145	0.02	0.29	0.07	5.06	0	—
160	cheese spread	180	105	0.133	0.02	0.24	0.06	4.31	0	—
	Yogurt low fat									
161	natural	8 [c]	5	Tr [c]	0.05	0.26	0.12	1.04	0.4	0.03
162	flavoured	8 [c]	5	Tr [c]	0.05	0.25	0.11	1.05	0.4	0.04
163	fruit	8	28	Tr	0.05	0.23	0.11	1.01	1.8	0.07
164	hazelnut	8	5	Tr	0.06	0.27	0.12	1.09	0.4	0.58

No	Food	Vitamin B_6 mg	Vitamin B_{12} µg	Folic acid Free µg	Folic acid Total µg	Panto-thenic acid mg	Biotin µg
	Cream						
142 143	single	0.03	0.2	3	4	0.30	1.4
145 146	double	0.02	0.1	2	2	0.19	0.8
148 149	whipping	0.02	0.2	2	3	0.21	0.9
150	sterilised, canned	0.01	Tr	Tr	Tr	0.28	1.3
	Cheese						
151	Camembert type	0.20[b]	1.2	—	60 *(35–95)*	1.4 *(0.4–3.6)*	6.0 *(1.2–17.8)*
152	Cheddar type	0.08 *(0.05–0.14)*	1.5	—	20 *(10–40)*	0.30 *(0.1–0.7)*	1.7 *(0.4–2.3)*
153	Danish Blue type	0.15 *(0.06–0.24)*	1.2 *(0.6–2.7)*	—	50 *(20–80)*	2.0 *(1.0–3.5)*	1.5 *(1.0–3.6)*
154	Edam type	0.08 *(0.05–0.12)*	1.4	—	20 *(5–35)*	0.30 *(0.1–1.3)*	1.5 *(0.7–5.1)*
155	Parmesan	0.10	1.5	—	20	0.30	1.7
156	Stilton	—	—	—	—	—	—
157	cottage cheese	0.01	(0.5)	2	9	—	—
158	cream cheese	(0.01)	(0.3)	(4)	(5)	—	—
159	processed cheese	—	—	1	2	—	—
160	cheese spread	—	—	—	—	—	—
	Yogurt low fat						
161	natural	0.04	Tr	1	2	—	—
162	flavoured	0.04	Tr	4	8	—	—
163	fruit	0.04	Tr	Tr	3	—	—
164	hazelnut	0.04	Tr	Tr	5	—	—

Notes

[a] Rind 0.5 mg.

[b] Rind 0.40 mg.

[c] Some brands are fortified; these values apply to unfortified yogurts.

No	Food	Description and number of samples	Water g	Lactose g	Other sugars g	Total nitrogen g

No	Food	Energy value		Protein (N × 6.38)	Fat	Carbo-hydrate	mg									
		kcal	kJ	g	g	g	Na	K	Ca	Mg	P	Fe	Cu	Zn	S	Cl

No	Food	Retinol µg	Carotene µg	Vitamin D µg	Thiamin mg	Riboflavin mg	Nicotinic acid mg	Potential nicotinic acid from tryptophan mg Trp ÷ 60	Vitamin C mg	Vitamin E mg

No	Food	Vitamin B_6 mg	Vitamin B_{12} µg	Folic acid Free µg	Folic acid Total µg	Pantothenic acid mg	Biotin µg	Notes

No	Food	Description and number of samples	Water g	Total nitrogen g
	Eggs			
165	whole, raw[a]	150 pooled samples of battery, deep litter and free range eggs	74.8	1.97
166	white, raw	34 eggs, English and Danish; vitamins from literature	88.3	1.44
167	yolk, raw	34 eggs, English and Danish; vitamins from literature	51.0	2.58
168	dried	6 packets; vitamins from literature	7.0	6.97
169	boiled	As raw, except for vitamin losses from literature	74.8	1.97
170	fried	6 eggs; vitamin losses from literature	63.3	2.26
171	poached	6 eggs, poached in water; vitamin losses from literature	74.7	1.99
172	omelette	Recipe, p340	68.8	1.70
173	scrambled	Recipe, p340	62.2	1.67
	Egg and cheese dishes			
174	**Cauliflower cheese**	Recipe, p340	78.4	0.90
175	**Cheese pudding**	Recipe, p340	68.2	1.63
176	**Cheese soufflé**	Recipe, p340	56.6	1.84
177	**Macaroni cheese**	Recipe, p340	67.4	1.19
178	**Pizza,** cheese and tomato	Recipe, p341	51.2	1.53
179	**Quiche Lorraine**	Recipe, p341	34.6	2.37
180	**Scotch egg**	Recipe, p341	53.1	1.86
181	**Welsh rarebit**	Recipe, p341	33.5	2.50

[a] An average egg is composed of 11 per cent shell, 58 per cent white and 31 per cent yolk

No	Food	Energy value		Protein	Fat	Carbo-hydrate	mg									
		kcal	kJ	g	g	g	Na	K	Ca	Mg	P	Fe	Cu	Zn	S	Cl
	Eggs															
165	whole, raw	147	612	12.3	10.9	Tr	140	140	52	12	220	2.0	0.10	1.5	180	160
166	white, raw	36	153	9.0	Tr	Tr	190	150	5	11	33	0.1	0.05	0.03	180	170
167	yolk, raw	339	1402	16.1	30.5	Tr	50	120	130	15	500	6.1	0.30	3.6	170	140
168	dried	564	2343	43.6	43.3	Tr	520	480	190	41	800	7.9	0.18	5.0	630	590
169	boiled	147	612	12.3	10.9	Tr	140	140	52	12	220	2.0	0.10	1.5	180	160
170	fried	232	961	14.1	19.5	Tr	220	180	64	14	260	2.5	0.12	1.8	210	200
171	poached	155	644	12.4	11.7	Tr	110	120	52	11	240	2.3	0.10	1.5	180	160
172	omelette	190	787	10.6	16.4	Tr	1030	120	47	18	190	1.7	0.09	1.3	160	1540
173	scrambled	246	1018	10.5	22.7	Tr	1050	130	60	17	190	1.7	0.09	1.3	150	1580
	Egg and cheese dishes															
174	**Cauliflower cheese**	113	471	5.7	8.0	4.9	250	250	160	16	120	0.4	0.03	0.8	—	410
175	**Cheese pudding**	170	707	10.2	10.8	8.4	460	150	230	19	210	0.8	0.06	1.3	—	730
176	**Cheese soufflé**	252	1049	11.5	19.0	9.4	420	150	230	19	230	1.1	0.07	1.4	—	670
177	**Macaroni cheese**	174	726	7.4	9.7	15.1	280	120	180	18	140	0.4	0.03	0.9	—	460
178	**Pizza,** cheese and tomato	234	982	9.4	11.5	24.8	340	180	240	19	170	1.1	0.13	1.2	—	570
179	**Quiche Lorraine**	391	1627	14.7	28.1	21.1	610	190	260	21	240	1.3	0.10	1.8	—	970
180	**Scotch egg**	279	1159	11.6	20.9	11.8	480	150	56	13	190	1.7	0.20	1.3	140	640
181	**Welsh rarebit**	365	1523	15.7	23.6	23.9	1030	130	420	29	290	1.1	0.07	2.1	150	1660

No	Food	Retinol µg	Carotene µg	Vitamin D µg	Thiamin mg	Riboflavin mg	Nicotinic acid mg	Potential nicotinic acid from tryptophan mg Trp ÷ 60	Vitamin C mg	Vitamin E mg
	Eggs									
165	whole, raw	140	Tr	1.75[a]	0.09	0.47	0.07	3.61	0	1.6
166	white, raw	0	0	0	0	0.43	0.09	2.64	0	0
167	yolk, raw	400	Tr	5.0[a]	0.30	0.54	0.02	4.73	0	4.6
168	dried	490[b]	Tr	6.0	0.35[b]	1.2	0.20	12.78	0	5.6
169	boiled	140	Tr	1.75	0.08	0.45	0.07	3.61	0	1.6
170	fried	140	Tr	1.75	0.07	0.42	0.07	4.14	0	1.6
171	poached	140	Tr	1.75	0.07	0.38	0.07	3.65	0	1.6
172	omelette	190	40	1.57	0.07	0.32	0.06	3.11	0	1.5
173	scrambled	130	80	1.54	0.07	0.33	0.07	3.05	Tr	1.6
	Egg and cheese dishes									
174	**Cauliflower cheese**	80	50	0.29	0.06	0.14	0.41	1.32	8	0.4
175	**Cheese pudding**	110	50	0.46	0.05	0.28	0.24	2.55	Tr	0.6
176	**Cheese soufflé**	200	40	1.39	0.07	0.26	0.26	2.90	Tr	1.3
177	**Macaroni cheese**	90	40	0.33	0.03	0.14	0.23	1.60	Tr	0.3
178	**Pizza,** cheese and tomato	70	230	0.06	0.11	0.14	1.01	2.06	3	0.7
179	**Quiche Lorraine**	160	50	0.88	0.14	0.24	1.05	3.37	Tr	0.9
180	**Scotch egg**	60	0	0.75	0.06	0.22	1.54	2.90	0	0.8
181	**Welsh rarebit**	210	130	0.19	0.07	0.21	0.67	3.55	Tr	0.4

No	Food	Vitamin B_6 mg	Vitamin B_{12} µg	Folic acid Free µg	Folic acid Total µg	Pantothenic acid mg	Biotin µg
	Eggs						
165	whole, raw	0.11	1.7[c]	25[d]	25	1.8	25
166	white, raw	Tr	0.1	1	1	0.3	Tr[e]
167	yolk, raw	0.30	4.9	48	52	4.6	60[e]
168	dried	0.40	7.0	—	—	6.2	—
169	boiled	0.10	1.7	22	22	1.6	25
170	fried	0.09	1.7	17	17	1.4	25
171	poached	0.09	1.7	16	16	1.4	25
172	omelette	0.08	1.5	15	15	1.3	22
173	scrambled	0.09	1.4	15	15	1.3	20
	Egg and cheese dishes						
174	**Cauliflower cheese**	0.11	0.3	—	13	0.4	2
175	**Cheese pudding**	0.06	0.9	—	8	0.5	8
176	**Cheese soufflé**	0.07	1.0	—	12	0.7	10
177	**Macaroni cheese**	0.03	0.3	2	3	0.2	1
178	**Pizza,** cheese and tomato	0.08	0.3	7	24	0.3	3
179	**Quiche Lorraine**	0.10	0.8	—	8	0.5	7
180	**Scotch egg**	0.07	1.2	8	8	0.8	12
181	**Welsh rarebit**	0.04	0.8	—	7	0.3	2

Notes

[a] If the hens have been fed a supplement, values may be considerably higher.

[b] There may be a considerable loss on storage.

[c] The value for battery eggs. Deep litter contain 2.6 µg and free range 2.9 µg.

[d] The value for battery eggs. Deep litter contain 32 µg and free range 39 µg.

[e] The natural antibiotic factor avidin, present in raw egg white, binds with the biotin in the yolk to form a complex unavailable to man. Avidin in cooked egg white is inactive.

No	Food	Description and number of samples	Water g	Total nitrogen g
140	**Butter** salted	Analytical (6 samples) and literature sources	15.4	0.07
182	**Cod liver oil**	3 samples	Tr	Tr
183	**Compound cooking fat**	7 samples	Tr	Tr
184	**Dripping** beef	Analysed as purchased	1.0	Tr
185	**Lard**	Analysed as purchased	1.0	Tr
186	**Low fat spread**	One brand	57.1	0
187	**Margarine** all kinds	Hard, soft and polyunsaturated	16.0	0.02
193	**Suet** block	Analysed as purchased	Tr	0.15
194	shredded	6 samples of the same brand	1.5	Tr
195	**Vegetable oils**	All kinds	Tr	Tr

No	Food	Energy value kcal	Energy value kJ	Protein g	Fat g	Carbo-hydrate g	Na (mg)	K (mg)	Ca (mg)	Mg (mg)	P (mg)	Fe (mg)	Cu (mg)	Zn (mg)	S (mg)	Cl (mg)
140	**Butter** salted	740	3041	0.4	82.0	Tr	870	15	15	2	24	0.2	0.03	0.15	9	1340
182	**Cod liver oil**	899	3696	Tr	99.9	0	Tr	Tr	Tr	Tr	Tr	Tr	Tr	Tr	Tr	Tr
183	**Compound cooking fat**	894	3674	Tr	99.3	0	Tr	Tr	Tr	Tr	Tr	Tr	Tr	Tr	Tr	Tr
184	**Dripping** beef	891	3663	Tr	99.0	0	5	4	1	Tr	13	0.2	—	—	9	2
185	**Lard**	891	3663	Tr	99.0	0	2	1	1	1	3	0.1	0.02	—	25	4
186	**Low fat spread**	366	1506	0	40.7	0	690	Tr	Tr	Tr	Tr	Tr	Tr	Tr	Tr	1035
187	**Margarine** all kinds	730	3000	0.1	81.0	0.1	800	5	4	1	12	0.3	0.04	—	12	1200
193	**Suet** block	895	3678	0.9	99.0	0	21	13	6	1	7	0.4	0.04	—	20	18
194	shredded	826	3402	Tr	86.7	12.1	Tr	Tr	Tr	Tr	Tr	Tr	Tr	Tr	Tr	Tr
195	**Vegetable oils**	899	3696	Tr	99.9	0	Tr	Tr	Tr	Tr	Tr	Tr	Tr	Tr	Tr	Tr

No	Food	Retinol µg	Carotene µg	Vitamin D µg	Thiamin mg	Riboflavin mg	Nicotinic acid mg	Potential nicotinic acid from tryptophan mg Trp ÷60	Vitamin C mg	Vitamin E mg
140	**Butter** salted	750	470	0.76	Tr	Tr	Tr	0.1	Tr	2.0
182	**Cod liver oil**	18 000	Tr	210	0	0	0	0	0	20.0
183	**Compound cooking fat**	0	0	0	0	0	0	0	0	Tr
184	**Dripping** beef	—	—	Tr	Tr	Tr	Tr	Tr	0	(0.3)
185	**Lard**	Tr	0	Tr	Tr	Tr	Tr	Tr	0	Tr
186	**Low fat spread**	900 *(800–1000)*	0	7.94 *(7.05–8.82)*	0	0	0	0	0	4.0[a]
187	**Margarine** all kinds	900 *(800–1000)*	0[b]	7.94 *(7.05–8.82)*	Tr	Tr	Tr	Tr	0	8.0[a]
193	**Suet** block	(52)	(73)	Tr	Tr	Tr	Tr	0.2	0	(1.5)
194	shredded	52	73	Tr	Tr	Tr	Tr	Tr	0	1.5
195	**Vegetable oils**	0	Tr[c]	0	Tr	Tr	Tr	Tr	0	see[d]

No	Food	Vitamin B_6 mg	Vitamin B_{12} µg	Folic acid Free µg	Folic acid Total µg	Pantothenic acid mg	Biotin µg
140	**Butter** salted	Tr	Tr	Tr	Tr	Tr	Tr
182	**Cod liver oil**	0	0	0	0	0	0
183	**Compound cooking fat**	0	0	0	0	0	0
184	**Dripping** beef	Tr	Tr	Tr	Tr	Tr	Tr
185	**Lard**	Tr	Tr	Tr	Tr	Tr	Tr
186	**Low fat spread**	0	0	0	0	0	0
187	**Margarine** all kinds	Tr	Tr	Tr	Tr	Tr	Tr
193	**Suet** block	Tr	Tr	Tr	Tr	Tr	Tr
194	shredded	Tr	Tr	Tr	Tr	Tr	Tr
195	**Vegetable oils**	Tr	0	Tr	Tr	Tr	Tr

Notes

[a] The vitamin E content will vary according to the blend of oils used. Polyunsaturated margarine contains about 25 mg α-tocopherol per 100 g.

[b] Some margarines may be fortified with carotene.

[c] Most vegetable oils contain only a trace of carotene, with the exception of unrefined palm oil, which contains about 30 000 µg β- and 24 000 µg α-carotene per 100 g.

[d] Active tocopherols present in vegetable oils are:

	Tocopherols mg/100g α	γ	δ	αT3
196 Coconut	0.5	0	0.6	0.5
197 Cottonseed	38.9	38.7	0	0
198 Maize	11.2	60.2	1.8	0
199 Olive	5.1	Tr	0	0
200 Palm	25.6	31.6	7.0	14.3
201 Peanut	13.0	21.4	2.1	0
202 Rapeseed	18.4	38.0	1.2	0
203 Safflowerseed	38.7	17.4	24.0	0
204 Soyabean	10.1	59.3	26.4	0
205 Sunflowerseed	48.7	5.1	0.8	0
206 Wheatgerm	133.0	26.0	27.1	2.6

(from Slover, 1971)

No	Food	Description and number of samples	Edible matter, proportion of weight purchased	Water g	Total nitrogen g
	Bacon				
208	**dressed carcase** raw	Average carcase weight 68 kg, 57% lean, 32% fat in whole carcase	0.89	48.8	2.08
210	**lean** average, raw	Average of five different cuts	—	67.0	3.23
211	**fat** average, raw	Average of five different cuts	—	12.8	0.76
212	cooked	Average of five different cuts	—	13.8	1.48
213	**collar joint** raw, lean and fat	12 samples, boneless; 70% lean	0.91	51.3	2.34
214	boiled, lean and fat	12 samples, 73% lean; soaked for 16 hours before cooking	0.64	49.0	3.26
215	lean only	12 samples; soaked for 16 hours before cooking	0.47	60.8	4.16
216	**gammon joint** raw, lean and fat	12 samples, boneless, 80% lean	0.93	60.8	2.82
217	boiled, lean and fat	12 samples 80% lean; soaked for 16 hours before cooking	0.63	53.9	3.95
218	lean only	12 samples; soaked for 16 hours before cooking	0.50	62.7	4.70
219	**gammon rashers** grilled, lean and fat	24 samples, 88% lean; rind removed before cooking	0.60	52.1	4.72
220	lean only	24 samples; rind removed before cooking	0.52	57.0	5.02
221	**rashers, raw** back, lean and fat	36 samples, 59% lean; rind removed	0.93	40.5	2.27
222	middle, lean and fat	36 samples, 59% lean; rind removed	0.90	40.8	2.28
223	streaky, lean and fat	36 samples, 61% lean; rind removed	0.85	41.8	2.34
225	**rashers, fried** average, lean only	Average of back, middle and streaky	—	39.2	5.24
226	back, lean and fat	36 samples, 68% lean; rind removed before cooking	0.55	29.7	3.99
227	middle, lean and fat	36 samples, 64% lean; rind removed before cooking	0.51	28.7	3.86
228	streaky, lean and fat	36 samples, 60% lean; rind removed before cooking	0.51	27.5	3.69
230	**rashers, grilled** average, lean only	Average of back, middle and streaky	—	46.0	4.88
231	back, lean and fat	36 samples, 72% lean; rind removed before cooking	0.55	36.0	4.04
232	middle, lean and fat	36 samples, 70% lean; rind removed before cooking	0.51	35.2	3.98
233	streaky, lean and fat	36 samples, 69% lean; rind removed before cooking	0.47	34.6	3.92

No	Food	Energy value		Protein (N × 6.25)	Fat	Carbo-hydrate	mg									
		kcal	kJ	g	g	g	Na	K	Ca	Mg	P	Fe	Cu	Zn	S	Cl
	Bacon															
208	**dressed carcase** raw	352	1453	13.0	33.3	0	1400	250	7	16	130	1.0	0.10	1.8	—	2090
210	**lean** average, raw	147	617	20.2	7.4	0[a]	1870[a]	350	9	22	180[a]	1.2	0.12	2.5	200	2800[a]
211	**fat** average, raw	747	3075	4.8	80.9	0	560	75	3	4	38[a]	0.7	0.06	0.6	—	810
212	cooked	692	2852	9.3	72.8	0	990	130	7	10	90	0.8	0.09	0.8	—	1520
213	**collar joint** raw, lean and fat	319	1318	14.6	28.9	0	1690	260	7	16	140	1.2	0.11	2.4	—	2560
214	boiled, lean and fat	325	1346	20.4	27.0	0	1100	170	13	15	140	1.6	0.18	3.9	—	1630
215	lean only	191	801	26.0	9.7	0	1350	210	15	19	170	1.9	0.22	5.1	270	2000
216	**gammon joint** raw, lean and fat	236	978	17.6	18.3	0	1180	310	7	20	160	0.9	0.11	1.7	—	1800
217	boiled, lean and fat	269	1119	24.7	18.9	0	960	210	9	18	150	1.3	0.15	2.7	—	1440
218	lean only	167	703	29.4	5.5	0	1110	250	10	21	180	1.5	0.17	3.3	280	1670
219	**gammon rashers** grilled, lean and fat	228	953	29.5	12.2	0	2140	480	9	31	260	1.4	0.17	3.2	—	3290
220	lean only	172	726	31.4	5.2	0	2210	520	10	33	270	1.5	0.18	3.5	310	3410
221	**rashers, raw** back	428	1766	14.2	41.2	0	1470	230	7	15	120	1.0	0.10	1.6	—	2140
222	middle	425	1756	14.3	40.9	0	1470	230	7	15	120	1.0	0.10	1.6	—	2150
223	streaky	414	1710	14.6	39.5	0	1500	240	8	15	120	1.0	0.10	1.7	—	2190
225	**rashers, fried** average, lean only	332	1383	32.8	22.3	0	2280	380	16	25	210	1.5	0.14	3.6	310	3510
226	back, lean and fat	465	1926	24.9	40.6	0	1910	300	13	20	170	1.3	0.12	2.6	—	2970
227	middle, lean and fat	477	1975	24.1	42.3	0	1870	300	13	20	170	1.3	0.12	2.5	—	2910
228	streaky, lean and fat	496	2050	23.1	44.8	0	1820	290	12	19	160	1.2	0.12	2.4	—	2840
230	**rashers, grilled** average, lean only	292	1218	30.5	18.9	0	2240	350	13	18	190	1.6	0.17	3.7	290	3250
231	back, lean and fat	405	1681	25.3	33.8	0	2020	290	12	16	160	1.5	0.16	3.0	—	2970
232	middle, lean and fat	416	1722	24.9	35.1	0	2000	290	12	16	160	1.5	0.16	2.9	—	2940
233	streaky, lean and fat	422	1749	24.5	36.0	0	1990	290	12	16	160	1.5	0.15	2.9	—	2930

[a] Sweetcure bacon contains 0.5 g sugars, 1200 mg Na, 280 mg P and 1500 mg Cl per 100 g lean and 140 mg P per 100 g fat

No	Food	Retinol µg	Carotene µg	Vitamin D µg	Thiamin mg	Riboflavin mg	Nicotinic acid mg	Potential nicotinic acid from tryptophan mg Trp ÷ 60	Vitamin C mg	Vitamin E mg
	Bacon									
208	**dressed carcase** raw	Tr	Tr	Tr	0.40	0.16	2.9	2.4	0	0.08
210	**lean** average, raw	Tr	Tr	Tr	0.65	0.25	4.7	3.8	0	0.06
211	**fat** average, raw[a]	Tr	Tr	Tr	—	—	—	0.9	0	0.11
212	cooked[a]	Tr	Tr	Tr	—	—	—	1.7	0	0.36
213	**collar joint** raw, lean and fat	Tr	Tr	Tr	0.41	0.21	2.7	2.7	0	0.07
214	boiled, lean and fat	Tr	Tr	Tr	0.27	0.22	2.6	3.8	0	(0.14)
215	lean only	Tr	Tr	Tr	0.37	0.30	3.6	4.9	0	(0.05)
216	**gammon joint** raw, lean and fat	Tr	Tr	Tr	0.62	0.17	4.1	3.3	0	0.07
217	boiled, lean and fat	Tr	Tr	Tr	0.44	0.15	3.4	4.6	0	(0.11)
218	lean only	Tr	Tr	Tr	0.55	0.19	4.2	5.5	0	(0.05)
219	**gammon rashers** grilled, lean and fat	Tr	Tr	Tr	0.88	0.24	6.3	5.5	0	(0.07)
220	lean only	Tr	Tr	Tr	1.00	0.27	7.1	5.9	0	(0.04)
221	**rashers, raw** back	Tr	Tr	Tr	0.35	0.14	3.1	2.7	0	0.07
222	middle	Tr	Tr	Tr	0.36	0.14	3.1	2.7	0	0.08
223	streaky	Tr	Tr	Tr	0.37	0.15	3.2	2.7	0	0.08
225	**rashers, fried** average, lean only	Tr	Tr	Tr	0.61	0.31	7.7	6.1	0	0.06
226	back, lean and fat	Tr	Tr	Tr	0.41	0.21	5.2	4.7	0	0.18
227	middle, lean and fat	Tr	Tr	Tr	0.39	0.20	5.0	4.5	0	0.20
228	streaky, lean and fat	Tr	Tr	Tr	0.37	0.19	4.6	4.3	0	0.21
230	**rashers, grilled** average, lean only	Tr	Tr	Tr	0.59	0.23	6.2	5.7	0	0.04
231	back, lean and fat	Tr	Tr	Tr	0.43	0.17	4.5	4.7	0	0.11
232	middle, lean and fat	Tr	Tr	Tr	0.41	0.16	4.4	4.6	0	0.11
233	streaky, lean and fat	Tr	Tr	Tr	0.40	0.16	4.2	4.6	0	0.12

No	Food	Vitamin B_6 mg	Vitamin B_{12} µg	Folic acid Free µg	Folic acid Total µg	Pantothenic acid mg	Biotin µg
	Bacon						
208	**dressed carcase** raw	0.30	Tr	1	2	0.3	1
210	**lean** average, raw	0.45	Tr	1	3	0.6	2
211	**fat** average, raw[a]	—	Tr	Tr	Tr	—	Tr
212	cooked[a]	—	Tr	Tr	Tr	—	Tr
213	**collar joint** raw, lean and fat	0.32	Tr	Tr	2	0.4	1
214	boiled, lean and fat	(0.24)	Tr	Tr	(Tr)	(0.4)	(2)
215	lean only	(0.33)	Tr	Tr	(1)	(0.5)	(3)
216	**gammon joint** raw, lean and fat	0.36	Tr	Tr	3	0.5	2
217	boiled, lean and fat	(0.26)	Tr	Tr	(Tr)	(0.4)	(2)
218	lean only	(0.33)	Tr	Tr	(1)	(0.5)	(3)
219	**gammon rashers** grilled, lean and fat	(0.33)	Tr	Tr	(2)	(0.6)	(3)
220	lean only	(0.37)	Tr	Tr	(2)	(0.7)	(3)
221	**rashers, raw** back	0.26	Tr	Tr	2	0.4	1
222	middle	0.27	Tr	Tr	2	0.4	1
223	streaky	0.28	Tr	Tr	2	0.4	1
225	**rashers, fried** average, lean only	0.45	Tr	Tr	2	0.5	4
226	back, lean and fat	0.30	Tr	Tr	1	0.3	3
227	middle, lean and fat	0.29	Tr	Tr	1	0.3	2
228	streaky, lean and fat	0.27	Tr	Tr	1	0.3	2
230	**rashers, grilled** average, lean only	0.37	Tr	Tr	2	0.7	3
231	back, lean and fat	0.27	Tr	Tr	1	0.5	2
232	middle, lean and fat	0.26	Tr	Tr	1	0.5	2
233	streaky, lean and fat	0.25	Tr	Tr	1	0.5	2

Notes

Vitamin E, vitamin B_6, vitamin B_{12}, folic acid, pantothenic acid and biotin were analysed on pooled samples of raw lean and cooked lean, and vitamin E also on pooled samples of raw fat and cooked fat. The values for the individual cuts have been calculated from these results.

[a] Bacon fat may contain small amounts of some vitamins. It is, however, extremely difficult to obtain satisfactory analytical values for inclusion in the tables.

No	Food	Description and number of samples	Edible matter, proportion of weight purchased	Water g	Total nitrogen g
	Beef				
235	**dressed carcase** raw	Average carcase weight 224 kg, 59% lean, 23% fat in whole carcase; includes kidney knob and channel fat	0.83	58.6	2.53
237	**lean** average, raw	Average of six different cuts	—	74.0	3.25
240	**fat** average, raw	Average of six different cuts	—	24.0	1.40
241	cooked	Average of six different cuts	—	25.2	1.91
242	**brisket** raw, lean and fat	18 samples, boned and rolled, 77% lean	0.95	62.2	2.68
243	boiled, lean and fat	18 samples, boned and rolled, 77% lean; salt added	0.61	48.4	4.41
244	**forerib** raw, lean and fat	18 samples, with bone, 72% lean	0.75	57.4	2.56
245	roast, lean and fat	18 samples, 72% lean; cooked on the bone	0.55	48.4	3.59
246	lean only	18 samples; cooked on the bone	0.40	59.1	4.46
247	**mince** raw	18 samples	1.00	64.5	3.01
248	stewed	18 samples; salt added	0.72	59.1	3.69
249	**rump steak** raw, lean and fat	18 samples, 86% lean	0.95	66.7	3.02
250	fried, lean and fat	18 samples, 87% lean	0.70	56.2	4.58
251	lean only	18 samples	0.60	61.1	4.92
252	grilled, lean and fat	18 samples, 89% lean	0.74	59.3	4.36
253	lean only	18 samples	0.66	63.8	4.58
254	**silverside** salted, boiled, lean and fat	17 samples, 85% lean; soaked for 18 hours before cooking	0.56	54.5	4.58
255	lean only	17 samples; soaked for 18 hours before cooking.	0.48	59.7	5.16
256	**sirloin** raw, lean and fat	18 samples, boneless, 72% lean	0.92	59.4	2.65
257	roast, lean and fat	18 samples, boneless, 80% lean	0.72	54.3	3.77
258	lean only	18 samples, boneless	0.58	62.0	4.41
259	**stewing steak** raw, lean and fat	18 samples, 85% lean	0.96	68.7	3.23
260	stewed, lean and fat	18 samples, 92% lean; salt added	0.60	57.1	4.94
261	**topside** raw, lean and fat	18 samples, 87% lean	0.94	68.4	3.13
262	roast, lean and fat	18 samples, 88% lean	0.74	60.2	4.26
263	lean only	18 samples	0.65	65.1	4.67

No	Food	Energy value kcal	Energy value kJ	Protein (N × 6.25) g	Fat g	Carbo-hydrate g	Na mg	K mg	Ca mg	Mg mg	P mg	Fe mg	Cu mg	Zn mg	S mg	Cl mg
	Beef															
235	**dressed carcase** raw	282	1168	15.8	24.3	0	55	280	7	17	150	1.9	0.13	3.3	—	55
237	**lean**, average, raw	123	517	20.3	4.6	0	61	350	7	20	180	2.1	0.14	4.3	190	59
240	**fat**, average, raw	637	2625	8.8	66.9	0	33	100	10	7	60	1.0	0.11	1.0	—	39
241	cooked	613	2526	11.9	62.8	0	50	160	14	11	90	1.4	0.14	1.4	—	56
242	**brisket** raw, lean and fat	252	1044	16.8	20.5	0	68	270	7	16	140	1.6	0.12	3.5	—	69
243	boiled, lean and fat	326	1354	27.6	23.9	0	73	200	12	18	150	2.8	0.13	6.3	—	92
244	**forerib** raw, lean and fat	290	1201	16.0	25.1	0	48	270	10	15	130	1.5	0.12	3.4	—	45
245	roast, lean and fat	349	1446	22.4	28.8	0	51	260	14	18	150	1.9	0.16	5.2	—	56
246	lean only	225	941	27.9	12.6	0	56	310	13	22	180	2.3	0.17	6.8	260	61
247	**mince** raw	221	919	18.8	16.2	0	86	290	15	17	160	2.7	0.15	4.3	180	86
248	stewed	229	955	23.1	15.2	0	320	290	18	20	170	3.1	0.24	5.8	220	470
249	**rump steak** raw, lean and fat	197	821	18.9	13.5	0	51	330	6	20	210	2.3	0.14	4.6	—	49
250	fried, lean and fat	246	1026	28.6	14.6	0	54	360	7	24	220	3.2	0.15	5.3	—	56
251	lean only	190	797	30.8	7.4	0	57	390	6	25	240	3.4	0.15	5.9	280	58
252	grilled, lean and fat	218	912	27.3	12.1	0	55	380	7	25	220	3.4	0.18	4.9	—	61
253	lean only	168	708	28.6	6.0	0	56	400	7	26	230	3.5	0.18	5.3	280	62
254	**silverside** salted, boiled, lean and fat	242	1012	28.6	14.2	0	910	200	11	18	140	2.8	0.25	5.5	—	1420
255	lean only	173	730	32.3	4.9	0	1000	230	10	20	150	3.2	0.27	6.2	310	1560
256	**sirloin** raw, lean and fat	272	1126	16.6	22.8	0	49	260	9	16	150	1.6	0.13	3.1	—	52
257	roast, lean and fat	284	1182	23.6	21.1	0	54	300	10	19	170	1.9	0.18	4.6	—	64
258	lean only	192	806	27.6	9.1	0	59	350	10	22	190	2.1	0.19	5.5	270	65
259	**stewing steak** raw, lean and fat	176	736	20.2	10.6	0	72	320	8	18	140	2.1	0.15	3.8	—	73
260	stewed, lean and fat	223	932	30.9	11.0	0	360	230	15	21	160	3.0	0.25	8.7	—	550
261	**topside** raw, lean and fat	179	748	19.6	11.2	0	43	340	5	19	170	1.9	0.13	3.3	—	44
262	roast, lean and fat	214	896	26.6	12.0	0	48	350	6	23	200	2.6	0.13	4.9	—	51
263	lean only	156	659	29.2	4.4	0	49	370	6	24	210	2.8	0.14	5.5	270	51

No	Food	Retinol µg	Carotene µg	Vitamin D µg	Thiamin mg	Riboflavin mg	Nicotinic acid mg	Potential nicotinic acid from tryptophan mg Trp ÷ 60	Vitamin C mg	Vitamin E mg
	Beef									
235	**dressed carcase** raw	(4)	Tr	Tr	0.05	0.20	3.8	3.4	0	0.19
237	**lean,** average, raw	Tr	Tr	Tr	0.07	0.24	5.2	4.3	0	0.15
240	**fat,** average, raw[a]	—	—	—	—	—	—	1.9	0	0.32
241	cooked[a]	—	—	—	—	—	—	2.6	0	0.55
242	**brisket** raw, lean and fat	Tr	Tr	Tr	0.05	0.16	3.7	3.6	0	0.19
243	boiled, lean and fat	Tr	Tr	Tr	0.04	0.30	4.3	5.9	0	0.35
244	**forerib** raw, lean and fat	Tr	Tr	Tr	0.04	0.14	3.6	3.4	0	0.20
245	roast, lean and fat	Tr	Tr	Tr	0.04	0.24	3.9	4.8	0	0.36
246	lean only	Tr	Tr	Tr	0.05	0.33	5.5	6.0	0	0.29
247	**mince** raw	Tr	Tr	Tr	0.06	0.31	4.0	4.0	0	(0.18)
248	stewed	Tr	Tr	Tr	0.05	0.33	4.4	4.9	0	(0.31)
249	**rump steak** raw, lean and fat	Tr	Tr	Tr	0.08	0.26	4.2	4.0	0	0.17
250	fried, lean and fat	Tr	Tr	Tr	0.08	0.35	5.5	6.1	0	0.33
251	lean only	Tr	Tr	Tr	0.09	0.40	6.3	6.6	0	0.29
252	grilled, lean and fat	Tr	Tr	Tr	0.08	0.32	5.7	5.8	0	0.32
253	lean only	Tr	Tr	Tr	0.09	0.36	6.4	6.1	0	0.29
254	**silverside** salted, boiled, lean and fat	Tr	Tr	Tr	0.03	0.27	3.3	6.1	0	0.33
255	lean only	Tr	Tr	Tr	0.04	0.32	3.9	6.9	0	0.29
256	**sirloin** raw, lean and fat	Tr	Tr	Tr	0.04	0.17	4.2	3.5	0	0.20
257	roast, lean and fat	Tr	Tr	Tr	0.06	0.25	4.8	5.0	0	0.34
258	lean only	Tr	Tr	Tr	0.07	0.31	6.0	5.9	0	0.29
259	**stewing steak** raw, lean and fat	Tr	Tr	Tr	0.06	0.23	4.2	4.3	0	0.18
260	stewed, lean and fat	Tr	Tr	Tr	0.03	0.33	3.6	6.6	0	0.31
261	**topside** raw, lean and fat	Tr	Tr	Tr	0.05	0.21	4.8	4.2	0	0.17
262	roast, lean and fat	Tr	Tr	Tr	0.07	0.31	5.7	5.7	0	0.32
263	lean only	Tr	Tr	Tr	0.08	0.35	6.5	6.2	0	0.29

No	Food	Vitamin B_6 mg	Vitamin B_{12} µg	Folic acid Free µg	Folic acid Total µg	Pantothenic acid mg	Biotin µg
	Beef						
235	**dressed carcase** raw	0.23	1	4	10	0.5	Tr
237	**lean** average, raw	0.32	2	4	10	0.7	Tr
240	**fat** average, raw[a]	—	Tr	—	—	—	Tr
241	cooked[a]	—	Tr	—	—	—	Tr
242	**brisket** raw, lean and fat	0.25	1	3	8	0.5	Tr
243	boiled, lean and fat	0.25	2	4	13	0.7	Tr
244	**forerib** raw, lean and fat	0.23	1	3	7	0.5	Tr
245	roast, lean and fat	0.24	1	3	13	0.6	Tr
246	lean only	0.33	2	5	17	0.9	Tr
247	**mince** raw	(0.27)	(2)	(3)	(9)	(0.6)	Tr
248	stewed	(0.30)	(2)	(4)	(16)	(0.8)	Tr
249	**rump steak** raw, lean and fat	0.27	2	3	9	0.6	Tr
250	fried, lean and fat	0.29	2	4	15	0.8	Tr
251	lean only	0.33	2	5	17	0.9	Tr
252	grilled, lean and fat	0.29	2	4	15	0.8	Tr
253	lean only	0.33	2	5	17	0.9	Tr
254	**silverside** salted boiled, lean and fat	0.28	2	4	15	0.8	Tr
255	lean only	0.33	2	5	17	0.9	Tr
256	**sirloin** raw, lean and fat	0.23	1	3	7	0.5	Tr
257	roast, lean and fat	0.26	2	4	14	0.7	Tr
258	lean only	0.33	2	5	17	0.9	Tr
259	**stewing steak** raw, lean and fat	0.27	2	3	9	0.6	Tr
260	stewed, lean and fat	0.30	2	4	16	0.8	Tr
261	**topside** raw, lean and fat	0.28	2	3	9	0.6	Tr
262	roast, lean and fat	0.29	2	4	15	0.8	Tr
263	lean only	0.33	2	5	17	0.9	Tr

Notes

Vitamin E, vitamin B_6, vitamin B_{12}, folic acid, pantothenic acid and biotin were analysed on pooled samples of raw lean and cooked lean, and vitamin E also on pooled samples of raw fat and cooked fat. The values for the individual cuts have been calculated from these results.

[a] Beef fat may contain small amounts of some vitamins. It is, however, extremely difficult to obtain satisfactory analytical values for inclusion in the tables.

No	Food	Description and number of samples	Edible matter, proportion of weight purchased	Water g	Total nitrogen g
	Lamb				
264	**dressed carcase** raw	Average carcase weight 17 kg, 55% lean, 28% fat in whole carcase; includes kidney knob and channel fat	0.84	54.4	2.34
266	**lean** average, raw	Average of six different cuts	—	70.1	3.33
269	**fat** average, raw	Average of six different cuts	—	21.2	0.99
270	cooked	Average of six different cuts	—	24.6	1.81
271	**breast** raw, lean and fat	15 samples, boneless, 59% lean	0.96	48.3	2.67
272	roast, lean and fat	15 samples, boneless, 60% lean	0.77	43.6	3.05
273	lean only	15 samples, boneless	0.45	57.8	4.09
274	**chops, loin** raw, lean and fat	15 samples, 60% lean	0.82	49.5	2.34
275	grilled, lean and fat	15 samples, 70% lean	0.54	46.6	3.76
276	lean and fat (weighed with bone)	Calculated from the previous item	0.54	36.3	2.93
277	lean only	15 samples	0.38	58.9	4.44
278	lean only (weighed with fat and bone)	Calculated from the previous item	0.38	32.4	2.44
279	**cutlets** raw, lean and fat	15 samples, 59% lean	0.76	48.7	2.35
280	grilled, lean and fat	15 samples, 67% lean	0.50	45.1	3.68
281	lean and fat (weighed with bone)	Calculated from the previous item	0.50	29.8	2.43
282	lean only	15 samples	0.33	58.9	4.44
283	lean only (weighed with fat and bone)	Calculated from the previous item	0.33	25.9	1.95
284	**leg** raw, lean and fat	15 samples, 80% lean	0.77	63.1	2.86
285	roast, lean and fat	15 samples, 82% lean	0.53	55.3	4.18
286	lean only	15 samples	0.43	61.8	4.71
287	**scrag and neck** raw, lean and fat	15 samples, 71% lean	0.60	55.7	2.50
288	stewed, lean and fat	15 samples, 84% lean	0.46	52.6	4.10
289	lean only	15 samples	0.38	55.9	4.44
290	lean only (weighed with fat and bone)	Calculated from the previous item	0.38	28.5	2.26

No	Food	Energy value kcal	Energy value kJ	Protein (N × 6.25) g	Fat g	Carbo-hydrate g	Na mg	K mg	Ca mg	Mg mg	P mg	Fe mg	Cu mg	Zn mg	S mg	Cl mg
	Lamb															
264	**dressed carcase** raw	333	1377	14.6	30.5	0	71	260	7	18	150	1.4	0.15	2.9	—	65
266	**lean** average, raw	162	679	20.8	8.8	0	88	350	7	24	190	1.6	0.17	4.0	210	76
269	**fat** average, raw	671	2762	6.2	71.8	0	36	96	7	6	54	0.7	0.15	0.8	—	37
270	cooked	616	2538	11.3	63.4	0	56	150	11	12	120	1.4	0.16	1.4	—	53
271	**breast** raw, lean and fat	378	1564	16.7	34.6	0	100	270	8	18	150	1.3	0.16	3.2	—	77
272	roast, lean and fat	410	1697	19.1	37.1	0	73	250	10	18	150	1.5	0.19	3.6	—	74
273	lean only	252	1049	25.6	16.6	0	86	330	10	24	200	1.7	0.23	5.3	240	89
274	**chops, loin** raw, lean and fat	377	1558	14.6	35.4	0	61	230	(7)[a]	17	140	1.2	0.16	2.1	—	60
275	grilled, lean and fat	355	1473	23.5	29.0	0	72	320	(9)[a]	24	210	1.9	0.17	3.4	—	83
276	lean and fat (weighed with bone)	277	1147	18.3	22.6	0	56	250	(7)[a]	19	160	1.5	0.13	2.7	—	65
277	lean only	222	928	27.8	12.3	0	75	380	(9)[a]	28	240	2.1	0.19	4.1	270	90
278	lean only (weighed with fat and bone)	122	512	15.3	6.8	0	41	210	(5)[a]	15	130	1.2	0.10	2.3	150	50
279	**cutlets** raw, lean and fat	386	1593	14.7	36.3	0	60	230	(7)[a]	16	130	1.2	0.15	2.1	—	60
280	grilled, lean and fat	370	1534	23.0	30.9	0	71	320	(9)[a]	23	200	1.9	0.18	3.3	—	82
281	lean and fat (weighed with bone)	244	1013	15.2	20.4	0	47	210	(6)[a]	15	130	1.3	0.12	2.2	—	54
282	lean only	222	928	27.8	12.3	0	75	380	(9)[a]	28	240	2.1	0.19	4.1	270	90
283	lean only (weighed with fat and bone)	97	407	12.2	5.4	0	33	170	(4)[a]	12	110	0.9	0.08	1.8	120	40
284	**leg** raw, lean and fat	240	996	17.9	18.7	0	52	310	6	22	170	1.7	0.14	2.8	—	54
285	roast, lean and fat	266	1106	26.1	17.9	0	65	310	8	25	200	2.5	0.28	4.6	—	62
286	lean only	191	800	29.4	8.1	0	67	340	8	28	220	2.7	0.31	5.3	290	64
287	**scrag and neck** raw, lean and fat	316	1309	15.6	28.2	0	71	260	(7)[a]	18	140	1.2	0.16	3.6	—	68
288	stewed, lean and fat	292	1216	25.6	21.1	0	240	190	(9)[a]	18	190	2.2	0.22	6.1	—	340
289	lean only	253	1054	27.8	15.7	0	250	200	(9)[a]	19	190	2.4	0.33	6.9	270	350
290	lean only (weighed with fat and bone)	128	536	14.1	8.0	0	130	100	(5)[a]	10	100	1.2	0.17	3.5	140	180

[a] The calcium content is extremely variable as scrapings of bone may easily be included in the edible portion. This is a minimum value

No	Food	Retinol µg	Carotene µg	Vitamin D µg	Thiamin mg	Riboflavin mg	Nicotinic acid mg	Potential nicotinic acid from tryptophan mg Trp ÷ 60	Vitamin C mg	Vitamin E mg
	Lamb									
264	**dressed carcase** raw	(4)	Tr	Tr	0.09	0.21	4.0	3.1	0	0.26
266	**lean** average, raw	Tr	Tr	Tr	0.14	0.28	6.0	4.4	0	0.10
269	**fat** average, raw[a]	—	—	—	—	—	—	1.3	0	0.30
270	cooked[a]	—	—	—	—	—	—	2.4	0	0.18
271	**breast** raw, lean and fat	Tr	Tr	Tr	0.08	0.17	3.8	3.6	0	0.18
272	roast, lean and fat	Tr	Tr	Tr	0.06	0.17	3.4	4.1	0	0.13
273	lean only	Tr	Tr	Tr	0.10	0.29	5.6	5.5	0	0.10
274	**chops, loin** raw, lean and fat	Tr	Tr	Tr	0.09	0.16	4.0	3.1	0	0.18
275	grilled, lean and fat	Tr	Tr	Tr	0.11	0.21	5.1	5.0	0	0.12
276	lean and fat (weighed with bone)	Tr	Tr	Tr	0.09	0.16	4.0	3.9	0	0.09
277	lean only	Tr	Tr	Tr	0.15	0.30	7.2	5.9	0	0.10
278	lean only (weighed with fat and bone)	Tr	Tr	Tr	0.08	0.17	4.0	3.3	0	0.06
279	**cutlets** raw, lean and fat	Tr	Tr	Tr	0.09	0.16	3.9	3.1	0	0.18
280	grilled, lean and fat	Tr	Tr	Tr	0.10	0.20	4.8	4.9	0	0.13
281	lean and fat (weighed with bone)	Tr	Tr	Tr	0.07	0.13	3.2	3.2	0	0.09
282	lean only	Tr	Tr	Tr	0.15	0.30	7.2	5.9	0	0.10
283	lean only (weighed with fat and bone)	Tr	Tr	Tr	0.07	0.13	3.2	2.6	0	0.04
284	**leg** raw, lean and fat	Tr	Tr	Tr	0.14	0.25	5.7	3.8	0	0.14
285	roast, lean and fat	Tr	Tr	Tr	0.12	0.31	5.4	5.6	0	0.11
286	lean only	Tr	Tr	Tr	0.14	0.38	6.6	6.3	0	0.10
287	**scrag and neck** raw, lean and fat	Tr	Tr	Tr	0.07	0.17	3.4	3.3	0	0.16
288	stewed, lean and fat	Tr	Tr	Tr	0.04	0.18	2.7	5.5	0	0.11
289	lean only	Tr	Tr	Tr	0.05	0.21	3.2	5.9	0	0.10
290	lean only (weighed with fat and bone)	Tr	Tr	Tr	0.03	0.11	1.6	3.0	0	0.05

No	Food	Vitamin B_6 mg	Vitamin B_{12} µg	Folic acid		Pantothenic acid mg	Biotin µg
				Free µg	Total µg		
	Lamb						
264	**dressed carcase** raw	0.17	2	Tr	4	0.5	1
266	**lean** average, raw	0.25	2	Tr	5	0.7	2
269	**fat** average, raw[a]	—	Tr	Tr	—	—	Tr
270	cooked[a]	—	Tr	Tr	—	—	Tr
271	**breast** raw, lean and fat	0.15	1	Tr	3	0.4	1
272	roast, lean and fat	0.13	1	Tr	3	0.4	1
273	lean only	0.22	2	Tr	4	0.7	2
274	**chops, loin** raw, lean and fat	0.15	1	Tr	3	0.4	1
275	grilled, lean and fat	0.15	2	Tr	3	0.5	1
276	lean and fat (weighed with bone)	0.12	2	Tr	2	0.4	1
277	lean only	0.22	2	Tr	4	0.7	2
278	lean only (weighed with fat and bone)	0.12	1	Tr	2	0.4	1
279	**cutlets** raw, lean and fat	0.15	1	Tr	3	0.4	1
280	grilled, lean and fat	0.15	2	Tr	3	0.5	1
281	lean and fat (weighed with bone)	0.10	1	Tr	2	0.3	1
282	lean only	0.22	2	Tr	4	0.7	2
283	lean only (weighed with fat and bone)	0.10	1	Tr	2	0.3	1
284	**leg** raw, lean and fat	0.20	2	Tr	4	0.6	1
285	roast, lean and fat	0.18	2	Tr	3	0.6	1
286	lean only	0.22	2	Tr	4	0.7	2
287	**scrag and neck** raw, lean and fat	0.18	2	Tr	4	0.5	1
288	stewed, lean and fat	0.19	2	Tr	4	0.6	1
289	lean only	0.22	2	Tr	4	0.7	2
290	lean only (weighed with fat and bone)	0.11	1	Tr	2	0.4	1

Notes

Vitamin E, vitamin B_6, vitamin B_{12}, folic acid, pantothenic acid and biotin were analysed on a pooled sample of raw lean and cooked lean, and vitamin E also on pooled samples of raw fat and cooked fat. The values for the individual cuts have been calculated from these results.

[a] Lamb fat may contain small amounts of some vitamins. It is, however, extremely difficult to obtain satisfactory analytical values for inclusion in the tables.

No	Food	Description and number of samples	Edible matter, proportion of weight purchased	Water g	Total nitrogen g
	***Lamb** contd*				
291	**shoulder** raw, lean and fat	15 samples, 68% lean	0.78	56.1	2.50
292	roast, lean and fat	15 samples, 73% lean	0.57	53.6	3.18
293	lean only	15 samples	0.42	64.8	3.80
	Pork				
294	**dressed carcase** raw	Average carcase weight 52 kg, 48% lean, 25% fat in whole carcase; includes kidney knob and channel fat, head, feet and skin	0.74	50.7	2.18
296	**lean** average, raw	Average of three different cuts	—	71.5	3.29
299	**fat** average, raw	Average of three different cuts	—	21.1	1.08
300	cooked	Average of three different cuts	—	20.9	2.37
301	**belly, rashers** raw, lean and fat	15 samples, 56% lean	0.87	48.7	2.44
302	grilled, lean and fat	15 samples, 56% lean	0.64	43.0	3.38
303	**chops, loin** raw, lean and fat	15 samples, 65% lean; without kidney	0.83	54.3	2.55
304	grilled, lean and fat	15 samples, 75% lean; without kidney	0.49	46.3	4.55
305	lean and fat (weighed with bone)	Calculated from the previous item	0.49	36.1	3.55
306	lean only	15 samples	0.37	56.1	5.17
307	lean only (weighed with fat and bone)	Calculated from the previous item	0.37	33.1	3.05
308	**leg** raw, lean and fat	15 samples, fillet end, 73% lean	0.85	59.5	2.66
309	roast, lean and fat	15 samples, fillet end, 76% lean	0.60	51.9	4.30
310	lean only	15 samples, fillet end	0.45	61.6	4.91
	Veal				
311	**cutlet** fried	Covered in egg and breadcrumbs	0.74	54.6	5.02
312	**fillet** raw	All lean	1.00	74.9	3.37
313	roast	All lean	0.75	55.1	5.05

No	Food	Energy value kcal	Energy value kJ	Protein (N × 6.25) g	Fat g	Carbo-hydrate g	Na mg	K mg	Ca mg	Mg mg	P mg	Fe mg	Cu mg	Zn mg	S mg	Cl mg
	***Lamb** contd*															
291	**shoulder** raw, lean and fat	314	1301	15.6	28.0	0	66	260	7	18	150	1.2	0.21	3.1	—	56
292	roast, lean and fat	316	1311	19.9	26.3	0	61	260	9	19	150	1.6	0.15	4.3	—	60
293	lean only	196	819	23.8	11.2	0	65	300	9	22	170	1.8	0.16	5.3	240	65
	Pork															
294	**dressed carcase** raw	338	1397	13.6	31.5	0	65	270	8	16	150	0.9	0.15	1.8	—	64
296	**lean** average, raw	147	615	20.7	7.1	0	76	370	8	22	200	0.9	0.15	2.4	200	71
299	**fat** average, raw	670	2757	6.8	71.4	0	38	87	7	5	49	0.7	0.10	0.4	—	44
300	cooked	619	2553	14.8	62.2	0	79	210	11	9	110	1.0	0.15	0.9	—	84
301	**belly, rashers** raw, lean and fat	381	1574	15.3	35.5	0	73	220	8	14	120	0.8	0.15	1.8	—	71
302	grilled, lean and fat	398	1646	21.1	34.8	0	95	310	11	19	170	1.0	0.16	2.6	—	92
303	**chops, loin** raw	329	1362	15.9	29.5	0	56	290	(8)[a]	17	160	0.8	0.13	1.6	—	54
304	grilled, lean and fat	332	1380	28.5	24.2	0	84	380	(11)[a]	26	230	1.2	0.17	2.9	—	79
305	lean and fat (weighed with bone)	258	1073	22.2	18.8	0	66	300	(9)[a]	20	180	0.9	0.13	2.3	—	62
306	lean only	226	945	32.3	10.7	0	84	420	(9)[a]	29	260	1.2	0.16	3.5	310	78
307	lean only (weighed with fat and bone)	133	558	19.1	6.3	0	50	250	(5)[a]	17	150	0.7	0.09	2.1	180	46
308	**leg** raw, lean and fat	269	1115	16.6	22.5	0	59	300	7	18	160	0.8	0.12	1.8	—	59
309	roast, lean and fat	286	1190	26.9	19.8	0	79	350	10	22	200	1.3	0.25	2.9	—	79
310	lean only	185	777	30.7	6.9	0	79	390	9	25	230	1.3	0.29	3.5	300	76
	Veal															
311	**cutlet** fried	215	904	31.4	8.1	4.4	110	420	10	33	280	(1.6)	—	—	330	120
312	**fillet** raw	109	459	21.1	2.7	0	110	360	8	25	260	1.2	—	—	220	68
313	roast	230	963	31.6	11.5	0	97	430	14	28	360	1.6	—	—	330	110

[a] The calcium content is extremely variable as scrapings of bone may easily be included in the edible portion. This is a minimum value

No	Food	Retinol µg	Carotene µg	Vitamin D µg	Thiamin mg	Riboflavin mg	Nicotinic acid mg	Potential nicotinic acid from tryptophan mg Trp ÷ 60	Vitamin C mg	Vitamin E mg
	***Lamb** contd*									
291	**shoulder** raw, lean and fat	Tr	Tr	Tr	0.10	0.18	3.6	3.3	0	0.17
292	roast, lean and fat	Tr	Tr	Tr	0.07	0.20	3.1	4.2	0	0.12
293	lean only	Tr	Tr	Tr	0.10	0.27	4.3	5.1	0	0.10
	Pork									
294	**dressed carcase** raw	(4)	Tr	Tr	0.58	0.19	4.1	2.5	0	0.01
296	**lean** average, raw	Tr	Tr	Tr	0.89	0.25	6.2	3.8	0	0
299	**fat** average, raw[a]	Tr	Tr	Tr	—	—	—	1.3	0	0.03
300	cooked[a]	Tr	Tr	Tr	—	—	—	2.8	0	0.12
301	**belly, rashers** raw, lean and fat	Tr	Tr	Tr	0.45	0.15	3.3	2.9	0	0.01
302	grilled, lean and fat	Tr	Tr	Tr	0.53	0.11	4.2	3.9	0	0.05
303	**chops, loin** raw, lean and fat	Tr	Tr	Tr	0.57	0.14	4.2	3.0	0	0.01
304	grilled, lean and fat	Tr	Tr	Tr	0.66	0.20	5.7	5.3	0	0.03
305	lean and fat (weighed with bone)	Tr	Tr	Tr	0.51	0.16	4.4	4.1	0	0.02
306	lean only	Tr	Tr	Tr	0.88	0.26	7.6	6.0	0	0
307	lean only (weighed with fat and bone)	Tr	Tr	Tr	0.52	0.15	4.5	3.6	0	0
308	**leg** raw, lean and fat	Tr	Tr	Tr	0.73	0.20	4.5	3.1	0	0.01
309	roast, lean and fat	Tr	Tr	Tr	0.65	0.27	5.0	5.0	0	0.03
310	lean only	Tr	Tr	Tr	0.85	0.35	6.6	5.7	0	0
	Veal									
311	**cutlet** fried	Tr	Tr	Tr	—	—	—	6.7	0	—
312	**fillet** raw	Tr	Tr	Tr	0.10	0.25	7.0	4.5	0	—
313	roast	Tr	Tr	Tr	0.06	0.27	7.0	6.7	0	—

No	Food	Vitamin B_6 mg	Vitamin B_{12} µg	Folic acid Free µg	Folic acid Total µg	Pantothenic acid mg	Biotin µg
	***Lamb** contd*						
291	**shoulder** raw, lean and fat	0.17	2	Tr	3	0.5	1
292	roast, lean and fat	0.16	2	Tr	3	0.5	1
293	lean only	0.22	2	Tr	4	0.7	2
	Pork						
294	**dressed carcase** raw	0.30	1	Tr	3	0.7	1
296	**lean** average, raw	0.45	3	Tr	5	1.1	3
299	**fat** average, raw[a]	—	Tr	Tr	—	—	Tr
300	cooked[a]	—	Tr	Tr	—	—	Tr
301	**belly, rashers** raw, lean and fat	0.25	2	Tr	3	0.6	2
302	grilled, lean and fat	0.23	1	Tr	4	0.7	2
303	**chops, loin** raw, lean and fat	0.29	2	Tr	3	0.7	2
304	grilled, lean and fat	0.31	1	Tr	6	1.0	2
305	lean and fat (weighed with bone)	0.24	1	Tr	5	0.8	2
306	lean only	0.41	2	Tr	7	1.3	3
307	lean only (weighed with fat and bone)	0.24	1	Tr	4	0.8	2
308	**leg** raw, lean and fat	0.33	2	Tr	4	0.8	2
309	roast, lean and fat	0.31	1	Tr	6	1.0	2
310	lean only	0.41	2	Tr	7	1.3	3
	Veal						
311	**cutlet** fried	—	1	Tr	—	—	Tr
312	**fillet** raw	0.30	1	Tr	5	0.6	Tr
313	roast	0.32	1	Tr	4	0.5	Tr

Notes

Vitamin E, vitamin B_6, vitamin B_{12}, folic acid, pantothenic acid and biotin were analysed on a pooled sample of raw lean and cooked lean, and vitamin E also on pooled samples of raw fat and cooked fat. The values for the individual cuts have been calculated from these results.

[a] Pork fat may contain small quantities of some vitamins. It is, however, extremely difficult to obtain satisfactory analytical values for inclusion in the tables.

No	Food	Description and number of samples	Edible matter, proportion of weight purchased	Water g	Total nitrogen g
	Poultry and game				
314	**Chicken** raw, meat only	15 samples, light and dark meat from dressed carcase	0.44	74.4	3.28
315	meat and skin	15 samples, dressed carcase excluding waste	0.64	64.4	2.82
316	light meat	15 samples	0.23	74.4	3.49
317	dark meat	15 samples	0.21	74.5	3.06
318	boiled, meat only	15 samples, light and dark meat from dressed carcase	0.39	63.4	4.67
319	light meat	15 samples	0.20	65.2	4.75
320	dark meat	15 samples	0.19	61.5	4.58
321	roast, meat only	15 samples, light and dark meat from dressed carcase	0.40	68.4	3.97
322	meat and skin	15 samples, dressed carcase excluding waste	0.55	61.9	3.61
323	light meat	15 samples	0.21	68.5	4.24
324	dark meat	15 samples	0.20	68.2	3.69
325	wing quarter (weighed with bone)	Meat only	0.37	34.2	1.99
326	leg quarter (weighed with bone)	Meat only	0.42	42.4	2.46
327	**Duck** raw, meat only	9 samples, meat from dressed carcase	0.28	75.0	3.15
328	meat, fat and skin	9 samples, dressed carcase excluding waste	0.67	43.9	1.80
329	roast, meat only	11 samples, meat from dressed carcase	0.21	64.2	4.05
330	meat, fat and skin	11 samples, dressed carcase excluding waste	0.40	49.6	3.14
331	**Goose** roast	Meat only	0.39	46.7	4.69
332	**Grouse** roast	Meat only	0.51	61.6	5.00
333	roast (weighed with bone)	Calculated from the previous item	0.51	40.6	3.30
334	**Partridge** roast	Meat only	0.39	54.5	5.87
335	roast (weighed with bone)	Calculated from the previous item	0.39	32.7	3.52

No	Food	Energy value		Protein (N × 6.25)	Fat	Carbo-hydrate	mg									
		kcal	kJ	g	g	g	Na	K	Ca	Mg	P	Fe	Cu	Zn	S	Cl
	Poultry and game															
314	**Chicken** raw, meat only	121	508	20.5	4.3	0	81	320	10	25	200	0.7	0.19	1.1	220	78
315	meat and skin	230	954	17.6	17.7	0	70	260	10	20	160	0.7	0.16	1.0	—	69
316	light meat	116	489	21.8	3.2	0	72	330	10	27	210[a]	0.5	0.14	0.7	210	70
317	dark meat	126	528	19.1	5.5	0	89	300	11	22	180	0.9	0.25	1.6	220	86
318	boiled, meat only	183	767	29.2	7.3	0	82	300	11	25	190	1.2	0.20	2.0	300	90
319	light meat	163	686	29.7	4.9	0	70	370	9	26	200[a]	0.6	0.17	1.0	290	82
320	dark meat	204	853	28.6	9.9	0	95	230	12	22	180	1.9	0.23	3.1	300	99
321	roast, meat only	148	621	24.8	5.4	0	81	310	9	24	210	0.8	0.12	1.5	260	87
322	meat and skin	216	902	22.6	14.0	0	72	270	9	21	170	0.8	0.12	1.4	—	77
323	light meat	142	599	26.5	4.0	0	71	330	9	26	220[a]	0.5	0.11	1.0	250	79
324	dark meat	155	648	23.1	6.9	0	91	290	9	22	190	1.0	0.13	2.1	260	95
325	wing quarter (weighed with bone)	74	311	12.4	2.7	0	41	160	5	12	110	0.4	0.06	0.8	130	44
326	leg quarter (weighed with bone)	92	388	15.4	3.4	0	50	190	6	15	130	0.5	0.07	0.9	160	54
327	**Duck** raw, meat only	122	513	19.7	4.8	0	110	290	12	19	200	2.4	0.34	1.9	210	98
328	meat, fat and skin	430	1772	11.3	42.7	0	77	210	11	14	130	2.4	0.27	1.3	—	69
329	roast, meat only	189	789	25.3	9.7	0	96	270	13	20	200	2.7	0.31	2.6	270	96
330	meat, fat and skin	339	1406	19.6	29.0	0	76	210	12	16	150	2.7	0.27	1.8	—	75
331	**Goose** roast	319	1327	29.3	22.4	0	150	410	10	31	270	4.6	(0.49)	—	320	160
332	**Grouse** roast	173	728	31.3	5.3	0	96	470	30	41	340	7.6	—	—	340	130
333	roast (weighed with bone)	114	480	20.6	3.5	0	63	310	20	27	220	5.0	—	—	220	88
334	**Partridge** roast	212	890	36.7	7.2	0	100	410	46	36	310	7.7	—	—	400	99
335	roast (weighed with bone)	127	533	22.0	4.3	0	60	240	28	22	190	4.6	—	—	240	59

[a] In frozen chickens treated with polyphosphates the light meat may contain up to 250 mg P per 100 g

No	Food	Retinol µg	Carotene µg	Vitamin D µg	Thiamin mg	Riboflavin mg	Nicotinic acid mg	Potential nicotinic acid from tryptophan mgTrp ÷ 60	Vitamin C mg	Vitamin E mg
	Poultry and game									
314	**Chicken** raw, meat only	Tr	Tr	Tr	0.10	0.16	7.8	3.8	0	0.10
315	meat and skin	Tr	Tr	Tr	0.08	0.14	6.0	3.3	0	—
316	light meat	Tr	Tr	Tr	0.10	0.10	9.9	4.1	0	0.08
317	dark meat	Tr	Tr	Tr	0.11	0.22	5.4	3.6	0	0.13
318	boiled, meat only	Tr	Tr	Tr	0.06	0.19	6.7	5.5	0	0.07
319	light meat	Tr	Tr	Tr	0.05	0.12	8.9	5.5	0	—
320	dark meat	Tr	Tr	Tr	0.07	0.28	4.3	5.3	0	—
321	roast, meat only	Tr	Tr	Tr	0.08	0.19	8.2	4.6	0	0.11
322	meat and skin	Tr	Tr	Tr	—	—	—	4.2	0	—
323	light meat	Tr	Tr	Tr	0.08	0.14	10.3	5.0	0	0.08
324	dark meat	Tr	Tr	Tr	0.09	0.24	6.1	4.3	0	0.15
325	wing quarter (weighed with bone)	Tr	Tr	Tr	0.04	0.10	4.1	2.3	0	0.06
326	leg quarter (weighed with bone)	Tr	Tr	Tr	0.05	0.12	5.1	2.9	0	0.07
327	**Duck** raw, meat only	—	—	—	0.36	0.45	5.3	4.2	0	0
328	meat, fat and skin	—	—	—	—	—	—	2.4	0	—
329	roast, meat only	—	—	—	0.26	0.47	5.1	5.4	0	0.02
330	meat, fat and skin	—	—	—	—	—	—	4.2	—	—
331	**Goose** roast	—	—	—	—	—	—	5.5	—	—
332	**Grouse** roast	—	—	—	(0.32)	(0.54)	(8.8)	5.8	—	—
333	roast (weighed with bone)	—	—	—	(0.21)	(0.36)	(5.8)	3.9	—	—
334	**Partridge** roast	—	—	—	—	—	—	6.9	—	—
335	roast (weighed with bone)	—	—	—	—	—	—	4.1	—	—

No	Food	Vitamin B_6 mg	Vitamin B_{12} µg	Folic acid Free µg	Folic acid Total µg	Pantothenic acid mg	Biotin µg	Notes
	Poultry and game							
314	**Chicken** raw, meat only	0.42	Tr	10	12	1.2	2	
315	meat and skin	0.30	Tr	7	8	0.9	2	
316	light meat	0.53	Tr	8	12	1.2	2	
317	dark meat	0.30	1	12	12	1.3	3	
318	boiled, meat only	0.35	Tr	5	8	1.1	4	
319	light meat	0.33	Tr	4	4	1.0	3	
320	dark meat	0.37	1	7	13	1.1	4	
321	roast, meat only	0.26	Tr	8	10	1.2	3	
322	meat and skin	—	Tr	—	—	—	—	
323	light meat	0.35	Tr	6	7	1.1	2	
324	dark meat	0.16	1	10	13	1.3	3	
325	wing quarter (weighed with bone)	0.13	Tr	4	5	0.6	2	
326	leg quarter (weighed with bone)	0.16	Tr	5	6	0.7	2	
327	**Duck** raw, meat only	0.34	3	7	25	1.6	6	
328	meat, fat and skin	—	—	—	—	—	—	
329	roast, meat only	0.25	3	7	10	1.5	4	
330	meat, fat and skin	—	—	—	—	—	—	
331	**Goose** roast	(0.43)	—	—	—	—	—	
332	**Grouse** roast	—	—	—	—	—	—	
333	roast (weighed with bone)	—	—	—	—	—	—	
334	**Partridge** roast	—	—	—	—	—	—	
335	roast (weighed with bone)	—	—	—	—	—	—	

No	Food	Description and number of samples	Edible matter, proportion of weight purchased	Water g	Total nitrogen g
	***Poultry and game** contd*				
336	**Pheasant** roast	Meat only	0.45	56.9	5.15
337	roast (weighed with bone)	Calculated from the previous item	0.45	35.8	3.24
338	**Pigeon** roast	Meat only	0.28	57.2	4.44
339	roast (weighed with bone)	Calculated from the previous item	0.28	25.2	1.95
340	**Turkey** raw, meat only	5 samples, light and dark meat from dressed carcase	0.57	75.5	3.51
341	meat and skin	5 samples, dressed carcase excluding waste	0.70	72.0	3.29
342	light meat	5 samples	0.32	75.2	3.71
343	dark meat	5 samples	0.25	75.9	3.24
344	roast, meat only	5 samples, light and dark meat from dressed carcase	0.46	68.0	4.61
345	meat and skin	5 samples, dressed carcase excluding waste	0.57	65.0	4.48
346	light meat	5 samples	0.25	68.4	4.76
347	dark meat	5 samples	0.21	67.7	4.44
	Other game				
348	**Hare** stewed	Meat only	0.44	60.7	4.78
349	stewed (weighed with bone)	Calculated from the previous item	0.44	44.3	3.48
350	**Rabbit** raw	9 samples, pieces of loin and leg, meat only	0.62	74.6	3.50
351	stewed	Meat only	0.35	63.9	4.37
352	stewed (weighed with bone)	Calculated from the previous item	0.35	32.5	2.23
353	**Venison** roast	Haunch, meat only	0.58	56.8	5.60

No	Food	Energy value		Protein (N × 6.25)	Fat	Carbo-hydrate	mg									
		kcal	kJ	g	g	g	Na	K	Ca	Mg	P	Fe	Cu	Zn	S	Cl
	***Poultry and game** contd*															
336	**Pheasant** roast	213	892	32.2	9.3	0	100	410	49	35	310	8.4	—	—	310	110
337	roast (weighed with bone)	134	563	20.3	5.9	0	66	260	31	22	190	5.3	—	—	190	68
338	**Pigeon** roast	230	961	27.8	13.2	0	110	410	16	34	400	19.4	—	—	300	99
339	roast (weighed with bone)	101	422	12.2	5.8	0	46	180	7	15	180	8.5	—	—	130	44
340	**Turkey** raw, meat only	107	454	21.9	2.2	0	54	300	8	23	190	0.8	0.13	1.7	220	48
341	meat and skin	145	606	20.6	6.9	0	49	270	9	19	170	0.8	0.12	1.6	—	43
342	light meat	103	435	23.2	1.1	0	43	320	6	25	200	0.5	0.11	1.2	210	42
343	dark meat	114	478	20.3	3.6	0	68	270	11	21	180	1.2	0.16	2.4	220	55
344	roast, meat only	140	590	28.8	2.7	0	57	310	9	27	220	0.9	0.15	2.4	290	52
345	meat and skin	171	717	28.0	6.5	0	52	280	9	24	200	0.9	0.14	2.1	—	47
346	light meat	132	558	29.8	1.4	0	45	340	7	29	230	0.5	0.14	1.5	280	42
347	dark meat	148	624	27.8	4.1	0	71	270	12	23	210	1.4	0.16	3.5	300	62
	Other game															
348	**Hare** stewed	192	804	29.9	8.0	0	40	210	21	22	250	10.8	—	—	320	74
349	stewed (weighed with bone)	139	585	21.8	5.8	0	29	150	15	16	180	7.9	—	—	230	54
350	**Rabbit** raw	124	520	21.9	4.0	0	67	360	22	25	220	1.0	0.54	1.4	200	74
351	stewed	179	749	27.3	7.7	0	32	210	11	22	200	1.9	—	—	250	43
352	stewed (weighed with bone)	91	381	13.9	3.9	0	16	110	6	11	100	1.0	—	—	130	22
353	**Venison** roast	198	832	35.0	6.4	0	86	360	29	33	290	7.8	—	—	320	89

No	Food	Retinol µg	Carotene µg	Vitamin D µg	Thiamin mg	Riboflavin mg	Nicotinic acid mg	Potential nicotinic acid from tryptophan mg Trp ÷ 60	Vitamin C mg	Vitamin E mg
	Poultry and game *contd*									
336	**Pheasant** roast	—	—	—	(0.04)	(0.15)	(6.1)	6.0	0	—
337	roast (weighed with bone)	—	—	—	(0.03)	(0.09)	(3.8)	3.8	0	—
338	**Pigeon** roast	—	—	—	—	—	(8.9)	5.2	0	Tr
339	roast (weighed with bone)	—	—	—	—	—	(3.9)	2.3	0	Tr
340	**Turkey** raw, meat only	Tr	Tr	Tr	0.09	0.16	7.9	4.1	0	Tr
341	meat and skin	—	—	—	—	—	—	3.8	0	—
342	light meat	Tr	Tr	Tr	0.08	0.11	9.9	4.3	0	Tr
343	dark meat	Tr	Tr	Tr	0.10	0.23	5.2	3.8	0	Tr
344	roast, meat only	Tr	Tr	Tr	0.07	0.21	8.5	5.4	0	Tr
345	meat and skin	—	—	—	—	—	—	5.2	0	—
346	light meat	Tr	Tr	Tr	0.07	0.14	10.0	5.6	0	0.02
347	dark meat	Tr	Tr	Tr	0.07	0.29	6.7	5.2	0	Tr
	Other game									
348	**Hare** stewed	—	—	—	—	—	—	5.6	0	—
349	stewed (weighed with bone)	—	—	—	—	—	—	4.1	0	—
350	**Rabbit** raw	—	—	—	0.10	0.19	8.4	4.1	0	0.13
351	stewed	—	—	—	0.07	0.28	8.5	5.1	0	—
352	stewed (weighed with bone)	—	—	—	0.04	0.14	4.3	2.6	0	—
353	**Venison** roast	—	—	—	0.22	—	—	6.5	0	—

No	Food	Vitamin B_6 mg	Vitamin B_{12} µg	Folic acid		Panto-thenic acid mg	Biotin µg	Notes
				Free µg	Total µg			
	***Poultry and game** contd*							
336	**Pheasant** roast	—	—	—	—	—	—	
337	roast (weighed with bone)	—	—	—	—	—	—	
338	**Pigeon** roast	—	—	—	—	—	—	
339	roast (weighed with bone)	—	—	—	—	—	—	
340	**Turkey** raw, meat only	0.46	2	11	15	0.8	2	
341	meat and skin	—	—	—	—	—	—	
342	light meat	0.59	1	7	8	0.8	1	
343	dark meat	0.30	3	17	25	0.9	2	
344	roast, meat only	0.32	2	12	15	0.8	2	
345	meat and skin	—	—	—	—	—	—	
346	light meat	0.31	1	9	13	0.7	1	
347	dark meat	0.32	3	16	17	0.9	2	
	Other game							
348	**Hare** stewed	—	—	—	—	—	—	
349	stewed (weighed with bone)	—	—	—	—	—	—	
350	**Rabbit** raw	0.50	10	4	5	0.8	1	
351	stewed	0.50	12	3	4	0.8	1	
352	stewed (weighed with bone)	0.26	6	2	2	0.4	Tr	
353	**Venison** roast	—	—	—	—	—	—	

No	Food	Description and number of samples	Edible matter, proportion of weight purchased	Water g	Total nitrogen g
	Offal				
354	**Brain, calf and lamb** raw	14 samples	1.00	79.4	1.64
355	**calf,** boiled	5 samples; soaked 2 hours, boiled 35 minutes	0.80	73.4	2.03
356	**lamb,** boiled	11 samples; soaked 2 hours, boiled 20 minutes	0.81	77.0	1.86
358	**Heart, lamb** raw	12 samples; fat and valves removed	0.73	75.6	2.73
359	**sheep,** roast	Ventricles only	0.53	57.3	4.18
360	**ox,** raw	18 samples; fat and valves removed	0.81	76.3	3.03
361	stewed	18 samples; fat and valves removed before cooking	0.50	61.5	5.02
362	**pig,** raw	Fat removed	—	79.2	2.74
364	**Kidney, lamb** raw	19 samples; core removed	0.93	78.9	2.64
365	fried	19 samples; core removed before cooking	0.59	66.5	3.94
366	**ox** raw	18 samples; core removed	0.82	79.8	2.51
367	stewed	18 samples; core removed before cooking, salt added	0.51	64.1	4.09
368	**pig** raw	20 samples; core removed	0.90	78.8	2.61
369	stewed	20 samples; core removed before cooking, salt added	0.56	66.3	3.91
371	**Liver, calf** raw	12 samples	1.00	69.7	3.21
372	fried	12 samples; coated in seasoned flour and fried	0.84	52.6	4.31
373	**chicken** raw	16 samples	1.00	72.9	3.05
374	fried	16 samples; coated in seasoned flour and fried	0.84	64.2	3.31
375	**lamb** raw	33 samples	1.00	67.3	3.22
376	fried	18 samples; coated in seasoned flour and fried	0.88	58.4	3.67
377	**ox** raw	33 samples	1.00	68.6	3.37
378	stewed	18 samples; coated in seasoned flour	0.82	62.6	3.96

No	Food	Energy value		Protein (N × 6.25)	Fat	Carbo-hydrate	mg									
		kcal	kJ	g	g	g	Na	K	Ca	Mg	P	Fe	Cu	Zn	S	Cl
	Offal															
354	**Brain, calf and lamb** raw	110	456	10.3	7.6	0	140	270	(12)[a]	15	340	1.6	0.30	1.2	130	150
355	**calf** boiled	152	630	12.7	11.2	0	210	190	(16)[a]	15	380	2.3	0.42	1.5	—	200
356	**lamb** boiled	126	523	11.6	8.8	0	210	190	(11)[a]	15	320	1.4	0.23	1.4	—	250
358	**Heart, lamb** raw	119	498	17.1	5.6	0	140	280	7	21	210	3.6	0.52	2.0	200	140
359	**sheep** roast	237	988	26.1	14.7	0	150	370	10	35	390	8.1	—	—	300	130
360	**ox** raw	108	455	18.9	3.6	0	95	320	5	25	230	4.9	0.43	2.0	190	95
361	stewed	179	752	31.4	5.9	0	180	210	7	29	270	7.7	0.73	3.5	310	210
362	**pig** raw	93	391	17.1	2.7	0	80	300	6	20	220	4.8	—	—	200	110
364	**Kidney, lamb** raw	90	380	16.5	2.7	0	220	270	10	17	260	7.4	0.42	2.4	180	270
365	fried	155	651	24.6	6.3	0	270	340	13	29	360	12.0	(0.65)	4.1	290	330
366	**ox** raw	86	363	15.7	2.6	0	180	230	10	15	230	5.7	0.42	1.9	170	200
367	stewed	172	720	25.6	7.7	0	400	180	16	19	300	8.0	0.66	3.0	270	520
368	**pig** raw	90	377	16.3	2.7	0	190	290	8	19	270	5.0	0.81	2.6	170	180
369	stewed	153	641	24.4	6.1	0	370	190	13	21	330	6.4	0.84	4.7	—	480
371	**Liver, calf** raw	153	642	20.1	7.3	1.9	93	330	7	20	360	8.0	11.0	7.8	240	89
372	fried	254	1063	26.9	13.2	7.3	170	410	15	26	470	7.5	12.0	6.2	300	210
373	**chicken** raw	135	567	19.1	6.3	0.6	85	300	8	21	320	9.5	0.52	3.4	220	100
374	fried	194	810	20.7	10.9	3.4	240	290	15	23	350	9.1	0.53	3.4	250	350
375	**lamb** raw	179	748	20.1	10.3	1.6	76	290	7	19	370	9.4	8.7	3.9	230	83
376	fried	232	970	22.9	14.0	3.9	190	300	12	22	400	10.0	9.9	4.4	270	250
377	**ox** raw	163	683	21.1	7.8	2.2	81	320	6	19	360	7.0	2.5	4.0	240	90
378	stewed	198	831	24.8	9.5	3.6	110	250	11	19	380	7.8	2.3	4.3	270	120

[a] The calcium content is extremely variable as scrapings of bone may easily be included in the edible portion. This is a minimum value

No	Food	Retinol µg	Carotene µg	Vitamin D µg	Thiamin mg	Riboflavin mg	Nicotinic acid mg	Potential nicotinic acid from tryptophan mg Trp ÷ 60	Vitamin C mg	Vitamin E mg
	Offal									
354	**Brain, calf and lamb** raw	Tr	Tr	Tr	0.07	0.24	3.0	2.2	23	1.2
355	**calf** boiled	Tr	Tr	Tr	0.08	0.19	2.2	2.7	17	2.3
356	**lamb** boiled	Tr	Tr	Tr	0.10	0.24	2.1	2.5	17	1.1
358	**Heart, lamb** raw	Tr	Tr	—	0.48	0.9	6.9	3.6	7	0.37
359	**sheep** roast	Tr	Tr	—	(0.45)	(1.5)	(9.1)	5.6	(11)	(0.70)
360	**ox** raw	Tr	Tr	—	0.45	0.8	6.3	4.0	7	0.45
361	stewed	Tr	Tr	—	0.21	1.1	4.7	6.7	6	0.72
362	**pig** raw	Tr	0	—	(0.48)	(0.9)	(6.9)	3.7	5	(0.37)
364	**Kidney, lamb** raw	100	—	—	0.49	1.8	8.3	3.5	7	0.45
365	fried	160	—	—	0.56	2.3	9.6	5.3	9	0.41
366	**ox** raw	150 *(40–270)*	—	—	0.37	2.1	6.0	3.4	10	0.18
367	stewed	250	—	—	0.25	2.1	4.8	5.5	10	0.42
368	**pig** raw	110 *(70–145)*	0	—	0.32	1.9	7.5	3.5	14	0.38
369	stewed	140	0	—	0.19	2.1	6.1	5.2	11	0.36
371	**Liver, calf** raw	14 600 *(8300–31 700)*	100	0.25	0.21	3.1	12.4	4.3	18	0.24
372	fried	17 400	100	0.25	0.27	4.2	15.6	5.8	13	0.50
373	**chicken** raw	9300 *(5700–12 800)*	0	0.21	0.36	2.7	10.2	4.1	23	0.25
374	fried	11 100	0	—	0.37	1.7	10.5	4.4	13	0.34
375	**lamb** raw	18 100 *(3000–55 000)*	60	0.50	0.27	3.3	14.2	4.3	10	0.46
376	fried	20 600	60	0.50	0.26	4.4	15.2	4.9	12	0.32
377	**ox** raw	16 500 *(11 600–24 000)*	1540	1.13	0.23	3.1	13.4	4.5	23	0.42
378	stewed	20 100	1540	1.13	0.18	3.6	10.3	5.3	15	0.44

No	Food	Vitamin B_6 mg	Vitamin B_{12} µg	Folic acid Free µg	Folic acid Total µg	Panto-thenic acid mg	Biotin µg	Notes
	Offal							
354	**Brain, calf and lamb** raw	0.10	9	2	6	2.0	2	
355	**calf,** boiled	0.12	7	1	3	1.4	3	
356	**lamb,** boiled	0.08	8	1	6	1.4	3	
358	**Heart, lamb,** raw	0.29	8	Tr	2	2.5	4	
359	**sheep** roast	(0.38)	(14)	(Tr)	(4)	(3.8)	(8)	
360	**ox** raw	0.23	13	2	4	2.4	2	
361	stewed	0.11	15	1	2	1.6	4	
362	**pig** raw	(0.29)	(8)	(Tr)	(2)	(2.5)	(4)	
364	**Kidney, lamb** raw	0.30	55	20	31	4.3	37	
365	fried	0.30	79	39	79	5.1	42	
366	**ox** raw	0.32	31	56	77	3.1	24	
367	stewed	0.30	31	49	75	3.0	49	
368	**pig** raw	0.25	14	5	42	3.0	32	
369	stewed	0.28	15	12	43	2.4	53	
371	**Liver, calf** raw	0.54	100	190	240	8.4	39	
372	fried	0.73	87	220	320	8.8	53	
373	**chicken** raw	0.40	56	290	590	6.1	210	
374	fried	0.45	49	160	500	5.5	170	
375	**lamb** raw	0.42	84	150	220	8.2	41	
376	fried	0.49	81	140	240	7.6	41	
377	**ox** raw	0.83	110	220	330	8.1	33	
378	stewed	0.52	110	180	290	5.7	50	

No	Food	Description and number of samples	Edible matter, proportion of weight purchased	Water g	Total nitrogen g
	***Offal** contd*				
379	**Liver, pig** raw	33 samples	1.00	69.5	3.41
380	stewed	18 samples; coated in seasoned flour	0.79	62.1	4.09
381	**Oxtail** raw	12 samples, lean only	0.38	68.6	3.20
382	stewed	12 samples, lean only; salt added	0.27	53.9	4.88
383	stewed (weighed with fat and bones)	Calculated from the previous item	0.27	20.5	1.85
384	**Sweetbread, lamb** raw	12 samples	1.00	75.5	2.44
385	fried	12 samples; soaked 2 hours, boiled for 1 hour then coated with egg and breadcrumbs and fried	0.60	59.9	3.10
387	**Tongue, lamb** raw	20 samples, fat and skin removed	0.57	67.9	2.45
388	**sheep** stewed	Fat and skin removed	0.33	56.9	2.91
389	**ox** pickled, raw	6 samples, fat and skin removed	0.60	62.4	2.51
390	boiled	Fat and skin removed	0.38	48.6	3.12
391	**Tripe** dressed	18 samples; lime treated before purchase	1.00	88.1	1.50
392	stewed	18 samples; lime treated before purchase, stewed in milk	0.56	78.5	2.37

No	Food	Energy value		Protein (N × 6.25)	Fat	Carbo-hydrate	mg									
		kcal	kJ	g	g	g	Na	K	Ca	Mg	P	Fe	Cu	Zn	S	Cl
	***Offal** contd*															
379	**Liver, pig** raw	154	647	21.3	6.8	2.1	87	320	6	21	370	21.0	2.7	6.9	230	95
380	stewed	189	793	25.6	8.1	3.6	130	250	11	22	390	17.0	2.5	8.2	280	150
381	**Oxtail** raw	171	714	20.0	10.1	0	110	270	9	20	160	2.7	0.20	5.6	190	110
382	stewed	243	1014	30.5	13.4	0	190	170	14	18	140	3.8	0.27	8.8	290	270
383	stewed (weighed with bones)	92	386	11.6	5.1	0	72	65	5	7	53	1.4	0.10	3.3	110	100
384	**Sweetbread, lamb** raw	131	549	15.3	7.8	0	75	420	8	21	400	1.7	0.20	1.9	140	120
385	fried	230	960	19.4	14.6	5.6	210	260	34	23	420	1.8	0.22	2.1	160	260
387	**Tongue, lamb** raw	193	800	15.3	14.6	0	420	250	6	33	170	2.2	0.64	2.7	190	550
388	**sheep** stewed	289	1197	18.2	24.0	0	80	110	11	13	200	3.4	—	—	190	80
389	**ox** pickled, raw	220	914	15.7	17.5	0	1210	300	7	19	150	4.9	0.37	3.5	190	1750
390	boiled	293	1216	19.5	23.9	0	1000	150	31	16	230	3.0	—	—	200	1450
391	**Tripe** dressed	60	252	9.4	2.5	0	46	8	75	8	37	0.5	0.09	1.5	80	8
392	stewed	100	418	14.8	4.5	Tr	73	100	150	15	90	0.7	0.14	2.3	140	58

No	Food	Retinol µg	Carotene µg	Vitamin D µg	Thiamin mg	Riboflavin mg	Nicotinic acid mg	Potential nicotinic acid from tryptophan mgTrp ÷ 60	Vitamin C mg	Vitamin E mg
	Offal *contd*									
379	**Liver, pig** raw	9200 *(5600–14 200)*	0	1.13	0.31	3.0	14.8	4.6	13	0.17
380	stewed	11 600	0	1.13	0.21	3.1	11.5	5.5	9	0.16
381	**Oxtail** raw	Tr	Tr	Tr	0.03	0.29	4.5	4.3	0	0.29
382	stewed	Tr	Tr	Tr	0.02	0.28	3.3	6.5	0	0.45
383	stewed (weighed with bones)	Tr	Tr	Tr	0.01	0.11	1.3	2.5	0	0.17
384	**Sweetbread, lamb** raw	Tr	Tr	Tr	0.03	0.25	3.7	3.3	18	0.44
385	fried	Tr	Tr	Tr	0.03	0.24	2.1	4.1	18	1.2
387	**Tongue, lamb** raw	Tr	Tr	Tr	0.17	0.49	4.9	3.3	7	0.21
388	**sheep** stewed	Tr	Tr	Tr	(0.13)	(0.45)	(3.7)	3.9	(6)	(0.32)
389	**ox** pickled, raw	Tr	Tr	Tr	0.10	0.38	6.4	3.4	3	0.28
390	boiled	Tr	Tr	Tr	(0.06)	(0.29)	(4.1)	4.2	(2)	(0.35)
391	**Tripe** dressed	Tr	Tr	Tr	Tr	0.01	0.06	2.0	(3)	0.08
392	stewed	Tr	Tr	Tr	Tr	0.08	0.02	3.2	(3)	0.09

No	Food	Vitamin B_6 mg	Vitamin B_{12} µg	Folic acid Free µg	Folic acid Total µg	Panto-thenic acid mg	Biotin µg	Notes
	Offal *contd*							
379	**Liver, pig** raw	0.68	25	59	110	6.5	27	
380	stewed	0.64	26	33	110	4.6	34	
381	**Oxtail** raw	0.27	3	3	7	1.0	1	
382	stewed	0.14	2	3	9	0.9	2	
383	stewed (weighed with bones)	0.05	1	1	3	0.3	1	
384	**Sweetbread, lamb** raw	0.03	6	11	13	1.0	3	
385	fried	0.02	4	6	14	0.8	5	
387	**Tongue, lamb** raw	0.17	7	1	4	1.0	1	
388	**sheep** stewed	(0.10)	(7)	(Tr)	(4)	(0.8)	(2)	
389	**ox** pickled, raw	0.18	5	3	6	0.8	2	
390	boiled	(0.09)	(4)	(2)	(5)	(0.5)	(3)	
391	**Tripe** dressed	Tr	Tr	Tr	2	Tr	Tr	
392	stewed	0.02	Tr	Tr	1	0.2	2	

No	Food	Description and number of samples	Edible matter, proportion of weight purchased	Water g	Total nitrogen g
	Meat products and dishes				
	Canned meats				
393	**Beef, corned**	18 samples	1.00	58.5	4.30
394	**Ham**	12 samples, 10 brands	1.00	72.5	2.95
395	**Ham and pork** chopped	12 samples, 5 brands	1.00	58.5	2.30
396	**Luncheon meat**	18 samples	1.00	51.5	2.02
397	**Stewed steak with gravy**	12 samples, 8 brands	1.00	70.0	2.37
398	**Tongue**	18 samples, lamb and ox	1.00	63.9	2.56
400	**Veal, jellied**	18 samples	1.00	68.8	4.00
	Offal products				
401	**Black pudding** fried	24 samples	—	44.0	2.06
402	**Faggots**	38 samples	1.00	47.1	1.78
403	**Haggis** boiled	8 samples	0.95	46.2	1.71
404	**Liver sausage**	24 samples	1.00	51.8	2.06
	Sausages				
405	**Frankfurters**	12 samples (cans and packets), 6 brands	1.00	59.5	1.52
406	**Polony**	24 samples	1.00	52.0	1.50
407	**Salami**	24 samples, 8 different countries of origin	1.00	28.0	3.09
408	**Sausages, beef** raw	20 samples	1.00	50.3	1.54
409	fried	20 samples	0.75	47.7	2.07
410	grilled	20 samples	0.75	47.9	2.08
411	**Sausages, pork** raw	18 samples	1.00	45.4	1.69
412	fried	18 samples	0.70	44.9	2.20
413	grilled	18 samples	0.72	45.1	2.13
414	**Saveloy**	60 samples	1.00	56.7	1.59

No	Food	Energy value kcal	Energy value kJ	Protein (N × 6.25) g	Fat g	Carbo-hydrate g	Na mg	K mg	Ca mg	Mg mg	P mg	Fe mg	Cu mg	Zn mg	S mg	Cl mg
	Meat products and dishes															
	Canned meats															
393	**Beef, corned**	217	905	26.9	12.1	0	950	140	14	15	120	2.9	0.24	5.6	240	1430
394	**Ham**	120	502	18.4	5.1	0	1250	280	9	18	280	1.2	0.22	2.3	180	1670
395	**Ham and pork** chopped	270	1118	14.4	23.6	0	1090	230	14	13	250	1.2	0.24	2.9	140	1210
396	**Luncheon meat**	313	1298	12.6	26.9	5.5	1050	140	15	8	200	1.1	0.33	2.2	120	1290
397	**Stewed steak with gravy**	176	730	14.8	12.5	1.0	380	240	14	14	98	2.1	0.19	3.3	130	550
398	**Tongue**	213	883	16.0	16.5	0	1050	97	32	14	140	2.5	0.29	2.3	210	1430
400	**Veal, jellied**	125	529	25.0	2.8	0	1190	240	15	19	180	1.5	0.34	3.3	230	1650
	Offal products															
401	**Black pudding** fried	305	1270	12.9	21.9	15.0	1210	140	35	16	110	20.0	0.37	1.3	110	1770
402	**Faggots**	268	1118	11.1	18.5	15.3	820	170	55	18	120	8.3	0.60	1.6	120	1160
403	**Haggis** boiled	310	1292	10.7	21.7	19.2	770	170	29	36	160	4.8	0.44	1.9	120	1200
404	**Liver sausage**	310	1283	12.9	26.9	4.3	860	170	26	12	230	6.4	0.63	2.3	130	1140
	Sausages															
405	**Frankfurters**	274	1135	9.5	25.0	3.0	980	98	34	9	130	1.5	0.24	1.4	90	1280
406	**Polony**	281	1168	9.4	21.1	14.2	870	120	42	13	130	1.3	0.32	1.2	90	1160
407	**Salami**	491	2031	19.3	45.2	1.9	1850	160	10	10	160	1.0	0.24	1.7	190	2460
408	**Sausages, beef** raw	299	1242	9.6	24.1	11.7	810	150	48	13	150	1.4	0.23	1.2	120	1100
409	fried	269	1124	12.9	18.0	14.9	1090	180	64	16	210	1.6	0.36	1.6	140	1470
410	grilled	265	1104	13.0	17.3	15.2	1100	190	73	17	210	1.7	0.30	1.7	140	1490
411	**Sausages, pork** raw	367	1520	10.6	32.1	9.5	760	160	41	11	160	1.1	0.27	1.2	120	1030
412	fried	317	1317	13.8	24.5	11.0	1050	200	55	15	210	1.5	0.37	1.7	160	1440
413	grilled	318	1320	13.3	24.6	11.5	1000	200	53	15	220	1.5	0.34	1.6	160	1340
414	**Saveloy**	262	1088	9.9	20.5	10.1	890	160	23	9	210	1.5	0.27	1.4	90	1030

No	Food	Retinol µg	Carotene µg	Vitamin D µg	Thiamin mg	Riboflavin mg	Nicotinic acid mg	Potential nicotinic acid from tryptophan mg Trp ÷ 60	Vitamin C mg	Vitamin E mg
	Meat products and dishes									
	Canned meats									
393	**Beef, corned**	Tr	Tr	Tr	Tr	0.23	2.5	6.5	0	0.78
394	**Ham**	Tr	Tr	Tr	0.52	0.25	3.9	3.0	0[a]	0.08
395	**Ham and pork** chopped	Tr	Tr	Tr	0.19	0.21	3.2	2.7	0[a]	0.11
396	**Luncheon meat**	Tr	Tr	Tr	0.07	0.12	1.8	2.7	0[a]	0.11
397	**Stewed steak with gravy**	Tr	Tr	Tr	Tr	0.13	2.4	2.8	0	0.59
398	**Tongue**	Tr	Tr	Tr	0.04	0.39	2.5	3.8	0	0.26
400	**Veal, jellied**	Tr	Tr	Tr	0.05	0.29	6.0	4.7	0	0.12
	Offal products									
401	**Black pudding** fried	Tr	Tr	Tr	0.09	0.07	1.0	2.8	0	0.24
402	**Faggots**	(1500)	Tr	(0.2)	0.14	0.49	3.0	2.1	Tr	—
403	**Haggis** boiled	(1800)	Tr	(0.05)	0.16	0.35	1.5	2.0	Tr	0.41
404	**Liver sausage**	(8300)	Tr	(0.6)	0.17	1.58	4.3	2.4	Tr	0.10
	Sausages									
405	**Frankfurters**	Tr	Tr	Tr	0.08	0.12	1.5	1.5	0	0.25
406	**Polony**	Tr	Tr	Tr	0.17	0.10	1.5	1.8	0	0.09
407	**Salami**	Tr	Tr	Tr	0.21	0.23	4.6	3.6	0	0.28
408	**Sausages, beef** raw	Tr	Tr	Tr	0.03	0.13	5.0	2.1	0	0.43
409	fried	Tr	Tr	Tr	Tr	0.14	6.9	2.8	0	0.28
410	grilled	Tr	Tr	Tr	Tr	0.14	5.4	2.8	0	0.22
411	**Sausages, pork** raw	Tr	Tr	Tr	0.04	0.12	3.4	2.3	0	0.24
412	fried	Tr	Tr	Tr	0.01	0.16	4.4	2.9	0	0.28
413	grilled	Tr	Tr	Tr	0.02	0.15	4.0	2.8	0	0.22
414	**Saveloy**	Tr	Tr	Tr	0.14	0.09	1.9	1.9	0	0.08

No	Food	Vitamin B_6 mg	Vitamin B_{12} µg	Folic acid Free µg	Folic acid Total µg	Pantothenic acid mg	Biotin µg	Notes
	Meat products and dishes							[a] Some brands have ascorbic acid added, and may contain from 12 to 60 mg per 100 g.
	Canned meats							
393	**Beef, corned**	0.06	2	1	2	0.4	2	
394	**Ham**	0.22	Tr	Tr	Tr	0.6	1	
395	**Ham and pork** chopped	0.05	1	Tr	1	0.4	2	
396	**Luncheon meat**	0.02	1	Tr	1	0.5	Tr	
397	**Stewed steak with gravy**	0.07	1	1	4	0.3	1	
398	**Tongue**	0.04	5	Tr	2	0.4	2	
400	**Veal, jellied**	0.14	2	3	3	0.3	3	
	Offal products							
401	**Black pudding** fried	0.04	1	2	5	0.6	2	
402	**Faggots**	0.17	5	19	22	1.1	4	
403	**Haggis** boiled	0.07	2	4	8	0.5	12	
404	**Liver sausage**	0.14	8	13	19	1.5	7	
	Sausages							
405	**Frankfurters**	0.03	1	Tr	1	0.4	2	
406	**Polony**	0.08	Tr	2	4	0.5	Tr	
407	**Salami**	0.15	1	2	3	0.8	3	
408	**Sausages, beef** raw	0.06	Tr	2	2	0.5	2	
409	fried	0.07	1	2	2	0.5	2	
410	grilled	0.07	1	2	4	0.5	2	
411	**Sausages, pork** raw	0.07	1	Tr	1	0.6	2	
412	fried	0.07	1	Tr	2	0.6	3	
413	grilled	0.06	1	1	3	0.6	3	
414	**Saveloy**	0.06	Tr	1	1	0.4	Tr	

No	Food	Description and number of samples	Edible matter, proportion of weight purchased	Water g	Total nitrogen g
	***Meat products and dishes** contd*				
415	**Beefburgers** frozen, raw	36 samples, 6 brands	1.00	56.3	2.43
416	fried	36 samples, 6 brands	0.76	53.0	3.27
417	**Brawn**	10 samples	1.00	72.0	1.99
418	**Meat paste**	67 samples, beef, chicken, ham and tongue, liver and bacon	1.00	67.1	2.43
419	**White pudding**	6 samples	1.00	22.8	1.12
	Meat and pastry products				
420	**Cornish pastie**	18 pasties, average 62% pastry, 38% filling	1.00	39.2	1.28
421	**Pork pie** individual	18 pies, average 55% pastry, 42% meat filling, 3% jelly	1.00	36.8	1.56
422	**Sausage roll** flaky pastry	Recipe p 341	—	23.0	1.21
423	short pastry	Recipe p 341	—	21.8	1.36
424	**Steak and kidney pie** pastry top only	Recipe p 341	—	48.8	2.47
425	individual	10 pies, purchased cooked; pastry top and bottom	1.00	42.6	1.46
	Cooked dishes				
426	**Beef steak pudding**	Recipe p 342	—	57.9	1.77
427	**Beef stew**	Recipe p 342	—	77.7	1.54
428	**Bolognese sauce**	Recipe p 342	—	75.0	1.27
429	**Curried meat**	Recipe p 342	—	68.2	1.55
430	**Hot pot**	Recipe p 342	—	72.1	1.48
431	**Irish stew**	Recipe p 342	—	76.0	0.83
432	**Irish stew** (weighed with bones)	Calculated from the previous item	—	69.4	0.76
433	**Moussaka**	Recipe p 343	—	65.7	1.49
434	**Shepherd's pie**	Recipe p 343	—	76.1	1.24

No	Food	Energy value kcal	Energy value kJ	Protein g	Fat g	Carbo-hydrate g	Na mg	K mg	Ca mg	Mg mg	P mg	Fe mg	Cu mg	Zn mg	S mg	Cl mg
	***Meat products and dishes** contd*															
415	**Beefburgers** frozen, raw	265	1102	15.2	20.5	5.3	600	270	23	17	190	2.5	0.25	3.2	160	800
416	fried	264	1099	20.4	17.3	7.0	880	340	33	23	250	3.1	0.28	4.2	220	1120
417	**Brawn**	153	636	12.4	11.5	0	750	85	38	7	59	1.0	0.19	1.3	100	1110
418	**Meat paste**	173	721	15.2	11.2	3.0	740	160	86	15	170	2.3	0.31	2.3	150	1060
419	**White pudding**	450	1876	7.0	31.8	36.3	370	190	38	61	230	2.1	0.43	1.6	110	600
	Meat and pastry products															
420	**Cornish pastie**	332	1388	8.0	20.4	31.1	590	190	60	18	110	1.5	0.35	1.0	100	860
421	**Pork pie** individual	376	1564	9.8	27.0	24.9	720	150	47	16	120	1.4	0.32	1.0	100	1030
422	**Sausage roll** flaky pastry	479	1991	7.2	36.2	33.1	550	110	70	13	97	1.3	0.17	0.7	—	810
423	short pastry	463	1929	8.1	31.8	38.4	580	120	82	15	110	1.5	0.18	0.7	—	850
424	**Steak and kidney pie** pastry top only	286	1195	15.2	18.3	16.2	680	240	37	20	140	2.8	0.21	2.4	—	1020
425	individual	323	1349	9.1	21.2	25.6	510	140	53	18	110	2.5	0.10	1.2	—	720
	Cooked dishes															
426	**Beef steak pudding**	223	934	10.8	12.1	18.9	360	180	110	15	140	1.5	0.11	1.8	—	430
427	**Beef stew**	119	498	9.6	7.5	3.6	400	200	19	14	73	1.2	0.10	1.8	—	590
428	**Bolognese sauce**	139	579	8.0	10.9	2.5	440	310	26	18	81	1.6	0.14	1.9	82	670
429	**Curried meat**	160	668	9.6	10.1	8.2	480	210	33	22	71	2.9	0.16	2.5	—	710
430	**Hot pot**	114	480	9.3	4.2	10.4	670	430	22	25	83	1.2	0.16	1.7	—	1040
431	**Irish stew**	124	520	5.2	7.3	10.1	360	340	12	18	60	0.6	0.12	1.1	—	570
432	**Irish stew** (weighed with bones)	114	475	4.7	6.7	9.2	330	310	11	16	55	0.5	0.11	1.0	—	520
433	**Moussaka**	195	811	9.3	13.4	9.8	320	350	88	21	130	1.3	0.13	1.8	99	510
434	**Shepherd's pie**	119	497	7.6	6.1	8.9	450	240	15	16	69	1.1	0.13	1.9	79	670

No	Food	Retinol µg	Carotene µg	Vitamin D µg	Thiamin mg	Riboflavin mg	Nicotinic acid mg	Potential nicotinic acid from tryptophan mg Trp ÷ 60	Vitamin C mg	Vitamin E mg
	***Meat products and dishes** contd*									
415	**Beefburgers** frozen, raw	Tr	Tr	Tr	0.04	0.21	3.7	2.8	0	0.27
416	fried	Tr	Tr	Tr	0.02	0.23	4.2	3.8	0	0.58
417	**Brawn**	Tr	Tr	Tr	0.05	0.08	1.0	2.3	0	0.06
418	**Meat paste**	Tr	Tr	Tr	0.03	0.26	3.8	2.8	0	0.17
419	**White pudding**	Tr	Tr	Tr	0.26	0.08	0.5	1.3	0	1.0
	Meat and pastry products									
420	**Cornish pastie**	Tr	Tr	Tr	0.10	0.06	1.6	1.7	0	1.3
421	**Pork pie** individual	Tr	Tr	Tr	0.16	0.09	1.8	2.1	0	0.43
422	**Sausage roll** flaky pastry	125	Tr	1.11	0.11	0.04	1.8	1.5	0	1.3
423	short pastry	100	Tr	0.86	0.12	0.04	2.0	1.7	0	1.0
424	**Steak and kidney pie** pastry top only	100	Tr	—	0.14	0.52	3.6	3.2	2	0.7
425	individual	—	Tr	—	0.12	0.15	1.7	1.7	0	—
	Cooked dishes									
426	**Beef steak pudding**	Tr	Tr	Tr	0.08	0.09	1.8	0.6	Tr	0.2
427	**Beef stew**	Tr	1600	Tr	0.04	0.10	1.7	2.1	Tr	0.15
428	**Bolognese sauce**	Tr	1940	Tr	0.06	0.12	1.6	1.7	5	1.3
429	**Curried meat**	Tr	Tr	Tr	0.02	0.09	0.9	2.0	2	0.83
430	**Hot pot**	Tr	1900	Tr	0.07	0.10	1.8	2.0	5	0.20
431	**Irish stew**	Tr	Tr	Tr	0.06	0.07	1.4	1.7	4	0.11
432	**Irish stew** (weighed with bones)	Tr	Tr	Tr	0.05	0.06	1.3	1.6	4	0.10
433	**Moussaka**	30	65	0.07	0.06	0.15	1.4	2.0	4	0.32
434	**Shepherd's pie**	14	Tr	0.14	0.04	0.12	1.7	1.5	2	0.28

No	Food	Vitamin B_6 mg	Vitamin B_{12} µg	Folic acid Free µg	Folic acid Total µg	Pantothenic acid mg	Biotin µg	Notes
	***Meat products and dishes** contd*							
415	**Beefburgers** frozen, raw	0.20	1	2	12	0.4	1	
416	fried	0.20	2	2	15	0.5	2	
417	**Brawn**	0.05	Tr	Tr	3	0.9	Tr	
418	**Meat paste**	0.08	3	7	9	0.3	3	
419	**White pudding**	0.06	1	2	6	0.8	18	
	Meat and pastry products							
420	**Cornish pastie**	0.12	1	3	3	0.6	1	
421	**Pork pie** individual	0.06	1	3	3	0.6	1	
422	**Sausage roll** flaky pastry	0.06	Tr	3	4	0.2	1	
423	short pastry	0.07	Tr	3	5	0.3	1	
424	**Steak and kidney pie** pastry top only	0.17	8	9	14	0.8	6	
425	individual	0.06	2	7	8	(0.3)	(1)	
	Cooked dishes							
426	**Beef steak pudding**	0.13	1	1	5	0.3	Tr	
427	**Beef stew**	0.13	1	Tr	5	0.3	Tr	
428	**Bolognese sauce**	0.16	1	Tr	10	0.4	Tr	
429	**Curried meat**	0.10	1	Tr	5	0.3	Tr	
430	**Hot pot**	0.21	1	1	8	0.4	Tr	
431	**Irish stew**	0.15	1	1	6	0.3	1	
432	**Irish stew** (weighed with bones)	0.14	1	1	5	0.3	1	
433	**Moussaka**	0.17	1	1	8	0.5	2	
434	**Shepherd's pie**	0.17	1	2	7	0.3	Tr	

No	Food	Description and number of samples	Edible matter, proportion of weight purchased	Water g	Total nitrogen g

No	Food	Energy value		Protein	Fat	Carbo-hydrate	mg									
		kcal	kJ	g	g	g	Na	K	Ca	Mg	P	Fe	Cu	Zn	S	Cl

No	Food	Retinol µg	Carotene µg	Vitamin D µg	Thiamin mg	Riboflavin mg	Nicotinic acid mg	Potential nicotinic acid from tryptophan mgTrp ÷ 60	Vitamin C mg	Vitamin E mg

No	Food	Vitamin B_6 mg	Vitamin B_{12} µg	Folic acid Free µg	Folic acid Total µg	Panto-thenic acid mg	Biotin µg	Notes

No	Food	Description and number of samples	Edible matter, proportion of weight purchased	Water g	Total nitrogen g
	White fish				
438	**Cod** raw, fresh fillets	Samples from 3 different shops	0.89	82.1	2.78
439	frozen steaks	12 packets, 3 brands	1.00	83.9	2.49
440	baked	Fillets; baked in the oven with added butter	0.69	76.6	3.43
441	baked (weighed with bones and skin)	Calculated from the previous item	0.69	65.1	2.92
442	fried in batter	Purchased cooked	1.00	60.9	3.14
443	grilled	Frozen steaks, 12 samples; butter and salt added	0.63	78.0	3.32
444	poached	Fillets; poached in milk, butter and salt added	0.75	77.7	3.35
445	poached (weighed with bones and skin)	Calculated from the previous item	0.75	67.6	2.91
446	steamed	Middle cuts	0.66	79.2	2.98
447	steamed (weighed with bones and skin)	Calculated from the previous item	0.66	64.1	2.42
448	**smoked** raw	Samples from 3 different shops	0.99	78.0	2.93
449	poached	Poached in milk, butter added	0.80	73.7	3.46
450	**dried** salt, boiled	Stockfish; soaked 24 hours and boiled	0.99	64.9	5.20
451	**Haddock, fresh** raw	Fillets	—	81.3	2.68
452	fried	Fish without heads, coated in crumbs; all except bones	1.15	65.1	3.42
453	fried (weighed with bones)	Calculated from the previous item	1.15	60.0	3.15
454	steamed	Middle cut	0.59	75.1	3.65
455	steamed (weighed with bones and skin)	Calculated from the previous item	0.59	57.1	2.77
456	**smoked** steamed	Flesh only	0.55	71.6	3.73
457	steamed (weighed with bones and skin)	Calculated from the previous item	0.55	46.5	2.42

No	Food	Energy value kcal	Energy value kJ	Protein (N × 6.25) g	Fat g	Carbo-hydrate g	Na mg	K mg	Ca mg	Mg mg	P mg	Fe mg	Cu mg	Zn mg	S mg	Cl mg
	White fish															
438	**Cod** raw, fresh fillets	76	322	17.4	0.7	0	77	320	16	23	170	0.3	0.06	0.4	200	110
439	frozen steaks	68	287	15.6	0.6	0	68	310	11	22	160	0.3	0.06	0.3	180	95
440	baked	96	408	21.4	1.2	0	340	350	22	26	190	0.4	0.07	0.5	230	520
441	baked (weighed with bones and skin)	82	348	18.3	1.0	0	290	300	19	22	160	0.4	0.06	0.4	200	440
442	fried in batter	199	834	19.6	10.3	7.5	100	370	80	24	200	0.5	(0.07)	—	—	150
443	grilled	95	402	20.8	1.3	0	91	380	10	26	200	0.4	0.07	0.5	240	130
444	poached	94	396	20.9	1.1	0	110	330	29	26	180	0.3	0.09	0.5	250	150
445	poached (weighed with bones and skin)	82	346	18.2	1.0	0	96	290	25	23	160	0.3	0.08	0.4	220	130
446	steamed	83	350	18.6	0.9	0	100	360	15	21	240	0.5	0.10	0.5	210	120
447	steamed (weighed with bones and skin	67	283	15.1	0.7	0	81	290	12	17	200	0.4	0.08	0.4	170	97
448	**smoked** raw	79	333	18.3	0.6	0	1170	390	14	25	190	0.4	0.17	0.4	210	1800
449	poached	101	426	21.6	1.6	0	1200	360	25	25	190	0.5	0.23	0.6	250	1800
450	**dried** salt, boiled	138	586	32.5	0.9	0	400	31	22	35	160	1.8	—	—	370	670
451	**Haddock, fresh** raw	73	308	16.8	0.6	0	120	300	18	23	170	0.6	0.19	0.3	220	160
452	fried	174	729	21.4	8.3	3.6	180	350	110	31	250	1.2	—	—	290	180
453	fried (weighed with bones)	160	669	19.7	7.6	3.3	160	320	100	28	230	1.1	—	—	260	170
454	steamed	98	417	22.8	0.8	0	120	320	55	28	230	0.7	0.13	(0.4)	300	(140)
455	steamed (weighed with bones and skin)	75	316	17.3	0.6	0	92	250	41	21	180	0.5	0.10	(0.3)	230	(110)
456	**smoked** steamed	101	429	23.3	0.9	0	1220	290	58	25	250	1.0	—	—	250	1900
457	steamed (weighed with bones and skin	66	279	15.1	0.6	0	790	190	37	17	160	0.7	—	—	160	1230

No	Food	Retinol µg	Carotene µg	Vitamin D µg	Thiamin mg	Riboflavin mg	Nicotinic acid mg	Potential nicotinic acid from tryptophan mg Trp ÷ 60	Vitamin C mg	Vitamin E mg
	White fish									
438	**Cod** raw, fresh fillets	Tr	Tr	Tr	0.08 *(0.05–0.18)*	0.07 *(0.02–0.16)*	1.7	3.2	Tr	0.44
439	frozen steaks	Tr	Tr	Tr	0.06	0.05	1.5	2.9	Tr	—
440	baked	Tr	Tr	Tr	0.07	0.07	1.7	4.0	Tr	0.59
441	baked (weighed with bones and skin)	Tr	Tr	Tr	0.06	0.06	1.3	3.4	Tr	0.50
442	fried in batter	Tr	Tr	Tr	—	—	—	3.7	Tr	—
443	grilled	Tr	Tr	Tr	0.08	0.06	1.9	3.9	Tr	—
444	poached	Tr	Tr	Tr	0.08	0.08	1.7	3.9	Tr	0.61
445	poached (weighed with bones and skin)	Tr	Tr	Tr	0.07	0.07	1.5	3.4	Tr	0.53
446	steamed	Tr	Tr	Tr	(0.09)	(0.09)	(2.1)	3.5	Tr	(0.54)
447	steamed (weighed with bones and skin)	Tr	Tr	Tr	(0.07)	(0.07)	(1.7)	2.8	Tr	(0.44)
448	**smoked** raw	Tr	Tr	Tr	0.08	0.07	1.4	3.4	Tr	—
449	poached	Tr	Tr	Tr	0.10	0.11	1.7	4.0	Tr	—
450	**dried** salt, boiled	Tr	Tr	Tr	(Tr)	(Tr)	—	6.1	Tr	—
451	**Haddock, fresh** raw	Tr	Tr	Tr	0.07 *(0.03–0.10)*	0.10 *(0.02–0.16)*	4.0	3.1	Tr	—
452	fried	Tr	Tr	Tr	—	—	—	4.0	Tr	—
453	fried (weighed with bones)	Tr	Tr	Tr	—	—	—	3.7	Tr	—
454	steamed	Tr	Tr	Tr	(0.08)	(0.13)	(5.1)	4.3	Tr	—
455	steamed (weighed with bones and skin)	Tr	Tr	Tr	(0.06)	(0.10)	(3.9)	3.2	Tr	—
456	**smoked** steamed	Tr	Tr	Tr	(0.10)	(0.11)	(1.7)	4.4	Tr	—
457	steamed (weighed with bones and skin)	Tr	Tr	Tr	(0.07)	(0.07)	(1.1)	2.8	Tr	—

No	Food	Vitamin B_6 mg	Vitamin B_{12} µg	Folic acid Free µg	Folic acid Total µg	Pantothenic acid mg	Biotin µg	Notes
	White fish							
438	**Cod** raw, fresh fillets	0.33	2	8	12	0.20	3	
439	frozen steaks	0.34	1	4	6	(0.20)	(3)	
440	baked	0.38	2	7	12	(0.20)	(3)	
441	baked (weighed with bones and skin)	0.32	2	6	10	(0.17)	(3)	
442	fried in batter	—	—	—	—	—	—	
443	grilled	0.41	2	6	10	(0.25)	(3)	
444	poached	0.37	2	5	14	(0.19)	(3)	
445	poached (weighed with bones and skin)	0.32	2	4	12	(0.16)	(3)	
446	steamed	(0.37)	(3)	(5)	(12)	(0.20)	(3)	
447	steamed (weighed with bones and skin)	(0.30)	(2)	(4)	(10)	(0.16)	(2)	
448	**smoked** raw	0.32	2	3	5	(0.20)	(3)	
449	poached	0.35	3	3	5	(0.20)	(3)	
450	**dried** salt, boiled	—	(Tr)	(Tr)	(Tr)	—	(Tr)	
451	**Haddock, fresh** raw	0.20	1	5	13	0.20	5	
452	fried	—	—	—	—	—	—	
453	fried (weighed with bones)	—	—	—	—	—	—	
454	steamed	(0.25)	(1)	(3)	(16)	(0.20)	(6)	
455	steamed (weighed with bones and skin)	(0.19)	(1)	(2)	(12)	(0.15)	(5)	
456	**smoked** steamed	(0.35)	(3)	(3)	(5)	(0.20)	(3)	
457	steamed (weighed with bones and skin)	(0.28)	(2)	(2)	(3)	(0.13)	(2)	

No	Food	Description and number of samples	Edible matter, proportion of weight purchased	Water g	Total nitrogen g
	***White fish** contd*				
458	**Halibut** raw	Literature sources	—	78.1	2.83
459	steamed	Middle cut	0.66	70.9	3.80
460	steamed (weighed with bones and skin)	Calculated from the previous item	0.66	53.8	2.88
461	**Lemon sole** raw	Literature sources	—	81.2	2.74
462	fried	Fish without head and fins, coated in crumbs; all except bones	0.91	60.4	2.57
463	fried (weighed with bones)	Calculated from the previous item	0.91	47.7	2.03
464	steamed	Flesh only from fish without head and fins	0.62	77.2	3.29
465	steamed (weighed with bones and skin)	Calculated from the previous item	0.62	54.9	2.34
466	**Plaice** raw	8 fish, purchased whole	0.42	79.5	2.86
467	fried in batter	6 samples, purchased cooked	1.00	52.4	2.52
468	fried in crumbs	Fillets from 8 fish, dipped in egg and breadcrumbs and fried; light skin included	0.53	59.9	2.88
469	steamed	Flesh only from fish without head and fins	0.49	78.0	3.02
470	steamed (weighed with bones and skin)	Calculated from the previous item	0.49	42.1	1.63
471	**Saithe** raw	Coley, coalfish; literature sources	—	81.0	(2.72)
472	steamed	Pieces from tail end	0.65	74.8	3.73
473	steamed (weighed with bones and skin)	Calculated from the previous item	0.65	63.5	3.17
475	**Whiting** fried	Fish without heads, coated in crumbs; all except bones	1.02	63.0	2.90
476	fried (weighed with bones)	Calculated from the previous item	1.02	56.8	2.61
477	steamed	Flesh only from fish without heads	0.57	76.9	3.35
478	steamed (weighed with bones)	Calculated from the previous item	0.57	52.2	2.28

No	Food	Energy value kcal	Energy value kJ	Protein (N × 6.25) g	Fat g	Carbo-hydrate g	Na mg	K mg	Ca mg	Mg mg	P mg	Fe mg	Cu mg	Zn mg	S mg	Cl mg
	***White fish** contd*															
458	**Halibut** raw	92	390	17.7	2.4	0	(84)	(260)	(10)	(17)	(190)	(0.5)	(0.05)	—	(190)	(60)
459	steamed	131	553	23.8	4.0	0	110	340	13	23	260	0.6	0.07	—	260	80
460	steamed (weighed with bones and skin)	99	417	18.0	3.0	0	84	260	10	18	190	0.5	0.05	—	190	61
461	**Lemon sole** raw	81	343	17.1	1.4	0	(95)	(230)	(17)	(17)	(200)	(0.5)	(0.10)	—	(200)	(97)
462	fried	216	904	16.1	13.0	9.3	140	250	95	22	240	1.1	0.16	—	190	120
463	fried (weighed with bones)	171	715	12.7	10.3	7.4	110	200	75	16	190	0.9	0.13	—	150	98
464	steamed	91	384	20.6	0.9	0	120	280	21	20	250	0.6	0.12	—	240	120
465	steamed (weighed with bones and skin)	64	270	14.6	0.6	0	82	200	15	14	180	0.4	0.09	—	170	83
466	**Plaice** raw	91	386	17.9	2.2	0	120	280	51	22	180	0.3	0.05	0.5	240	170
467	fried in batter	279	1165	15.8	18.0	14.4	220	230	93	21	170	1.0	0.17	1.0	210	280
468	fried in crumbs	228	951	18.0	13.7	8.6	220	280	67	24	180	0.8	0.20	0.7	240	310
469	steamed	93	392	18.9	1.9	0	120	280	38	24	250	0.6	—	—	250	110
470	steamed (weighed with bones and skin)	50	210	10.2	1.0	0	65	150	20	13	130	0.3	—	—	130	61
471	**Saithe** raw	(73)	(308)	(17.0)	(0.5)	0	(73)	(260)	(14)	(23)	(190)	(0.5)	—	—	(190)	(200)
472	steamed	99	418	23.3	0.6	0	97	350	19	31	250	0.6	—	—	270	83
473	steamed (weighed with bones and skin)	84	355	19.8	0.5	0	83	300	16	26	210	0.5	—	—	230	71
475	**Whiting** fried	191	801	18.1	10.3	7.0	200	320	48	33	260	0.7	—	—	270	190
476	fried (weighed with bones)	173	722	16.3	9.3	6.3	180	290	43	29	230	0.6	—	—	240	180
477	steamed	92	389	20.9	0.9	0	130	300	42	28	190	1.0	—	—	310	93
478	steamed (weighed with bones)	63	265	14.3	0.6	0	86	200	29	19	130	0.7	—	—	210	63

No	Food	Retinol µg	Carotene µg	Vitamin D µg	Thiamin mg	Riboflavin mg	Nicotinic acid mg	Potential nicotinic acid from tryptophan mgTrp ÷60	Vitamin C mg	Vitamin E mg
	***White fish** contd*									
458	**Halibut** raw	Tr[a]	Tr	Tr[a]	0.08 *(0.03–0.12)*	0.10 *(0.04–0.18)*	5.0	3.3	Tr	0.90
459	steamed	Tr[a]	Tr	Tr[a]	(0.08)	(0.11)	(5.2)	4.4	Tr	(1.0)
460	steamed (weighed with bones and skin)	Tr[a]	Tr	Tr[a]	(0.06)	(0.08)	(4.0)	3.4	Tr	(0.76)
461	**Lemon sole** raw	Tr	Tr	Tr	0.09	0.08	3.5	3.2	Tr	—
462	fried	Tr	Tr	Tr	—	—	—	3.0	Tr	—
463	fried (weighed with bones)	Tr	Tr	Tr	—	—	—	2.4	Tr	—
464	steamed	Tr	Tr	Tr	(0.09)	(0.09)	(3.6)	3.8	Tr	—
465	steamed (weighed with bones and skin)	Tr	Tr	Tr	(0.06)	(0.06)	(2.6)	2.7	Tr	—
466	**Plaice** raw	Tr	Tr	Tr	0.30 *(0.02–0.46)*	0.10 *(0.09–0.33)*	3.2	3.3	Tr	—
467	fried in batter	Tr	Tr	Tr	0.20	0.15	2.0	2.9	Tr	—
468	fried in crumbs	Tr	Tr	Tr	0.23	0.18	2.9	3.4	Tr	—
469	steamed	Tr	Tr	Tr	(0.30)	(0.11)	(3.2)	3.5	Tr	—
470	steamed (weighed with bones and skin)	Tr	Tr	Tr	(0.16)	(0.06)	(1.7)	1.9	Tr	—
471	**Saithe** raw	Tr	Tr	Tr	0.10	0.20	3.4	3.2	Tr	0.36
472	steamed	Tr	Tr	Tr	(0.12)	(0.26)	(4.0)	4.4	Tr	(0.47)
473	steamed (weighed with bones and skin)	Tr	Tr	Tr	(0.10)	(0.22)	(3.4)	3.7	Tr	(0.40)
475	**Whiting** fried	Tr	Tr	Tr	—	—	—	3.4	Tr	—
476	fried (weighed with bones)	Tr	Tr	Tr	—	—	—	3.1	Tr	—
477	steamed	Tr	Tr	Tr	—	—	—	3.9	Tr	—
478	steamed (weighed with bones)	Tr	Tr	Tr	—	—	—	2.7	Tr	—

No	Food	Vitamin B_6 mg	Vitamin B_{12} µg	Folic acid		Pantothenic acid mg	Biotin µg	Notes
				Free µg	Total µg			
	***White fish** contd*							[a] These are values for Atlantic halibut. Pacific halibut have been reported to contain 120 µg retinol and 1 µg vitamin D per 100 g.
458	**Halibut** raw	0.20	1	4	12	0.30	5	
459	steamed	(0.23)	(1)	(2)	(14)	(0.28)	(5)	
460	steamed (weighed with bones and skin)	(0.17)	(1)	(2)	(11)	(0.21)	(4)	
461	**Lemon sole** raw	—	1	5	11	0.30	(5)	
462	fried	—	—	—	—	—	—	
463	fried (weighed with bones)	—	—	—	—	—	—	
464	steamed	—	(1)	(3)	(13)	(0.31)	(5)	
465	steamed (weighed with bones and skin)	—	(1)	(2)	(9)	(0.22)	(4)	
466	**Plaice** raw	0.43	2	5	10	0.80	—	
467	fried in batter	—	—	—	—	—	—	
468	fried in crumbs	0.36	1	8	17	—	—	
469	steamed	(0.47)	(2)	(3)	(11)	(0.70)	—	
470	steamed (weighed with bones and skin)	(0.25)	(1)	(2)	(6)	(0.38)	—	
471	**Saithe** raw	0.47	4	—	—	0.38	7	
472	steamed	(0.62)	(5)	—	—	(0.40)	(8)	
473	steamed (weighed with bones and skin)	(0.53)	(4)	—	—	(0.34)	(7)	
475	**Whiting** fried	—	—	—	—	—	—	
476	fried (weighed with bones)	—	—	—	—	—	—	
477	steamed	—	—	—	—	—	—	
478	steamed (weighed with bones)	—	—	—	—	—	—	

No	Food	Description and number of samples	Edible matter, proportion of weight purchased	Water g	Total nitrogen g
	Fatty fish				
480	**Eel** raw	Yellow eels, flesh only	0.67	71.3 [a]	2.66
481	stewed	Yellow eels, flesh only; stewed in water	—	(61.3)	(3.30)
482	**Herring** raw	12 fish, sampled in November, flesh only	0.55	63.9 [b]	2.69
483	fried	Flesh, skin and roes; covered in oatmeal	0.77	58.7	3.69
484	fried (weighed with bones)	Calculated from the previous item	0.77	51.6	3.24
485	grilled	12 fish, flesh only	0.53	65.5	3.26
486	grilled (weighed with bones)	Calculated from the previous item	0.53	44.5	2.22
487	**Bloater** grilled	Flesh only from fish without heads or roes	0.65	55.6	3.76
488	grilled (weighed with bones)	Calculated from the previous item	0.65	41.1	2.78
489	**Kipper** baked	Flesh only	0.45	58.7	4.08
490	baked (weighed with bones)	Calculated from the previous item	0.45	31.6	2.20
491	**Mackerel** raw	Literature sources	—	64.0 [b]	3.04
492	fried	Flesh only from fish without heads	0.61	65.6	3.44
493	fried (weighed with bones)	Calculated from the previous item	0.61	47.8	2.51
494	**Pilchards** canned in tomato sauce	6 cans, 4 brands (South African), total contents of can	1.00	70.0	3.01
495	**Salmon** raw	Atlantic salmon; literature sources	—	68.0	(2.94)
496	steamed	Shoulder cut, flesh only	0.73	65.4	3.21
497	steamed (weighed with bones and skin)	Calculated from the previous item	0.73	53.0	2.60

[a] The water content varies according to stage of maturity: elvers contain 81.8 g and silver eels 57.1 g water per 100 g

[b] The values for water vary throughout the year from about 75 g per 100 g in February–April to 60 g per 100 g in July–October

No	Food	Energy value kcal	Energy value kJ	Protein (N × 6.25) g	Fat g	Carbo-hydrate g	Na mg	K mg	Ca mg	Mg mg	P mg	Fe mg	Cu mg	Zn mg	S mg	Cl mg
	Fatty fish															
480	**Eel** raw	168	700	16.6	11.3[a]	0	89	270	19	19	220	0.7	0.05	0.5	190	57
481	stewed	(201)	(839)	(20.6)	(13.2)	0	(84)	(250)	(21)	(20)	(230)	(0.9)	(0.06)	(0.6)	(230)	(53)
482	**Herring** raw	234	970	16.8	18.5[b]	0	67	340	33	29	210	0.8	0.12	0.5	190	76
483	fried	234	975	23.1	15.1	1.5	100	420	39	35	340	1.0	—	—	260	130
484	fried (weighed with bones)	206	858	20.3	13.3	1.3	89	370	34	31	300	1.7	—	—	230	110
485	grilled	199	828	20.4	13.0	0	170	370	33	32	240	1.0	0.11	0.5	230	220
486	grilled (weighed with bones)	135	562	13.9	8.8	0	120	250	22	22	160	0.7	0.07	0.4	160	150
487	**Bloater** grilled	251	1043	23.5	17.4	0	700	450	120	45	360	2.2	—	—	310	1130
488	grilled (weighed with bones)	186	773	17.4	12.9	0	520	330	91	33	260	1.6	—	—	230	840
489	**Kipper** baked	205	855	25.5	11.4	0	990	520	65	48	430	1.4	—	—	280	1520
490	baked (weighed with bones)	111	464	13.8	6.2	0	540	280	35	26	230	0.8	—	—	150	820
491	**Mackerel** raw	223	926	19.0	16.3[b]	0	(130)	(360)	(24)	(30)	(240)	(1.0)	0.19	0.5	(180)	(97)
492	fried	188	784	21.5	11.3	0	150	420	28	35	280	1.2	0.20	—	210	110
493	fried (weighed with bones)	138	574	15.7	8.3	0	110	310	21	25	200	0.9	0.15	—	150	83
494	**Pilchards** canned in tomato sauce	126	531	18.8	5.4	0.7	370	420	300	39	350	2.7	0.19	1.6	—	580
495	**Salmon** raw	(182)	(757)	(18.4)	(12.0)	0	(98)	(310)	(27)	(26)	(280)	(0.7)	0.20	0.8	(170)	(59)
496	steamed	197	823	20.1	13.0	0	110	330	29	29	300	0.8	—	—	190	64
497	steamed (weighed with bones and skin)	160	666	16.3	10.5	0	87	270	23	23	250	0.6	—	—	150	52

[a] The fat content varies according to stage of maturity: elvers contain 2.2g and silver eels 27.8g fat per 100g

[b] The values for fat vary throughout the year from about 5g per 100g in February–April to 20g per 100g in July–October

No	Food	Retinol µg	Carotene µg	Vitamin D µg	Thiamin mg	Riboflavin mg	Nicotinic acid mg	Potential nicotinic acid from tryptophan mgTrp ÷ 60	Vitamin C mg	Vitamin E mg
	Fatty fish									
480	**Eel** raw	1200 *(260–2500[a])*	Tr	—[b]	0.20	0.35 *(0.05–0.50)*	3.5	3.1	Tr	—
481	stewed	(1900)	Tr	—[b]	(0.13)	(0.40)	(2.8)	3.9	Tr	—
482	**Herring** raw	45 *(6–120)*	Tr	22.5 *(7.5–42.5)*	Tr *(Tr–0.13)*	0.18 *(0.09–0.33)*	4.1 *(2–6)*	3.1	Tr	0.21
483	fried	(49)	Tr	(25.0)	Tr	(0.18)	(4.0)	4.3	Tr	(0.30)
484	fried (weighed with bones)	(43)	Tr	(22.0)	Tr	(0.16)	(3.5)	3.8	Tr	(0.26)
485	grilled	(49)	Tr	(25.0)	Tr	0.18	4.0	3.8	Tr	0.30
486	grilled (weighed with bones)	(33)	Tr	(17.0)	Tr	0.12	2.7	2.6	Tr	0.20
487	**Bloater** grilled	(49)	Tr	(25.0)	Tr	(0.18)	(4.0)	4.4	Tr	(0.30)
488	grilled (weighed with bones)	(36)	Tr	(18.5)	Tr	(0.13)	(3.0)	3.2	Tr	(0.22)
489	**Kipper** baked	(49)	Tr	(25.0)	Tr	(0.18)	(4.0)	4.8	Tr	(0.30)
490	baked (weighed with bones)	(26)	Tr	(13.5)	Tr	(0.10)	(2.2)	2.6	Tr	(0.16)
491	**Mackerel** raw	45 *(15–60)*	Tr	17.5 *(2.5–25)*	0.09 *(0.02–0.20)*	0.35 *(0.16–0.66)*	8.0	3.6	Tr	—
492	fried	(52)	Tr	(21.1)	(0.09)	(0.38)	(8.7)	4.0	Tr	—
493	fried (weighed with bones)	(38)	Tr	(15.4)	(0.07)	(0.28)	(6.4)	2.9	Tr	—
494	**Pilchards** canned in tomato sauce	Tr[c]	Tr	8	0.02	0.29	7.6	3.5	Tr	0.70
495	**Salmon** raw	Tr[d]	Tr	Tr[d]	0.20	0.15 *(0.06–0.22)*	7.0	3.4	Tr	—
496	steamed	Tr[d]	Tr	Tr[d]	(0.20)	(0.11)	(7.0)	3.8	Tr	—
497	steamed (weighed with bones and skin)	Tr[d]	Tr	Tr[d]	(0.16)	(0.09)	(5.7)	3.0	Tr	—

No	Food	Vitamin B_6 mg	Vitamin B_{12} µg	Folic acid Free µg	Folic acid Total µg	Panto-thenic acid mg	Biotin µg
	Fatty fish						
480	**Eel** raw	0.30	1	—	—	0.15	—
481	stewed	(0.24)	(1)	—	—	(0.17)	—
482	**Herring** raw	0.45	6	3	5	1.0	10
483	fried	(0.57)	(11)	(3)	(10)	(0.88)	(10)
484	fried (weighed with bones)	(0.50)	(10)	(3)	(9)	(0.77)	(9)
485	grilled	0.57	11	3	10	(0.88)	(10)
486	grilled (weighed with bones)	0.39	8	2	7	(0.60)	(7)
487	**Bloater** grilled	(0.57)	(11)	(3)	(10)	(0.88)	(10)
488	grilled (weighed with bones)	(0.42)	(8)	(2)	(7)	(0.65)	(7)
489	**Kipper** baked	(0.57)	(11)	(3)	(10)	(0.88)	(10)
490	baked (weighed with bones)	(0.31)	(6)	(2)	(5)	(0.48)	(5)
491	**Mackerel** raw	0.70	10	—	—	1.0	7
492	fried	(0.84)	(12)	—	—	(0.96)	(8)
493	fried (weighed with bones)	(0.61)	(9)	—	—	(0.70)	(6)
494	**Pilchards** canned in tomato sauce	—	12	—	—	—	—
495	**Salmon** raw	0.75	5	4	26	2.0	5
496	steamed	(0.83)	(6)	(2)	(29)	(1.8)	(4)
497	steamed (weighed with bones and skin)	(0.67)	(5)	(2)	(23)	(1.5)	(3)

Notes

[a] The retinol content of eels increases with maturity.

[b] Whole body oil is a rich source of vitamin D, and it contains about 120 µg per 100 g oil.

[c] New analysis shows only a trace to be present.

[d] These are values for Atlantic salmon. Pacific salmon may contain 90 (20–150) µg retinol and 12.5 (5–20) µg vitamin D per 100 g.

No	Food	Description and number of samples	Edible matter, proportion of weight purchased	Water g	Total nitrogen g
	***Fatty fish** contd*				
498	**Salmon** canned	10 cans, red salmon; backbone and skin removed	0.94	70.4	3.24
499	smoked	4 samples	1.00	64.9	4.06
500	**Sardines** canned in oil, fish only	10 cans, 6 brands; fish after draining off oil	0.83	58.4	3.79
501	fish plus oil	10 cans, 6 brands; total contents of can	1.00	48.5	3.15
502	canned in tomato sauce	10 cans, 4 brands; total contents of can	1.00	65.0	2.84
504	**Sprats** fried	Fish without heads; fried in deep fat	0.59	33.7	3.98
505	fried (weighed with bones)	Calculated from the previous item	0.59	29.6	3.50
506	**Trout, brown** steamed	Flesh only from whole fish	0.54	70.6	3.76
507	steamed (weighed with bones)	Calculated from the previous item	0.54	46.5	2.48
508	**Tuna** canned in oil	6 cans, 2 brands, skipjack tuna	1.00	54.6	3.65
509	**Whitebait** fried	Whole fish; rolled in flour and fried	0.77	23.5	3.12
	Cartilaginous fish				
511	**Dogfish** fried in batter	7 samples, purchased cooked (rock salmon)	0.92	54.2	3.42
512	fried (weighed with waste)	Calculated from the previous item	0.92	49.9	3.15
514	**Skate** fried in batter	6 samples, purchased cooked	0.82	61.8	3.67
515	fried (weighed with waste)	Calculated from the previous item	0.82	50.7	3.01

No	Food	Energy value kcal	kJ	Protein (N × 6.25) g	Fat g	Carbo-hydrate g	Na mg	K mg	Ca mg	Mg mg	P mg	Fe mg	Cu mg	Zn mg	S mg	Cl mg
	***Fatty fish** contd*															
498	**Salmon** canned	155	649	20.3	8.2	0	570	300	93	30	240	1.4	0.09	0.9	220	880
499	smoked	142	598	25.4	4.5	0	1880	420	19	32	250	0.6	0.09	0.4	—	2850
500	**Sardines** canned in oil, fish only	217	906	23.7	13.6	0	650	430	550	52	520	2.9	0.19	3.0	310	1000
501	fish plus oil	334	1382	19.7	28.3	0	540	360	460	43	430	2.4	0.17	2.5	260	830
502	canned in tomato sauce	177	740	17.8	11.6	0.5	700	410	460	51	400	4.6	0.23	2.7	230	1110
504	**Sprats** fried	441	1826	24.9	37.9	0	130	410	710	46	640	4.5	—	—	280	180
505	fried (weighed with bones)	388	1608	21.9	33.4	0	120	360	620	40	560	4.0	—	—	250	160
506	**Trout, brown** steamed	135	566	23.5	4.5	0	88[a]	370	36	31	270	1.0	—	—	220	70[a]
507	steamed (weighed with bones)	89	375	15.5	3.0	0	58	250	24	20	180	0.7	—	—	140	46
508	**Tuna** canned in oil	289	1202	22.8	22.0	0	420	280	7	28	190	1.1	0.09	0.8	—	690
509	**Whitebait** fried	525	2174	19.5	47.5	5.3	230	110	860	50	860	5.1	—	—	270	330
	Cartilaginous fish															
511	**Dogfish** fried in batter	265	1103	16.7[b]	18.8	7.7	290	310	42	23	220	1.1	0.13	0.5	200	340
512	fried (weighed with waste)	244	1016	15.4[b]	17.3	7.1	270	290	39	21	200	1.0	0.12	0.4	181	310
514	**Skate** fried in batter	199	830	17.9[b]	12.1	4.9	140	240	50	27	180	1.0	0.09	0.9	250	170
515	fried (weighed with waste)	163	680	14.7[b]	9.9	4.0	110	200	40	22	150	0.8	0.07	0.7	210	140

[a] Sea trout contains 210mg Na and 260mg Cl per 100g

[b] (Total N — non-protein N) × 6.25

No	Food	Retinol µg	Carotene µg	Vitamin D µg	Thiamin mg	Riboflavin mg	Nicotinic acid mg	Potential nicotinic acid from tryptophan mg Trp ÷ 60	Vitamin C mg	Vitamin E mg
	***Fatty fish** contd*									
498	**Salmon** canned	90 *(20–150)*	Tr	12.5 *(5–20)*	0.04	0.18	7.0	3.8	Tr	1.5
499	smoked	Tr[a]	Tr	Tr[a]	0.16	0.17	8.8	4.7	Tr	—
500	**Sardines** canned in oil, fish only	Tr[b]	Tr	7.5	0.04	0.36	8.2	4.4	Tr	0.30
501	fish plus oil	Tr[b]	Tr	6.2	0.03	0.30	6.8	3.7	Tr	1.1
502	canned in tomato sauce	Tr[b]	Tr	7.5	0.02	0.28	5.5	3.3	Tr	0.51[c]
504	**Sprats** fried	—	Tr	—	—	—	—	4.6	Tr	—
505	fried (weighed with bones)	—	Tr	—	—	—	—	4.1	Tr	—
506	**Trout, brown** steamed	—	Tr	—	—	—	—	4.4	Tr	—
507	steamed (weighed with bones)	—	Tr	—	—	—	—	2.9	Tr	—
508	**Tuna** canned in oil	—	Tr	5.8	0.04	0.11	12.9	4.3	Tr	6.3[d]
509	**Whitebait** fried	—	Tr	—	—	—	—	3.6	Tr	—
	Cartilaginous fish									
511	**Dogfish** fried in batter	—	Tr	—	0.06	0.10	5.6	—	Tr	2.1
512	fried (weighed with waste)	—	Tr	—	0.06	0.09	5.2	—	Tr	1.9
514	**Skate** fried in batter	—	Tr	—	0.03	0.10	2.4	—	Tr	1.2
515	fried (weighed with waste)	—	Tr	—	0.02	0.08	2.0	—	Tr	1.0

No	Food	Vitamin B_6 mg	Vitamin B_{12} µg	Folic acid: Free µg	Folic acid: Total µg	Panto-thenic acid mg	Biotin µg
	***Fatty fish** contd*						
498	**Salmon** canned	0.45	4	4	12	0.50	(5)
499	smoked	—	—	—	—	—	—
500	**Sardines** canned in oil, fish only	0.48	28	3	8	0.50	5
501	fish plus oil	0.40	23	2	7	0.42	4
502	canned in tomato sauce	0.35	14	7	13	(0.50)	(5)
504	**Sprats** fried	—	—	—	—	—	—
505	fried (weighed with bones)	—	—	—	—	—	—
506	**Trout, brown** steamed	—	—	—	—	—	—
507	steamed (weighed with bones)	—	—	—	—	—	—
508	**Tuna** canned in oil	0.44	5	7	15	0.42	3
509	**Whitebait** fried	—	—	—	—	—	—
	Cartilaginous fish						
511	**Dogfish** fried in batter	—	—	—	—	—	—
512	fried (weighed with waste)	—	—	—	—	—	—
514	**Skate** fried in batter	—	—	—	—	—	—
515	fried (weighed with waste)	—	—	—	—	—	—

Notes

[a] These are values for Atlantic salmon. Pacific salmon may contain 90µg retinol and 12.5µg vitamin D per 100g.

[b] New analysis shows only a trace to be present.

[c] Also contains 0.25 mg γ-tocopherol per 100g.

[d] Also contains 2.9 mg γ-tocopherol per 100g.

No	Food	Description and number of samples	Edible matter, proportion of weight purchased	Water g	Total nitrogen g
	Crustacea				
518	**Crab** boiled	Boiled in fresh water	0.16	72.5	3.21
519	boiled (weighed with shell)	Calculated from the previous item	0.16	14.5	0.64
520	canned	6 cans, 2 brands	1.00	79.2	2.90
521	**Lobster** boiled	Boiled in fresh water	0.29	72.4	3.54
522	boiled (weighed with shell)	Calculated from the previous item	0.29	26.1	1.27
523	**Prawns** boiled	Purchased cooked, probably in sea or salt water	0.38	70.0	3.62
524	boiled (weighed with shell)	Calculated from the previous item	0.38	26.6	1.38
525	**Scampi** fried	5 packets, frozen; prepared in breadcrumbs	0.78	39.4	1.95
527	**Shrimps** boiled	Purchased cooked, probably in sea or salt water	0.33	62.5	3.80
528	boiled (weighed with shell)	Calculated from the previous item	0.33	20.6	1.26
529	canned	10 cans, 3 brands; drained shrimps	0.65	74.9	3.33
	Molluscs				
531	**Cockles** boiled	Purchased cooked (in sea or salt water) without shells	1.00	78.9	1.80
532	**Mussels** raw	Purchased alive	0.32	84.1	1.93
533	boiled	Boiled in fresh water	0.20	79.0	2.75
534	boiled (weighed with shell)	Calculated from the previous item	0.20	23.7	0.83
535	**Oysters** raw	Purchased alive	0.12	85.7	1.72
536	raw (weighed with shell)	Calculated from the previous item	0.12	10.3	0.21
538	**Scallops** steamed	Purchased without shells	0.56	73.1	3.71

No	Food	Energy value		Protein (N × 6.25)	Fat	Carbo-hydrate	mg									
		kcal	kJ	g	g	g	Na	K	Ca	Mg	P	Fe	Cu	Zn	S	Cl
	Crustacea															
518	**Crab** boiled	127	534	20.1	5.2	0	370	270	29	48	350	1.3	4.8	5.5	470	570
519	boiled (weighed with shell)	25	105	4.0	1.0	0	73	54	6	10	70	0.3	1.0	1.1	93	110
520	canned	81	341	18.1	0.9	0	550	100	120	32	140	2.8	0.42	5.0	—	830
521	**Lobster** boiled	119	502	22.1	3.4	0	330	260	62	34	280	0.8	1.7	1.8	510	530
522	boiled (weighed with shell)	42	179	7.9	1.2	0	120	93	22	12	100	0.3	0.65	0.6	190	190
523	**Prawns** boiled	107	451	22.6	1.8	0	1590	260	150	42	350	1.1	(0.70)	(1.6)	370	2550
524	boiled (weighed with shell)	41	172	8.6	0.7	0	610	99	55	16	130	0.4	(0.27)	(0.6)	140	970
525	**Scampi** fried	316	1321	12.2	17.6	28.9	380	390	99	30	310	1.1	0.22	0.6	—	740
527	**Shrimps** boiled	117	493	23.8	2.4	0	3840	400	320	110	270	1.8	0.80	(5.3)	340	5850
528	boiled (weighed with shell)	39	164	7.9	0.8	0	1260	130	110	35	89	0.6	0.26	(1.7)	110	1930
529	canned	94	398	20.8	1.2	0	980	100	110	49	150	5.1	0.23	2.4	—	1510
	Molluscs															
531	**Cockles** boiled	48	203	11.3	0.3	Tr	3520	43	130	51	200	26.0[a]	0.28	1.2	320	5220
532	**Mussels** raw	66	276	12.1	1.9	Tr	290	320	88	23	240	5.8	0.36	1.6	370	460
533	boiled	87	366	17.2	2.0	Tr	210	92	200	25	330	7.7	0.48	2.1	350	320
534	boiled (weighed with shell)	26	111	5.2	0.6	Tr	63	28	59	8	99	2.3	0.16	0.6	100	95
535	**Oysters** raw	51	217	10.8	0.9	Tr	510	260	190	42	270	6.0	7.6	45.0[b]	250	820
536	raw (weighed with shell)	6	26	1.3	0.1	Tr	61	31	22	5	32	0.7	0.91	5.4	30	98
538	**Scallops** steamed	105	446	23.2	1.4	Tr	270	480	120	38	340	3.0	—	—	570	410

[a] The iron content of cockles can be as high as 40mg per 100g

[b] The zinc content of oysters may vary from 6 to 100mg per 100g

No	Food	Retinol µg	Carotene µg	Vitamin D µg	Thiamin mg	Riboflavin mg	Nicotinic acid mg	Potential nicotinic acid from tryptophan mg Trp ÷ 60	Vitamin C mg	Vitamin E mg
	Crustacea									
518	**Crab** boiled	Tr	Tr	Tr	0.10	0.15	2.5	3.8	Tr	—
519	boiled (weighed with shell)	Tr	Tr	Tr	0.02	0.03	0.5	0.8	Tr	—
520	canned	Tr	Tr	Tr	Tr	0.05	1.1	3.4	Tr	—
521	**Lobster** boiled	Tr	Tr	Tr	0.08	0.05	1.5	4.1	Tr	1.5
522	boiled (weighed with shell)	Tr	Tr	Tr	0.03	0.02	0.5	1.5	Tr	0.50
523	**Prawns** boiled	Tr	Tr	Tr	—	—	—	4.2	Tr	—
524	boiled (weighed with shell)	Tr	Tr	Tr	—	—	—	1.6	Tr	—
525	**Scampi** fried	Tr	Tr	Tr	0.08	0.05	1.3	2.3	Tr	—
527	**Shrimps** boiled	Tr	Tr	Tr	0.03	0.03	3.0	4.4	Tr	—
528	boiled (weighed with shell)	Tr	Tr	Tr	0.01	0.01	1.0	1.5	Tr	—
529	canned	Tr	Tr	Tr	0.01	0.02	0.8	3.9	Tr	—
	Molluscs									
531	**Cockles** boiled	—	Tr	Tr	—	—	—	2.4	Tr	—
532	**Mussels** raw	—	Tr	Tr	—	—	—	2.6	Tr	0.90
533	boiled	—	Tr	Tr	—	—	—	3.7	Tr	(1.2)
534	boiled (weighed with shell)	—	Tr	Tr	—	—	—	1.1	Tr	(0.36)
535	**Oysters** raw	75	Tr	Tr	0.10	0.20	1.5	2.3	Tr[a]	0.85
536	raw (weighed with shell)	9	Tr	Tr	0.01	0.02	0.2	0.3	Tr	0.10
538	**Scallops** steamed	—	Tr	Tr	—	—	—	5.0	Tr	—

No	Food	Vitamin B_6 mg	Vitamin B_{12} µg	Folic acid		Panto-thenic acid mg	Biotin µg	Notes
				Free µg	Total µg			
	Crustacea							[a] Some species of oysters contain vitamin C; Pacific oysters (*Ostrea gigas*) have been shown to contain 22 mg, and Olympia oysters (*Ostrea lurida*) 38 mg per 100 g.
518	**Crab** boiled	0.35	Tr	3	20	0.60	Tr	
519	boiled (weighed with shell)	0.07	Tr	1	4	0.12	Tr	
520	canned	—	Tr	—	—	—	Tr	
521	**Lobster** boiled	—	1	7	17	1.63	5	
522	boiled (weighed with shell)	—	Tr	3	6	0.59	2	
523	**Prawns** boiled	—	—	—	—	—	—	
524	boiled (weighed with shell)	—	—	—	—	—	—	
525	**Scampi** fried	—	—	—	—	—	—	
527	**Shrimps** boiled	0.10	1	—	—	0.30	1	
528	boiled (weighed with shell)	0.03	Tr	—	—	0.10	Tr	
529	canned	0.03	2	4	15	0.35	(1)	
	Molluscs							
531	**Cockles** boiled	—	Tr	—	—	—	—	
532	**Mussels** raw	—	—	—	—	—	—	
533	boiled	—	—	—	—	—	—	
534	boiled (weighed with shell)	—	—	—	—	—	—	
535	**Oysters** raw	0.03	15	10	—	0.50	10	
536	raw (weighed with shell)	Tr	2	1	—	0.06	1	
538	**Scallops** steamed	—	—	17	17	0.14	Tr	

No	Food	Description and number of samples	Edible matter, proportion of weight purchased	Water g	Total nitrogen g
	***Molluscs** contd*				
539	**Whelks** boiled	Purchased cooked, probably boiled in sea or salt water	0.15	77.5	2.96
540	boiled (weighed with shell)	Calculated from the previous item	0.15	11.6	0.44
541	**Winkles** boiled	Purchased cooked, probably boiled in sea water	0.19	79.1	2.45
542	boiled (weighed with shell)	Calculated from the previous item	0.19	15.1	0.47
	Fish products and dishes				
543	**Fish cakes** frozen	14 packets, 4 brands, white fish	1.00	70.2	1.68
544	fried	14 packets, 4 brands, white fish	1.02	63.3	1.45
545	**Fish fingers** frozen	11 packets, 3 brands; in breadcrumbs	1.00	63.9	2.02
546	fried	11 packets, 3 brands; in breadcrumbs	0.90	55.6	2.16
547	**Fish paste**	30 samples, sardine, crab, lobster, salmon	1.00	67.1	2.45
548	**Fish pie**	Recipe p 343	—	74.3	1.13
549	**Kedgeree**	Recipe p 343	—	68.4	2.03
550	**Roe, cod** hard, raw	Literature sources	1.00	70.0	3.89
551	fried	Parboiled, sliced and fried in crumbs	0.93	62.0	3.35
552	**herring** soft, raw	Literature sources	1.00	82.0	2.42 [a]
553	fried	Rolled in flour and fried	0.80	52.3	3.85 [b]

[a] Includes 0.29g purine nitrogen per 100g

[b] Includes 0.48g purine nitrogen per 100g

No	Food	Energy value		Protein	Fat	Carbo-hydrate	mg									
		kcal	kJ	g	g	g	Na	K	Ca	Mg	P	Fe	Cu	Zn	S	Cl
	***Molluscs** contd*															
539	**Whelks** boiled	91	385	18.5	1.9	Tr	270	320	54	160	230	6.2	(7.2)	7.2	450	590
540	boiled (weighed with shell)	14	59	2.8	0.3	Tr	40	47	8	24	34	0.9	(1.1)	1.1	67	88
541	**Winkles** boiled	74	312	15.3	1.4	Tr	1140	150	140	360	220	15.0	(1.3)	(5.7)	380	1800
542	boiled (weighed with shell)	14	60	2.9	0.3	Tr	220	29	26	68	42	2.9	(0.25)	(1.1)	72	340
	Fish products and dishes															
543	**Fish cakes** frozen	112	477	10.5	0.8	16.8	480	280	76	20	120	1.0	0.10	0.5	—	820
544	fried	188	785	9.1	10.5	15.1	500	260	70	18	110	1.0	0.13	0.4	—	730
545	**Fish fingers** frozen	178	749	12.6	7.5	16.1	320	240	43	18	190	0.7	0.06	0.4	—	380
546	fried	233	975	13.5	12.7	17.2	350	260	45	19	220	0.7	0.08	0.4	—	400
547	**Fish paste**	169	704	15.3	10.4	3.7	600	300	280	33	310	9.0	0.37	1.4	—	940
548	**Fish pie**	128	540	7.1	5.7	13.0	210	310	40	18	92	0.4	0.10	0.4	—	330
549	**Kedgeree**	151	633	13.1	7.1	9.2	790	160	36	16	160	0.9	0.07	0.6	150	1220
550	**Roe, cod** hard, raw	113	476	24.3	1.7	0	—	—	—	—	—	—	—	—	—	—
551	fried	202	844	20.9	11.9	3.0	130	260	17	11	500	1.6	—	—	240	190
552	**herring** soft, raw	80	337	13.3[a]	3.0	0	—	—	—	—	—	—	—	—	—	—
553	fried	244	1019	21.1[a]	15.8	4.7	87	240	16	8	920	1.5	—	—	240	120

[a] (Total N − purine N) × 6.25

No	Food	Retinol µg	Carotene µg	Vitamin D µg	Thiamin mg	Riboflavin mg	Nicotinic acid mg	Potential nicotinic acid from tryptophan mg Trp ÷ 60	Vitamin C mg	Vitamin E mg
	***Molluscs** contd*									
539	**Whelks** boiled	—	Tr	Tr	—	—	—	4.0	Tr	0.80
540	boiled (weighed with shell)	—	Tr	Tr	—	—	—	0.6	Tr	0.10
541	**Winkles** boiled	—	Tr	Tr	—	—	—	3.3	Tr	—
542	boiled (weighed with shell)	—	Tr	Tr	—	—	—	0.6	Tr	—
	Fish products and dishes									
543	**Fish cakes** frozen	Tr	Tr	Tr	0.06	0.06	1.3	2.0	Tr	—
544	fried	Tr	Tr	Tr	0.06	0.06	1.1	1.7	Tr	—
545	**Fish fingers** frozen	Tr	Tr	Tr	0.09	0.06	1.1	2.4	Tr	—
546	fried	Tr	Tr	Tr	0.08	0.07	1.4	2.5	Tr	—
547	**Fish paste**	—	Tr	—	0.02	0.20	4.1	2.9	Tr	0.85[a]
548	**Fish pie**	30	4	0.17	0.07	0.09	1.0	1.4	2	0.42
549	**Kedgeree**	80	0	0.79	0.07	0.14	0.8	2.8	0	1.0
550	**Roe cod** hard, raw	140[b]	Tr	2.0	1.5	1.0	1.5	4.5	30	6.4
551	fried	(150)[b]	Tr	(2.2)	(1.3)	(0.9)	(1.3)	3.9	(26)	(6.9)
552	**herring** soft, raw	—	—	—	0.20	0.50	2.0	2.5	5	—
553	fried	—	—	—	(0.20)	(0.50)	(2.0)	3.9	(5)	—

No	Food	Vitamin B_6 mg	Vitamin B_{12} µg	Folic acid Free µg	Folic acid Total µg	Pantothenic acid mg	Biotin µg
	***Molluscs** contd*						
539	**Whelks** boiled	—	—	—	—	—	—
540	boiled (weighed with shell)	—	—	—	—	—	—
541	**Winkles** boiled	—	—	—	—	—	—
542	boiled (weighed with shell)	—	—	—	—	—	—
	Fish products and dishes						
543	**Fish cakes** frozen	—	—	—	—	—	—
544	fried	—	—	—	—	—	—
545	**Fish fingers** frozen	0.21	1	5	15	—	—
546	fried	0.21	2	6	16	—	—
547	**Fish paste**	—	—	—	—	—	—
548	**Fish pie**	0.22	1	2	6	0.3	1
549	**Kedgeree**	0.19	2	7	9	0.5	7
550	**Roe, cod** hard, raw	0.32	10	—	—	3.0	13
551	fried	(0.28)	(11)	—	—	(2.6)	(15)
552	**herring** soft, raw	—	5	—	—	0.49	—
553	fried	—	(6)	—	—	(0.49)	—

Notes

[a] Also contains 0.17 mg γ-tocopherol per 100 g.

[b] 90 per cent is present as retinaldehyde.

No	Food	Description and number of samples	Edible matter, proportion of weight purchased	Water g	Total nitrogen g

No	Food	Energy value		Protein	Fat	Carbo-hydrate	mg									
		kcal	kJ	g	g	g	Na	K	Ca	Mg	P	Fe	Cu	Zn	S	Cl

No	Food	Retinol µg	Carotene µg	Vitamin D µg	Thiamin mg	Riboflavin mg	Nicotinic acid mg	Potential nicotinic acid from tryptophan mgTrp ÷60	Vitamin C mg	Vitamin E mg

No	Food	Vitamin B_6 mg	Vitamin B_{12} µg	Folic acid		Panto-thenic acid mg	Biotin µg	Notes
				Free µg	Total µg			

No	Food	Description and number of samples	Edible matter, proportion of weight purchased	Water g	Sugars g	Starch g	Dietary fibre g	Total nitrogen g
554	**Ackee** canned	8 cans, drained contents only	—	76.7	0.8	Tr	2.7	0.46
555	**Artichokes, globe** boiled	Base of leaves and soft inside parts; boiled 35 minutes	0.41	84.4	—	0	—	0.18
556	boiled (weighed as served)	Calculated from the previous item	0.41	36.3	—	0	—	0.08
557	**Artichokes, Jerusalem** boiled	Flesh only; boiled 20 minutes	0.85	80.2	—	0	—	0.25
558	**Asparagus** boiled	Soft tips only; boiled 25 minutes	0.20	92.4	1.1	0	1.5	0.54
559	boiled (weighed as served)	Calculated from the previous item	0.20	46.2	0.6	0	0.8	0.27
560	**Aubergine** raw	Eggplant; flesh only	0.77	93.4	2.9	0.2	2.5	0.11
	Beans							
561	**French** boiled	Pods and beans; cut up and boiled 30 minutes	1.00	95.5	0.8	0.3	3.2	0.12
562	**runner** raw	Samples from 6 shops; pod ends and sides trimmed	0.79	89.0	2.8	1.1	2.9	0.36
563	boiled	Trimmed pods broken up and boiled 20 minutes	0.99	90.7	1.3	1.4	3.4	0.31
564	**broad** boiled	Whole beans without pods; boiled 30 minutes	0.31	83.7	0.6	6.5	4.2	0.66
565	**butter** raw	Whole beans	1.00	11.6	3.6	46.2	21.6	3.06
566	boiled	Soaked 24 hours, boiled 2 hours	2.50	70.5	1.5	15.6	5.1	1.13
567	**haricot** raw	Whole beans	1.00	11.3	2.8	42.7	25.4	3.42
568	boiled	Soaked 24 hours, boiled 2 hours	2.60	69.6	0.8	15.8	7.4	1.06
569	**baked** canned in tomato sauce	11 cans, 4 brands	1.00	73.6	5.2	5.1	7.3	0.82

No	Food	Energy value		Protein (N × 6.25)	Fat	Carbo-hydrate	mg									
		kcal	kJ	g	g	g	Na	K	Ca	Mg	P	Fe	Cu	Zn	S	Cl
554	**Ackee** canned	151	625	2.9	15.2	0.8	240	270	35	40	47	0.7	0.27	0.6	—	340
555	**Artichokes, globe** boiled	15	62	1.1	Tr	2.7 [a]	15	330	44	27	40	0.5	0.09	—	16	84
556	boiled (weighed as served)	7	28	0.5	Tr	1.2 [a]	6	140	19	12	17	0.2	0.04	—	7	36
557	**Artichokes, Jerusalem** boiled	18	78	1.6	Tr	3.2 [a]	3	420	30	11	33	0.4	0.12	0.1	22	58
558	**Asparagus** boiled	18	75	3.4	Tr	1.1	2	240	26	10	85	0.9	0.20	0.3	47	31
559	boiled (weighed as served)	9	39	1.7	Tr	0.6	1	120	13	5	42	0.5	0.10	0.1	23	16
560	**Aubergine** raw	14	62	0.7	Tr	3.1	3	240	10	10	12	0.4	0.08	—	9	61
	Beans															
561	**French** boiled	7	31	0.8	Tr	1.1	3	100	39	10	15	0.6	0.10	0.3	8	11
562	**runner** raw	26	114	2.3	0.2	3.9	2	280	27	27	47	0.8	0.07	0.4	—	18
563	boiled	19	83	1.9	0.2	2.7	1	150	22	17	41	0.7	0.05	0.3	—	5
564	**broad** boiled	48	206	4.1	0.6	7.1	20	230	21	28	99	1.0	0.43	—	27	14
565	**butter** raw	273	1162	19.1	1.1	49.8	62	1700	85	164	320	5.9	1.22	2.8	110	47
566	boiled	95	405	7.1	0.3	17.1	16	400	19	33	87	1.7	0.16	1.0	47	2
567	**haricot** raw	271	1151	21.4	1.6	45.5	43	1160	180	180	310	6.7	0.61	2.8	170	2
568	boiled	93	396	6.6	0.5	16.6	15	320	65	45	120	2.5	0.14	1.0	46	1
569	**baked** canned in tomato sauce	64	270	5.1	0.5	10.3	480	300	45	31	91	1.4	0.21	0.7	(43)	800

[a] This vegetable contains inulin; 50 per cent total carbohydrate taken to be available

No	Food	Retinol µg	Carotene µg	Vitamin D µg	Thiamin mg	Riboflavin mg	Nicotinic acid mg	Potential nicotinic acid from tryptophan mg Trp ÷ 60	Vitamin C mg	Vitamin E mg
554	**Ackee** canned	0	—	0	0.03	0.07	0.6	0.5	30	—
555	**Artichokes, globe** boiled	0	90	0	0.07	0.03	0.9	0.2	8	—
556	boiled (weighed as served)	0	40	0	0.03	0.01	0.4	0.1	3	—
557	**Artichokes, Jerusalem** boiled	0	(Tr)	0	0.10	Tr	—	0.3	2	0.2
558	**Asparagus** boiled	0	500 *(400–800)*	0	0.10	0.08	0.8	0.6	20 *(15–30)*	2.5
559	boiled (weighed as served)	0	250	0	0.05	0.04	0.4	0.3	10	1.3
560	**Aubergine** raw	0	Tr	0	0.05	0.03	0.8	0.1	5	—
	Beans									
561	**French** boiled	0	400 *(300–500)*	0	0.04	0.07	0.3	0.2	5	0.2
562	**runner** raw	0	400 *(300–500)*	0	0.05	0.10	0.9	0.4	20 *(15–40)*	0.2 [a]
563	boiled	0	400	0	0.03	0.07	0.5	0.3	5 [b]	0.2 [a]
564	**broad** boiled	0	250	0	(0.10)	0.04	3.0	0.7	15 [c]	Tr [d]
565	**butter** raw	0	Tr	0	0.45	0.13	2.5	3.1	0	—
566	boiled	0	Tr	0	—	—	—	1.1	0	—
567	**haricot** raw	0	Tr	0	0.45	0.13	2.5	3.4	0	—
568	boiled	0	Tr	0	—	—	—	1.1	0	—
569	**baked** canned in tomato sauce	0	—	0	0.07	0.05	0.5	0.8	(Tr)	0.6 [e]

No	Food	Vitamin B_6 mg	Vitamin B_{12} µg	Folic acid Free µg	Folic acid Total µg	Pantothenic acid mg	Biotin µg
554	**Ackee** canned	0.06	0	8	41	—	—
555	**Artichokes, globe** boiled	0.07	0	—	(30)	0.21	4.1
556	boiled (weighed as served)	0.03	0	—	(13)	0.09	1.8
557	**Artichokes, Jerusalem** boiled	—	0	—	—	—	—
558	**Asparagus** boiled	0.04	0	(5)	(30)	0.13	0.4
559	boiled (weighed as served)	0.02	0	(3)	(15)	0.07	0.2
560	**Aubergine** raw	0.08	0	8	20	0.22	—
	Beans						
561	**French** boiled	0.06	0	3	28	0.07	0.8
562	**runner** raw	0.07	0	57	60	0.05	0.7
563	boiled	0.04	0	3	28	0.04	0.5
564	**broad** boiled	—	0	—	—	3.8	2.1
565	**butter** raw	0.58	0	25	110	1.0	—
566	boiled	—	0	—	—	—	—
567	**haricot** raw	0.56	0	—	—	0.7	—
568	boiled	—	0	—	—	—	—
569	**baked** canned in tomato sauce	0.12	0	4	29	—	—

Notes

[a] Also contains 0.3mg γ-tocopherol per 100g.

[b] Canned green beans contain 2mg per 100g.

[c] Canned broad beans contain 6mg per 100g.

[d] Also contains 2.5mg γ-tocopherol per 100g.

[e] Also contains 0.8mg γ-tocopherol per 100g.

No	Food	Description and number of samples	Edible matter, proportion of weight purchased	Water g	Sugars g	Starch g	Dietary fibre g	Total nitrogen g
	Beans *contd*							
570	**mung** green gram, raw	Literature sources	1.00	12.0	1.2	34.4	(22.0)	3.52
571	cooked dahl	Recipe p343	—	72.5	0.8	10.6	(6.4)	1.02
572	**red kidney** raw	Literature sources	1.00	11.0	(3.0)	(42.0)	(25.0)	3.54
573	**Beansprouts** canned	10 cans, drained contents	0.55	95.4	0.4	0.4	3.0	0.25
574	**Beetroot** raw	Flesh only, no skin	0.82	87.1	6.0	0	3.1	0.21
575	boiled [a]	Flesh only, no skin; boiled 2 hours	0.80	82.7	9.9	0	2.5	0.29
576	**Broccoli tops** raw	10 samples; predominantly leaves, thick stems removed	0.70	89.0	2.5	Tr	3.6	0.52
577	boiled	10 samples; predominantly leaves, thick stems removed; boiled 15 minutes	0.84	89.9	1.5	0.1	4.1	0.49
578	**Brussels sprouts** raw	10 samples; inner leaves only	0.63	88.1	2.6	0.1	4.2	0.64
579	boiled	10 samples; inner leaves only; boiled 15 minutes	0.72	91.5	1.6	0.1	2.9	0.45
580	**Cabbage, red** raw	Inner leaves	0.70	89.7	3.5	Tr	3.4	0.27
581	**Savoy** raw	Inner leaves	0.53	89.9	3.3	Tr	3.1	0.53
582	boiled	Inner leaves; boiled 30 minutes	0.65	95.7	1.1	Tr	2.5	0.21
583	**spring** boiled	Inner leaves; boiled 30 minutes	0.59	96.6	0.8	Tr	2.2	0.18
584	**white** raw	7 cabbages; whole cabbage as purchased	1.00	90.3	3.7	0.1	2.7	0.31
585	**winter** raw	20 cabbages, January King; inner leaves	0.57	88.3	2.7	0.1	3.4	0.45
586	boiled	20 cabbages, January King; inner leaves, boiled 15 minutes	0.61	93.0	2.2	0.1	2.8	0.27

[a] Weighed cold

No	Food	Energy value		Protein (N × 6.25)	Fat	Carbo-hydrate	mg									
		kcal	kJ	g	g	g	Na	K	Ca	Mg	P	Fe	Cu	Zn	S	Cl
	Beans *contd*															
570	**mung** green gram, raw	231	981	22.0	1.0	35.6	28	850	100	170	330	8.0	0.97	—	190	12
571	cooked, dahl	106	447	6.4	4.2	11.4	820	270	34	51	100	2.6	0.29	—	61	1260
572	**red kidney** raw	272	1159	22.1	1.7	45.0	(40)	(1160)	140	(180)	410	6.7	(0.61)	(2.8)	(170)	(2)
573	**Beansprouts** canned	9	40	1.6	Tr	0.8	80	36	13	10	20	1.0	0.09	0.8	—	120
574	**Beetroot** raw	28	118	1.3	Tr	6.0	84	300	25	15	32	0.4	0.07	0.4	—	59
575	boiled	44	189	1.8	Tr	9.9	64	350	30	17	36	0.4	0.08	0.4	22	76
576	**Broccoli tops** raw	23	96	3.3	Tr	2.5	12	340	100	18	67	1.5	0.07	0.6	—	55
577	boiled	18	78	3.1	Tr	1.6	6	220	76	12	60	1.0	0.08	0.4	—	37
578	**Brussels sprouts** raw	26	111	4.0	Tr	2.7	4	380	32	19	65	0.7	0.06	0.5	—	28
579	boiled	18	75	2.8	Tr	1.7	2	240	25	13	51	0.5	0.05	0.4	78	16
580	**Cabbage, red** raw	20	85	1.7	Tr	3.5	32	300	53	17	32	0.6	0.09	0.3	68	45
581	**Savoy**, raw	26	109	3.3	Tr	3.3	23	260	75	20	68	0.9	(0.07)	0.3	88	22
582	boiled	9	40	1.3	Tr	1.1	8	120	53	7	27	0.7	0.07	0.2	30	9
583	**spring** boiled	7	32	1.1	Tr	0.8	12	110	30	6	32	0.5	0.07	0.2	27	6
584	**white** raw	22	93	1.9	Tr	3.8	7	280	44	13	36	0.4	(0.03)	0.3	—	23
585	**winter** raw	22	92	2.8	Tr	2.8	7	390	57	17	54	0.6	0.06	0.4	—	31
586	boiled	15	66	1.7	Tr	2.3	4	160	38	8	34	0.4	(0.03)	0.2	—	13

No	Food	Retinol µg	Carotene µg	Vitamin D µg	Thiamin mg	Riboflavin mg	Nicotinic acid mg	Potential nicotinic acid from tryptophan mgTrp ÷ 60	Vitamin C mg	Vitamin E mg
	Beans *contd*									
570	**mung** green gram, raw	0	24	0	0.45	0.20	2.0	3.5	Tr	—
571	cooked, dahl	60	44	0.06	0.09	0.04	0.4	1.0	Tr	—
572	**red kidney** raw	0	Tr	0	0.54	0.18	2.0	3.5	Tr	—
573	**Beansprouts** canned	0	Tr	0	0.02	0.03	0.2	0.3	1 [a]	—
574	**Beetroot** raw	0	Tr	0	0.03	0.05	0.1	0.2	6	0
575	boiled	0	Tr	0	0.02	0.04	0.1	0.3	5	0
576	**Broccoli tops** raw	0	2500 *(900–7000)*	0	0.10	0.30	1.0	0.6	110 *(70–160)*	1.3
577	boiled	0	2500 *(900–7000)*	0	0.06	0.20	0.6	0.6	34 *(20–70)*	1.1
578	**Brussels sprouts** raw	0	400 *(120–550)*	0	0.10	(0.15)	0.7	0.8	90 *(70–140)*	1.0
579	boiled	0	400 *(120–550)*	0	0.06	(0.10)	0.4	0.5	40 *(30–90)*	0.9
580	**Cabbage, red** raw	0	(20)	0	0.06	0.05	0.3	0.3	55	0.2
581	**Savoy** raw	0	300 [b]	0	0.06	0.05	0.3	0.5	60 *(50–80)*	0.2 [c]
582	boiled	0	300 [b]	0	0.03	0 03	0.2	0.2	15 *(10–40)*	0.2 [c]
803	**spring** boiled	0	500	0	0.03	0.03	0.2	0.2	25 *(10–50)*	0.2
584	**white** raw	0	(Tr)	0	0.06	0.05	0.3	0.3	40 [d]	0.2
585	**winter** raw	0	300 [b]	0	0.06	0.05	0.3	0.5	55 *(40–70)*	0.2 [c]
586	boiled	0	300 [b]	0	0.03	0.03	0.2	0.3	20 *(10–40)*	0.2 [c]

No	Food	Vitamin B_6 mg	Vitamin B_{12} µg	Folic acid Free µg	Folic acid Total µg	Pantothenic acid mg	Biotin µg
	Beans *contd*						
570	**mung** green gram, raw	(0.50)	0	25	140	—	—
571	cooked, dahl	(0.09)	0	1	20	—	—
572	**red kidney** raw	0.44	0	24	130	0.50	—
573	**Beansprouts** canned	0.03	0	4	12	—	—
574	**Beetroot** raw	0.05	0	70	90	0.12	Tr
575	boiled	0.03	0	20	(50)	0.10	Tr
576	**Broccoli tops** raw	0.21	0	89	130	1.0	(0.5)
577	boiled	0.13	0	7	110	0.70	(0.3)
578	**Brussels sprouts** raw	0.28	0	84	110	0.40	0.4
579	boiled	0.17	0	7	87	0.28	0.3
580	**Cabbage, red** raw	0.21	0	(60)	(90)	0.32	0.1
581	**Savoy** raw	0.16	0	(60)	(90)	0.21	0.1
582	boiled	0.10	0	(2)	(35)	0.15	Tr
583	**spring** boiled	0.10	0	(18)	(50)	0.15	Tr
584	**white** raw	0.16	0	19	26	0.21	0.1
585	**winter** raw	0.16	0	60	90	0.21	0.1
586	boiled	0.10	0	2	35	0.15	Tr

Notes

[a] Fresh beansprouts contain about 30mg per 100g.

[b] This is an average figure. The amount of carotene in leafy vegetables depends on the amount of chlorophyll, and the outer green leaves may contain 50 times as much as inner white ones.

[c] The value for inner leaves. Outer leaves contain 7.0mg α-tocopherol per 100g.

[d] About 20 per cent is lost on shredding.

No	Food	Description and number of samples	Edible matter, proportion of weight purchased	Water g	Sugars g	Starch g	Dietary fibre g	Total nitrogen g
587	**Carrots, old** raw	Flesh only	0.96	89.9	5.4	0	2.9	0.11
588	boiled	Flesh only; cut up and boiled 45 minutes	0.87	91.5	4.2	0.1	3.1	0.10
589	**young** boiled	Purchased with leaves; flesh only, boiled 25 minutes	0.50	91.1	4.4	0.1	3.0	0.14
590	canned	6 samples; drained contents	0.63	91.2	4.4	Tr	3.7	0.11
591	**Cauliflower** raw	16 cauliflowers; flower and stalk	0.62	92.7	1.5	Tr	2.1	0.30
592	boiled	16 cauliflowers; flower and stalk, boiled 20 minutes	0.60	94.5	0.8	Tr	1.8	0.26
593	**Celeriac** boiled	Flesh only; boiled 30 minutes	0.79	90.2	1.5	0.5	4.9	0.26
594	**Celery** raw	Stem only	0.73	93.5	1.2	0.1	1.8	0.15
595	boiled	Stem only; boiled 30 minutes	0.72	95.7	0.7	0	2.2	0.10
596	**Chicory** raw	Stem and young leaves	0.79	96.2	—	0	—	0.12
597	**Cucumber** raw	Flesh only	0.77	96.4	1.8	0	0.4	0.10
598	**Endive** raw	Leaves only	0.63	93.7	1.0	0	2.2	0.28
599	**Horseradish** raw	Flesh of root	0.45	74.7	7.3	3.7	8.3	0.72
600	**Laverbread**	6 samples; cooked puréed seaweed coated in oatmeal	1.00	87.7	Tr	1.6	3.1	0.51
601	**Leeks** raw	Bulb only	0.36	86.0	6.0	0	3.1	0.31
602	boiled	Bulb only; boiled 30 minutes	0.44	90.8	4.6	0	3.9	0.28
603	**Lentils** raw	As purchased	1.00	12.2	2.4	50.8	11.7	3.80
604	split, boiled	As purchased; boiled 20 minutes	3.27	72.1	0.8	16.2	3.7	1.22
605	masur dahl, cooked	Recipe p 344	—	78.4	0.7	10.7	2.4	0.78

No	Food	Energy value kcal	Energy value kJ	Protein (N × 6.25) g	Fat g	Carbo-hydrate g	Na mg	K mg	Ca mg	Mg mg	P mg	Fe mg	Cu mg	Zn mg	S mg	Cl mg
587	**Carrots, old** raw	23	98	0.7	Tr	5.4	95	220	48	12	21	0.6	0.08	0.4	7	69
588	boiled	19	79	0.6	Tr	4.3	50	87	37	6	17	0.4	0.08	0.3	5	31
589	**young** boiled	20	87	0.9	Tr	4.5	23	240	29	8	30	0.4	0.08	0.3	9	28
590	canned	19	82	0.7	Tr	4.4	280	84	27	5	15	1.3	0.04	0.3	—	450
591	**Cauliflower** raw	13	56	1.9	Tr	1.5	8	350	21	14	45	0.5	(0.03)	0.3	—	31
592	boiled	9	40	1.6	Tr	0.8	4	180	18	8	32	0.4	(0.03)	0.2	—	14
593	**Celeriac** boiled	14	59	1.6	Tr	2.0	28	400	47	12	71	0.8	0.13	—	13	23
594	**Celery** raw	8	36	0.9	Tr	1.3	140	280	52	10	32	0.6	0.11	0.1	15	180
595	boiled	5	21	0.6	Tr	0.7	67	130	52	9	19	0.4	0.11	0.1	8	100
596	**Chicory** raw	9	38	0.8	Tr	1.5[a]	7	180	18	13	21	0.7	0.14	0.2	13	25
597	**Cucumber** raw	10	43	0.6	0.1	1.8	13	140	23	9	24	0.3	0.09	0.1	11	25
598	**Endive** raw	11	47	1.8	Tr	1.0	10	380	44	10	67	2.8	0.09	—	26	71
599	**Horseradish** raw	59	253	4.5	Tr	11.0	8	580	120	36	70	2.0	0.14	—	210	19
600	**Laverbread**	52	217	3.2	3.7	1.6	560	220	20	31	51	3.5	0.12	0.8	—	820
601	**Leeks** raw	31	128	1.9	Tr	6.0	9	310	63	10	43	1.1	0.10	(0.1)	—	43
602	boiled	24	104	1.8	Tr	4.6	6	280	61	13	28	2.0	0.09	(0.1)	49	43
603	**Lentils** raw	304	1293	23.8	1.0	53.2	36	670	39	77	240	7.6	0.58	3.1	120	64
604	split, boiled	99	420	7.6	0.5	17.0	12	210	13	25	77	2.4	0.19	1.0	39	20
605	masur dahl, cooked	90	380	4.9	3.1	11.4	320	150	11	17	52	1.7	0.12	0.6	28	490

[a] This vegetable contains inulin ; 50 per cent total carbohydrate taken to be available

No	Food	Retinol µg	Carotene µg	Vitamin D µg	Thiamin mg	Riboflavin mg	Nicotinic acid mg	Potential nicotinic acid from tryptophan mg Trp ÷ 60	Vitamin C mg	Vitamin E mg
587	**Carrots, old** raw	0	12 000 *(10 000–14 000)*	0	0.06	0.05	0.6	0.1	6 *(4–10)*	0.5
588	boiled	0	12 000 *(10 000–14 000)*	0	0.05	0.04	0.4	0.1	4 *(2–6)*	0.5
589	**young** boiled	0	6000 *(5000–7000)*	0	0.05	0.04	0.4	0.1	4 *(2–6)*	0.5
590	canned	0	7000	0	0.04	0.02	0.3	0.1	3	(0.5)
591	**Cauliflower** raw	0	30 *(6–50)*	0	0.10	0.10	0.6	0.5	60 *(50–90)*	0.2 [a]
592	boiled	0	30 *(6–50)*	0	0.06	0.06	0.4	0.4	20 *(15–40)*	0.1 [b]
593	**Celeriac** boiled	0	0	0	0.04	0.04	0.5	0.3	4	—
594	**Celery** raw	0	Tr	0	0.03	0.03	0.3	0.2	7	0.2
595	boiled	0	Tr	0	0.02	0.02	0.2	0.1	5	0.2
596	**Chicory** raw	0	Tr	0	0.05	0.05	0.5	0.1	4	—
597	**Cucumber** raw	0	Tr	0	0.04	0.04	0.2	0.1	8	Tr
598	**Endive** raw	0	2000 *(1000–6000)*	0	0.06	0.10	0.4	0.3	12	—
599	**Horseradish** raw	0	0	0	0.05	0.03	0.5	0.7	120	—
600	**Laverbread**	0	—	0	0.03	0.10	0.6	0.5	5	1.1
601	**Leeks** raw	0	40 [c]	0	0.10	0.05	0.6	0.3	18 *(15–30)*	0.8
602	boiled	0	40 [c]	0	0.07	0.03	0.4	0.3	15 *(10–25)*	0.8
603	**Lentils** raw	0	60	0	0.50	0.20	2.0	3.8	Tr	—
604	split, boiled	0	20	0	0.11	0.04	0.4	1.2	Tr	—
605	masur dahl, cooked	27	30	0.03	0.07	0.03	0.3	0.8	Tr	—

No	Food	Vitamin B_6 mg	Vitamin B_{12} µg	Folic acid Free µg	Folic acid Total µg	Panto-thenic acid mg	Biotin µg
587	**Carrots, old** raw	0.15	0	12	15	0.25	0.6
588	boiled	0.09	0	1	8	0.18	0.4
589	**young** boiled	0.09	0	1	8	0.18	0.4
590	canned	0.02	0	1	7	0.10	0.4
591	**Cauliflower** raw	0.20	0	30	39	0.60	1.5
592	boiled	0.12	0	2	49	0.42	1.0
593	**Celeriac** boiled	0.10	0	—	—	—	—
594	**Celery** raw	0.10	0	6	12	0.40	0.1
595	boiled	0.06	0	(1)	(6)	0.28	Tr
596	**Chicory** raw	0.05	0	33	52	—	—
597	**Cucumber** raw	0.04	0	14	16	0.30	(0.4)
598	**Endive** raw	—	0	62	330	—	—
599	**Horseradish** raw	0.15	0	—	—	—	—
600	**Laverbread**	—	0	8	47	—	—
601	**Leeks** raw	0.25	0	—	—	0.12	1.4
602	boiled	0.15	0	7	—	0.10	1.0
603	**Lentils** raw	0.60	0	25	35	1.36	—
604	split, boiled	0.11	0	1	5	0.31	—
605	masur dahl, cooked	0.07	0	1	4	0.20	—

Notes

[a] Also contains 0.2 mg γ-tocopherol per 100g.

[b] Also contains 0.1 mg γ-tocopherol per 100g.

[c] Bulb only. The leaves contain about 2000 µg.

No	Food	Description and number of samples	Edible matter, proportion of weight purchased	Water g	Sugars g	Starch and dextrins g	Dietary fibre g	Total nitrogen g
606	**Lettuce** raw	29 lettuces; inner leaves of long and headed forms	0.70	95.9	1.2	Tr	1.5	0.16
607	**Marrow** raw	10 marrows; flesh only	0.50	93.5	3.0	0.7	(1.8)	0.10
608	boiled	Flesh only; boiled 25 minutes	0.64	97.8	1.3	0.1	0.6	0.06
609	**Mushrooms** raw	Flesh and stem	0.75	91.5	0	0	2.5	0.64[a]
610	fried	Flesh and stem; fried in dripping	0.61	64.2	0	0	(4.0)	0.90[a]
611	**Mustard and cress** raw	Leaves and stems	1.00	92.5	0.9	0	3.7	0.26
612	**Okra** raw	Literature sources (ladies' fingers)	0.88	90.0	2.3	Tr	(3.2)	0.32
613	**Onions** raw	Flesh only	0.97	92.8	5.2	0	1.3	0.15
614	boiled	Flesh only; boiled 30 minutes	0.85	96.6	2.7	0	1.3	0.09
615	fried	Flesh only; cut up and fried in dripping	0.49	42.0	10.1	0	(4.5)	0.29
616	spring, raw	Flesh of bulb	0.31	86.8	8.5	0	3.1	0.15
617	**Parsley** raw	Leaves	0.53	78.7	Tr	0	9.1	0.83
618	**Parsnips** raw	Flesh only	0.74	82.5	8.8	2.5	4.0	0.27
619	boiled	Flesh only; boiled 30 minutes	0.78	83.2	2.7	10.8	2.5	0.20
620	**Peas, fresh** raw	Whole peas, no pods	0.37	78.5	4.0	6.6	5.2	0.92
621	boiled	Whole peas, no pods; boiled 20 minutes	0.37	80.0	1.8	5.9	5.2	0.80
622	**frozen** raw	15 packets	1.00	79.1	4.1	3.4	7.8	0.91
623	boiled	15 packets; boiled 5 minutes	—	80.7	1.0	3.3	12.0	0.87
624	**canned** garden	10 cans; drained contents	0.63	81.6	3.6	3.4	6.3	0.74
625	processed	10 cans; drained contents	0.54	71.5	1.3	12.4	7.9	0.99

[a] 60 per cent of this nitrogen is present as urea

No	Food	Energy value kcal	Energy value kJ	Protein (N × 6.25) g	Fat g	Carbo-hydrate g	Na mg	K mg	Ca mg	Mg mg	P mg	Fe mg	Cu mg	Zn mg	S mg	Cl mg
606	**Lettuce** raw	12	51	1.0	0.4	1.2	9	240	23	8	27	0.9	(0.03)	0.2	—	53
607	**Marrow** raw	16	69	0.6	Tr	3.7	1	210	17	12	20	0.2	(0.03)	0.2	—	30
608	boiled	7	29	0.4	Tr	1.4	1	84	14	7	13	0.2	0.03	0.2	6	14
609	**Mushrooms** raw	13	53	1.8[a]	0.6	0	9	470	3	13	140	1.0	0.64	0.1	34	85
610	fried	210	863	2.2[a]	22.3	0	11	570	4	16	170	1.3	0.78	0.1	74	100
611	**Mustard and cress** raw	10	47	1.6	Tr	0.9	19	340	66	27	66	(1.0)	0.12	—	170	89
612	**Okra** raw	17	71	2.0	Tr	2.3	7	190	70	60	60	1.0	0.19	—	30	41
613	**Onions** raw	23	99	0.9	Tr	5.2	10	140	31	8	30	0.3	0.08	0.1	51	20
614	boiled	13	53	0.6	Tr	2.7	7	78	24	5	16	0.3	0.07	0.1	24	5
615	fried	345	1424	1.8	33.3	10.1	20	270	61	15	59	0.6	0.16	0.1	88	38
616	spring, raw	35	151	0.9	Tr	8.5	13	230	140	11	24	1.2	0.13	—	50	36
617	**Parsley** raw	21	88	5.2	Tr	Tr	33	1080	330	52	130	8.0	0.52	0.9	—	160
618	**Parsnips** raw	49	210	1.7	Tr	11.3	17	340	55	22	69	0.6	0.10	0.1	17	41
619	boiled	56	238	1.3	Tr[b]	13.5	4	290	36	13	32	0.5	0.10	0.1	15	33
620	**Peas, fresh** raw	67	283	5.8	0.4	10.6	1	340	15	30	100	1.9	0.23	0.7	50	38
621	boiled	52	223	5.0	0.4	7.7	Tr	170	13	21	83	1.2	0.15	0.5	44	8
622	**frozen** raw	53	227	5.7	0.4	7.2	3	190	33	27	90	1.5	0.22	0.9	—	20
623	boiled	41	175	5.4	0.4	4.3	2	130	31	23	84	1.4	0.19	0.7	—	12
624	**canned** garden	47	201	4.6	0.3	7.0	230	130	24	17	73	1.6	0.16	0.7	—	350
625	processed	80	339	6.2	0.4	13.7	330	170	27	24	91	1.5	0.22	0.8	—	510

[a] (Total N − urea N) × 6.25

[b] Roast parsnips contain 6.5 g fat per 100 g

No	Food	Retinol µg	Carotene µg	Vitamin D µg	Thiamin mg	Riboflavin mg	Nicotinic acid mg	Potential nicotinic acid from tryptophan mg Trp 60 ÷	Vitamin C mg	Vitamin E mg
606	**Lettuce** raw	0	1000[a]	0	0.07	0.08	0.3	0.1	15 *(10–30)*	0.5[b]
607	**Marrow** raw	0	30	0	Tr	(Tr)	0.3	0.1	5	Tr
608	boiled	0	30	0	Tr	(Tr)	0.2	0.1	2	Tr
609	**Mushrooms** raw	0	0	0	0.10	0.40	4.0	0.6	3	Tr
610	fried	0	0	0	(0.07)	0.35	3.5	0.9	1	Tr
611	**Mustard and cress** raw	0	(500)	0	—	—	—	0.3	40	0.7
612	**Okra** raw	0	90	0	0.10	0.10	1.0	0.3	25	—
613	**Onions** raw	0	0	0	0.03	0.05	0.2	0.2	10 *(3–15)*	Tr
614	boiled	0	0	0	0.02	0.04	0.1	0.1	6	Tr
615	fried	0	0	0	—	—	—	0.4	—	—
616	spring, raw	0	Tr	0	(0.03)	(0.05)	(0.2)	0.2	25 *(20–30)*	Tr
617	**Parsley** raw	0	7000 *(3000–10 000)*	0	0.15	0.30	1.0	0.8	150 *(100–200)*	1.8
618	**Parsnips** raw	0	Tr	0	0.10	0.08	1.0	0.3	15 *(5–30)*	1.0
619	boiled	0	Tr	0	0.07	0.06	0.7	0.2	10 *(5–20)*	1.0
620	**Peas, fresh** raw	0	300 *(250–400)*	0	0.32	0.15	2.5	0.9	25 *(15–35)*	Tr[d]
621	boiled	0	300	0	0.25	0.11	1.5	0.8	15[c]	Tr[d]
622	**frozen** raw	0	300	0	0.32	0.10	2.1	0.9	17	Tr[e]
623	boiled	0	300	0	0.24	0.07	1.5	0.9	13	Tr[f]
624	**canned** garden	0	300	0	0.13	0.10	2.1	0.7	8	Tr[d]
625	processed	0	300	0	0.10	0.04	0.5	1.0	Tr	Tr[g]

No	Food	Vitamin B_6 mg	Vitamin B_{12} µg	Folic acid Free µg	Folic acid Total µg	Pantothenic acid mg	Biotin µg
606	**Lettuce** raw	0.07	0	19	34	0.20	0.7
607	**Marrow** raw	0.06	0	13	13	0.10	0.4
608	boiled	0.03	0	(1)	(6)	0.07	—
609	**Mushrooms** raw	0.10	0	20	23	2.0	—
610	fried	(0.06)	0	17	(20)	(1.4)	—
611	**Mustard and cress** raw	—	0	—	—	—	—
612	**Okra** raw	0.08	0	25	100	0.26	—
613	**Onions** raw	0.10	0	15	16	0.14	0.9
614	boiled	0.06	0	Tr	8	0.10	0.6
615	fried	—	0	—	—	—	—
616	spring, raw	(0.10)	0	40	40	(0.14)	(0.9)
617	**Parsley** raw	0.20	0	—	—	0.30	0.4
618	**Parsnips** raw	0.10	0	57	67	0.50	0.1
619	boiled	0.06	0	(6)	(30)	0.35	Tr
620	**Peas, fresh** raw	0.16	0	—	—	0.75	0.5
621	boiled	0.10	0	—	—	0.32	0.4
622	**frozen** raw	0.10	0	2	78	(0.75)	0.5
623	boiled	0.07	0	5	78	(0.32)	0.4
624	**canned** garden	0.06	0	8	52	0.15	(0.4)
625	processed	0.03	0	1	3	(0.08)	(Tr)

Notes

[a] This is an average figure. The outer green leaves may contain 50 times as much carotene as the inner white ones.

[b] Also contains 0.7mg γ-tocopherol per 100g.

[c] Cooked air-dried peas contain 11mg per 100g and cooked accelerated-freeze-dried peas 16mg per 100g.

[d] Also contains 0.8mg γ-tocopherol per 100g.

[e] Also contains 1.0mg γ-tocopherol per 100g.

[f] Also contains 0.9mg γ-tocopherol per 100g.

[g] Also contains 1.2mg γ-tocopherol per 100g.

No	Food	Description and number of samples	Edible matter, proportion of weight purchased	Water g	Sugars g	Starch g	Dietary fibre g	Total nitrogen g
	Peas *contd*							
626	**dried** raw	Whole peas	1.00	13.3	2.4	47.6	16.7	3.45
627	boiled	Whole peas; soaked 24 hours, boiled 2 hours	2.70	70.3	0.9	18.2	4.8	1.11
628	**split** dried, raw	Peas as purchased	1.00	12.1	1.9	54.7	11.9	3.54
629	boiled	Soaked 24 hours, boiled 2 hours	2.50	67.3	0.9	21.0	5.1	1.33
630	**chick** Bengal gram, raw	Literature sources	1.00	9.9	(10.0)	(40.0)	(15.0)	3.23
631	cooked, dahl	Whole peas; recipe p 344	—	65.8	(5.2)	(16.8)	(6.0)	1.28
632	channa dahl	10 samples; made from split peas	—	74.2	1.5	8.9	5.2	0.85
633	**red** pigeon, raw	Literature sources	1.00	10.0	(9.0)	(45.0)	(15.0)	3.20
634	**Peppers, green** raw	30 peppers; flesh only	0.86	93.5	2.2	Tr	0.9	0.15
635	boiled	30 peppers; flesh only, boiled 15 minutes	0.72	93.7	1.7	0.1	0.9	0.15
636	**Plantain** green, raw	Literature sources	0.61	67.0	0.8	27.5	(5.8)	0.16
637	boiled	10 samples; boiled 30 minutes	0.71	63.9	0.9	30.2	6.4	0.16
638	ripe, fried	8 samples; fried in oil	0.43	34.7	11.5	36.0	5.8	0.24
639	**Potatoes, old** raw	Flesh only	0.86	75.8	0.5	20.3	2.1	0.34
640	boiled	Flesh only, boiled 30 minutes	0.86	80.5	0.4	19.3	1.0	0.23
641	mashed	Boiled and mashed with margarine and milk	0.94	76.9	0.6	17.4	0.9	0.24
642	baked	Baked in skins; flesh only	0.68	71.0	0.6	24.4	2.5	0.41
643	baked (weighed with skins)	Calculated from the previous item	0.68	57.5	0.5	19.8	2.0	0.33
644	roast	Flesh only; roasted in shallow fat	0.66	64.3	—	—	—	0.45
645	chips	Fried in deep fat	0.49	47.0	—	—	—	0.61

No	Food	Energy value kcal	Energy value kJ	Protein (N×6.25) g	Fat g	Carbo-hydrate g	Na mg	K mg	Ca mg	Mg mg	P mg	Fe mg	Cu mg	Zn mg	S mg	Cl mg
	Peas *contd*															
626	**dried** raw	286	1215	21.6	1.3	50.0	38	990	61	116	300	4.7	0.49	3.5	130	60
627	boiled	103	438	6.9	0.4	19.1	13	270	24	30	110	1.4	0.17	1.0	39	9
628	**split** dried, raw	310	1318	22.1	1.0	56.6	38	910	33	130	270	5.4	0.58	(4.0)	170	56
629	boiled	118	503	8.3	0.3	21.9	14	270	11	30	120	1.7	0.25	(1.2)	46	10
630	**chick** Bengal gram, raw	320	1362	20.2	5.7	50.0	40	800	140	160	300	6.4	0.76	—	180	60
631	cooked, dahl	144	610	8.0	3.3	22.0	850	400	64	67	130	3.1	0.33	—	84	1310
632	channa dahl	97	407	5.3	4.5	9.5	480	260	30	31	92	1.8	0.20	0.8	—	700
633	**red** pigeon, raw	301	1278	20.0	2.0	54.0	29	1100	100	130	300	5.0	1.25	—	180	5
634	**Peppers green** raw	15	65	0.9	0.4	2.2	2	210	9	11	25	0.4	0.07	0.2	—	18
635	boiled	14	59	0.9	0.4	1.8	2	170	9	10	22	0.4	0.06	0.2	—	15
636	**Plantain** green, raw	112	477	1.0	0.2	28.3	(1)	(350)	7	33	35	0.5	0.16	0.1	15	(80)
637	boiled	122	518	1.0	0.1	31.1	4	330	9	34	34	0.4	0.10	0.2	—	50
638	ripe, fried	267	1126	1.5	9.2	47.5	3	610	6	54	66	0.8	0.20	0.4	—	110
639	**Potatoes, old** raw	87	372	2.1	0.1	20.8	7	570	8	24	40	0.5	0.15	0.3	35	79
640	boiled	80	343	1.4	0.1	19.7	3	330	4	15	29	0.3	0.11	0.2	22	41
641	mashed	119	499	1.5	5.0	18.0	24	300	12	14	32	0.3	0.10	0.3	24	71
642	baked	105	448	2.6	0.1	25.0	8	680	9	29	48	0.8	0.18	0.3	42	94
643	baked (weighed with skins)	85	364	2.1	0.1	20.3	6	550	8	24	39	0.6	0.15	0.2	34	76
644	roast	157	662	2.8	4.8	27.3	9	750	10	32	53	0.7	0.20	0.4	56	100
645	chips	253	1065	3.8	10.9[a]	37.3	12	1020	14	43	72	0.9	0.27	0.6	45	140

[a] The fat content may vary from 7 to 15g per 100g

No	Food	Retinol µg	Carotene µg	Vitamin D µg	Thiamin mg	Riboflavin mg	Nicotinic acid mg	Potential nicotinic acid from tryptophan mgTrp ÷ 60	Vitamin C mg	Vitamin E mg
	Peas *contd*									
626	**dried** raw	0	250	0	0.60	0.30	3.0	3.5	Tr	Tr
627	boiled	0	80	0	0.11	0.07	1.0	1.1	Tr	Tr
628	**split** dried, raw	0	150	0	0.70	0.20	3.2	3.5	Tr	Tr
629	boiled	0	(50)	0	0.11	0.06	1.0	1.3	Tr	Tr
630	**chick** Bengal gram, raw	0	190	0	0.50	0.15	1.5	2.7	3	—
631	cooked, dahl	52	210	0.05	0.14	0.05	0.5	1.1	3	—
632	channa dahl	—	—	—	0.08	0.03	0.4	0.7	Tr	—
633	**red** pigeon, raw	0	30	0	0.50	0.15	2.3	1.6	Tr	—
634	**Peppers, green** raw	0	200 *(60–1000)*	0	Tr	0.03	0.7	0.2	100 *(60–170)*	0.8
635	boiled	0	200 *(60–1000)*	0	0.01	0.02	0.6	0.2	60	0.8
636	**Plantain** green, raw	0	60	0	0.05	0.05	0.7	0.2	20	—
637	boiled	0	60	0	Tr	0.01	0.3	0.2	3	—
638	ripe, fried	0	(120)	0	0.11	0.02	0.6	0.2	12	—
639	**Potatoes, old** raw	0	Tr	0	0.11	0.04	1.2	0.5	*(8–20[a])*	0.1
640	boiled	0	Tr	0	0.08	0.03	0.8	0.3	*(4–14[b])*	0.1
641	mashed	Tr	Tr	Tr	0.08	0.04	0.8	0.4	*(4–12[b])*	0.1
642	baked	0	Tr	0	0.10	0.04	1.2	0.6	*(5–16[b])*	0.1
643	baked (weighed with skins)	0	Tr	0	0.08	0.03	1.0	0.5	*(4–13[b])*	0.1
644	roast	0	Tr	0	0.10	0.04	1.2	0.7	*(5–16[b])*	0.1
645	chips	0	Tr	0	0.10	0.04	1.2	0.9	*(5–16[b])*	0.1

No	Food	Vitamin B_6 mg	Vitamin B_{12} µg	Folic acid Free µg	Folic acid Total µg	Pantothenic acid mg	Biotin µg
	Peas *contd*						
626	**dried** raw	0.13	0	21	33	2.0	—
627	boiled	—	0	Tr	—	—	—
628	**split** dried, raw	0.13	0	21	33	2.0	—
629	boiled	—	0	Tr	—	—	—
630	**chick** Bengal gram, raw	—	0	30	180	—	—
631	cooked, dahl	—	0	2	37	—	—
632	channa dahl	—	0	4	30	—	—
633	**red** pigeon, raw	—	0	19	100	—	—
634	**Peppers, green** raw	0.17	0	5	11	0.23	—
635	boiled	0.14	0	1	11	0.16	—
636	**Plantain** green, raw	(0.50)	0	2	16	0.37	—
637	boiled	(0.30)	0	5	18	0.26	—
638	ripe, fried	(1.0)	0	16	37	0.73	—
639	**Potatoes, old** raw	0.25	0	10	14	0.30	0.1
640	boiled	0.18	0	3	10	0.20	Tr
641	mashed	0.18	0	(3)	(10)	0.20	Tr
642	baked	0.18	0	3	10	0.20	Tr
643	baked (weighed with skins)	0.14	0	2	8	0.16	Tr
644	roast	0.18	0	3	7	0.20	Tr
645	chips	0.18	0	(3)	(10)	0.20	Tr

Notes

[a] Raw potatoes	Vitamin C mg per 100g
Maincrop, freshly dug	30
stored 1–3 months	20
stored 4–5 months	15
stored 6–7 months	10
stored 8–9 months	8

[b] Method of cooking	Vitamin C % of value in raw potato
Boiled, peeled; Mashed	50–70
Boiled, unpeeled; Baked; Roast; Steamed	60–80
Fried in deep fat (chips)	65–75

No	Food	Description and number of samples	Edible matter, proportion of weight purchased	Water g	Sugars g	Starch g	Dietary fibre g	Total nitrogen g
	Potatoes *contd*							
646	**old** chips, frozen	11 packets, as purchased	1.00	73.1	0.4	19.2	1.9	0.35
647	frozen, fried	11 packets; fried in shallow oil	0.74	48.3	0.5	28.5	3.2	0.48
648	**new** boiled	Flesh only; boiled 15 minutes	0.96	78.8	0.7	17.6	2.0	0.25
649	canned	10 cans; drained contents	0.63	84.2	0.4	12.2	2.5	0.19
650	**instant** powder	20 packets	1.00	7.2	2.2	71.0	16.5	1.45
651	made up	Calculated from the powder	4.50	79.4	0.5	15.6	3.6	0.32
652	**crisps**	26 packets, mixed plain and flavoured	1.00	2.7	0.7	48.6	11.9	1.00
653	**Pumpkin** raw	Flesh only	0.81	94.7	2.7	0.7	0.5	0.10
654	**Radishes** raw	Flesh and skin; purchased with leaves	0.50	93.3	2.8	0	1.0	0.16
655	**Salsify** boiled	Flesh only; boiled 45 minutes	0.63	81.2	—	0	—	0.30
656	**Seakale** boiled	Stem only; boiled 20 minutes	0.74	95.6	0.6	0	1.2	0.23
657	**Spinach** boiled	Leaves; boiled 15 minutes without added water	0.42	85.1	1.2	0.2	6.3	0.81
658	**Spring greens** boiled	Leaves; boiled 30 minutes	1.00	93.6	0.9	0	3.8	0.27
659	**Swedes** raw	Flesh only	0.86	91.4	4.2	0.1	2.7	0.18
660	boiled	Flesh only; boiled 45 minutes	0.82	91.6	3.7	0.1	2.8	0.14
661	**Sweetcorn, on-the-cob** raw	16 cobs; kernels only	0.66	65.2	1.7	22.0	3.7	0.66
662	boiled	16 cobs; boiled 15 minutes; kernels only	0.64	65.1	1.7	21.1	4.7	0.65
663	**canned** kernels	10 cans; whole contents	1.00	73.4	8.9	7.2	5.7	0.47

No	Food	Energy value kcal	Energy value kJ	Protein (N × 6.25) g	Fat g	Carbo-hydrate g	Na mg	K mg	Ca mg	Mg mg	P mg	Fe mg	Cu mg	Zn mg	S mg	Cl mg
	Potatoes *contd*															
646	**old** chips, frozen	109	462	2.2	3.0	19.6	25	420	8	21	61	0.7	0.11	0.3	—	45
647	frozen, fried	291	1214	3.0	18.9	29.0	34	540	11	27	77	1.0	0.15	0.4	—	71
648	**new** boiled	76	324	1.6	0.1	18.3	41	330	5	20	33	0.4	0.15	0.3	24	46
649	canned	53	226	1.2	0.1	12.6	260	230	11	10	31	0.7	0.08	0.3	—	440
650	**instant** powder	318	1356	9.1	0.8	73.2	1190	1550	89	69	220	2.4	0.37	1.1	—	1750
651	made up	70	299	2.0	0.2	16.1	260	340	20	15	48	0.5	0.08	0.2	—	380
652	**crisps**	533	2224	6.3	35.9	49.3	550	1190	37	56	130	2.1	0.22	0.8	—	890
653	**Pumpkin** raw	15	65	0.6	Tr	3.4	1	310	39	8	19	0.4	0.08	(0.2)	10	37
654	**Radishes** raw	15	62	1.0	Tr	2.8	59	240	44	11	27	1.9	0.13	0.1	38	19
655	**Salsify** boiled	18	77	1.9	Tr	2.8[a]	8	180	60	14	53	1.2	0.12	—	25	46
656	**Seakale** boiled	8	33	1.4	Tr	0.6	4	50	48	11	34	0.6	0.07	—	52	12
657	**Spinach** boiled	30	128	5.1	0.5	1.4	120	490	600	59	93	4.0	0.26	0.4	86	56
658	**Spring greens** boiled	10	43	1.7	Tr	0.9	10	120	86	9	31	1.3	0.08	0.4	29	16
659	**Swedes** raw	21	88	1.1	Tr	4.3	52	140	56	11	19	0.4	0.05	—	39	31
660	boiled	18	76	0.9	Tr	3.8	14	100	42	7	18	0.3	0.04	—	31	9
661	**Sweetcorn, on-the-cob** raw	127	538	4.1	2.4	23.7	1	300	4	46	130	1.1	0.16	1.2	—	11
662	boiled	123	520	4.1	2.3	22.8	1	280	4	45	120	0.9	0.15	1.0	—	14
663	**canned** kernels	76	325	2.9	(0.5)	16.1	310	200	3	23	67	0.6	0.05	0.6	—	460

[a] This vegetable contains inulin ; 50 per cent total carbohydrate taken to be available

No	Food	Retinol µg	Carotene µg	Vitamin D µg	Thiamin mg	Riboflavin mg	Nicotinic acid mg	Potential nicotinic acid from tryptophan mgTrp ÷60	Vitamin C mg	Vitamin E mg
	Potatoes *contd*									
646	**old** chips, frozen	0	Tr	0	0.08	0.01	1.6	0.5	6	—
647	frozen, fried	0	Tr	0	0.09	0.02	2.1	0.7	4	—
648	**new** boiled	0	Tr	0	0.11	0.03	1.2	0.4	18[a]	0.1
649	canned	0	Tr	0	0.02	0.03	0.7	0.3	17	0.1
650	**instant** powder	0	Tr	0	0.04	0.14	5.6	2.2	12[b]	—
651	made up	0	Tr	0	0.01	0.03	1.2	0.5	3[b]	—
652	**crisps**	0	Tr	0	0.19	0.07	4.6	1.5	17	6.1
653	**Pumpkin** raw	0	1500 *(700–2000)*	0	0.04	0.04	0.4	0.1	5	Tr
654	**Radishes** raw	0	Tr	0	0.04	0.02	0.2	0.2	25 *(10–35)*	0
655	**Salsify** boiled	0	—	0	0.03	—	—	0.3	4	—
656	**Seakale** boiled	0	—	0	0.06	—	—	0.2	18	—
657	**Spinach** boiled	0	6000 *(4000–10 000)*	0	0.07	0.15	0.4	1.4	25 *(10–60)*	2.0
658	**Spring greens** boiled	0	4000 *(1000–10 000)*	0	0.06	0.20	0.5	0.3	30 *(20–70)*	(1.1)
659	**Swedes** raw	0	Tr	0	0.06	0.04	1.2	0.2	25 *(15–40)*	0
660	boiled	0	Tr	0	0.04	0.03	0.8	0.2	17 *(8–25)*	0
661	**Sweetcorn, on-the-cob** raw	0	(240)	0	0.15	0.08	1.8	0.4	12	0.8[c]
662	boiled	0	(240)	0	0.20	0.08	1.7	0.4	9	0.5[d]
663	**canned** kernels	0	(210)	0	0.05	0.08	1.2	0.3	5	(0.5)[d]

No	Food	Vitamin B_6 mg	Vitamin B_{12} µg	Folic acid Free µg	Folic acid Total µg	Pantothenic acid mg	Biotin µg
	Potatoes *contd*						
646	**old** chips, frozen	0.28	0	3	12	—	Tr
647	frozen, fried	0.39	0	5	11	—	Tr
648	**new** boiled	0.20	0	3	10	0.20	Tr
649	canned	0.16	0	3	11	—	Tr
650	**instant** powder	(0.82)	0	10	24	(0.91)	(0.5)
651	made up	(0.18)	0	2	5	(0.20)	(0.1)
652	**crisps**	0.89	0	4	20	—	—
653	**Pumpkin** raw	0.06	0	(13)	(13)	0.40	(0.4)
654	**Radishes** raw	0.10	0	18	24	0.18	—
655	**Salsify** boiled	—	0	—	—	—	—
656	**Seakale** boiled	—	0	—	—	—	—
657	**Spinach** boiled	0.18	0	30	(140)	0.21	0.1
658	**Spring greens** boiled	(0.16)	0	(7)	(110)	0.30	(0.4)
659	**Swedes** raw	0.20	0	23	27	0.11	0.1
660	boiled	0.12	0	9	21	0.07	Tr
661	**Sweetcorn, on-the-cob** raw	0.19	0	43	52	0.54	—
662	boiled	0.16	0	18	33	0.38	—
663	**canned** kernels	(0.16)	0	9	32	0.22	—

Notes

[a] Raw new potatoes contain 30mg per 100g.

[b] Some brands are fortified and may contain up to 10 times these values.

[c] Also contains 1.6mg γ-tocopherol per 100g.

[d] Also contains 1.0mg γ-tocopherol per 100g.

No	Food	Description and number of samples	Edible matter, proportion of weight purchased	Water g	Sugars g	Starch g	Dietary fibre g	Total nitrogen g
664	**Sweet potatoes** raw	Literature sources	0.86	70.0	(9.7)	(11.8)	(2.5)	0.19
665	boiled	Flesh only; boiled 30 minutes	0.88	72.0	9.1	11.0	2.3	0.17
666	**Tomatoes** raw	Flesh, skin and seeds	1.00	93.4	2.8	Tr	1.5	0.14
667	fried	Flesh, skin and seeds; fried in dripping	0.87	86.5	3.3	Tr	3.0	0.16
668	canned	10 cans; drained contents[a]	0.60	94.0	2.0	Tr	0.9	0.17
669	**Turnips** raw	Flesh only	0.84	93.3	3.8	0	2.8	0.12
670	boiled	Flesh only; boiled 30 minutes	0.80	94.5	2.3	0	2.2	0.11
671	**Turnip tops** boiled	Leaves; boiled 20 minutes	0.45	92.8	0	0.1	3.9	0.43
672	**Watercress** raw	Leaves and part of stem	0.77	91.1	0.6	0.1	3.3	0.46
673	**Yam** raw	Literature sources	0.86	73.0	1.0	31.4	(4.1)	0.32
674	boiled	6 samples; flesh only, boiled 30 minutes	0.91	65.8	0.2	29.6	3.9	0.25

[a] The values given are also applicable to the total contents of the can

No	Food	Energy value kcal	Energy value kJ	Protein (N × 6.25) g	Fat g	Carbo-hydrate g	Na mg	K mg	Ca mg	Mg mg	P mg	Fe mg	Cu mg	Zn mg	S mg	Cl mg
664	**Sweet potatoes** raw	91	387	1.2	0.6	21.5	(19)	(320)	(22)	(13)	(47)	(0.7)	(0.16)	—	(16)	(64)
665	boiled	85	363	1.1	0.6	20.1	18	300	21	12	44	0.6	0.15	—	15	60
666	**Tomatoes** raw	14	60	0.9	Tr	2.8	3	290	13	11	21	0.4	0.10	0.2	11	51
667	fried	69	288	1.0	5.9	3.3	3	340	15	13	25	0.5	0.12	0.2	9	59
668	canned	12	51	1.1	Tr	2.0	29	270	9	11	22	0.9	0.11	0.3	—	78
669	**Turnips** raw	20	86	0.8	0.3	3.8	58	240	59	7	28	0.4	0.07	—	22	70
670	boiled	14	60	0.7	0.3	2.3	28	160	55	7	19	0.4	0.04	—	21	31
671	**Turnip tops** boiled	11	48	2.7	Tr	0.1	7	78	98	10	45	3.1	0.09	0.4	39	15
672	**Watercress** raw	14	61	2.9	Tr	0.7	60	310	220	17	52	1.6	0.14	0.2	130	160
673	**Yam** raw	131	560	2.0	0.2	32.4	—	(500)	10	(40)	(40)	0.3	0.16	0.4	—	—
674	boiled	119	508	1.6	0.1	29.8	17	300	9	14	33	0.3	0.15	0.4	—	40

No	Food	Retinol µg	Carotene µg	Vitamin D µg	Thiamin mg	Riboflavin mg	Nicotinic acid mg	Potential nicotinic acid from tryptophan mg Trp ÷ 60	Vitamin C mg	Vitamin E mg
664	**Sweet potatoes** raw	0	4000 [a]	0	0.10	0.06	0.8	0.4	25	(4.0)
665	boiled	0	4000 [a]	0	0.08	0.04	0.6	0.3	15	(4.0)
666	**Tomatoes** raw	0	600 *(200–1000)*	0	0.06	0.04	0.7	0.1	20 *(10–30)*	1.2 [b]
667	fried	0	—	0	—	—	—	0.1	(10)	—
668	canned	0	500 *(300–600)*	0	0.06	0.03	0.7	0.1	18	1.2 [b]
669	**Turnips** raw	0	0	0	0.04	0.05	0.6	0.2	25 *(15–40)*	0
670	boiled	0	0	0	0.03	0.04	0.4	0.2	17 *(8–25)*	0
671	**Turnip tops** boiled	0	6000 *(4000–12 000)*	0	0.06	0.20	0.5	0.4	40 *(20–70)*	(1.0)
672	**Watercress** raw	0	3000 *(1500–3500)*	0	0.10	0.10	0.6	0.5	60 *(40–80)*	1.0
673	**Yam** raw	0	12	0	0.10	0.03	0.4	0.4	10	—
674	boiled	0	12	0	0.05	0.01	0.5	0.3	2	—

No	Food	Vitamin B_6 mg	Vitamin B_{12} µg	Folic acid Free µg	Folic acid Total µg	Pantothenic acid mg	Biotin µg
664	**Sweet potatoes** raw	0.22	0	35	52	0.94	—
665	boiled	0.13	0	4	25	0.66	—
666	**Tomatoes** raw	0.11	0	15	28	0.33	1.5
667	fried	—	0	—	—	—	—
668	canned	0.11	0	11	25	0.20	1.5
669	**Turnips** raw	0.11	0	17	20	0.20	0.1
670	boiled	0.06	0	(1)	(10)	0.14	Tr
671	**Turnip tops** boiled	0.16	0	(7)	(110)	0.30	(0.4)
672	**Watercress** raw	0.13	0	(200)	—	0.10	0.4
673	**Yam** raw	—	0	—	—	0.63	—
674	boiled	—	0	2	6	0.44	—

Notes

[a] There is considerable variation according to variety; some yellow sweet potatoes contain 12 000µg but the white varieties contain only a trace.

[b] Also contains 0.2mg γ-tocopherol per 100g.

No	Food	Description and number of samples	Edible matter, proportion of weight purchased	Water g	Sugars g	Starch g	Dietary fibre g	Total nitrogen g

No	Food	Energy value		Protein (N × 6.25)	Fat	Carbo-hydrate	mg									
		kcal	kJ	g	g	g	Na	K	Ca	Mg	P	Fe	Cu	Zn	S	Cl

No	Food	Reitnol µg	Carotene µg	Vitamin D µg	Thiamin mg	Riboflavin mg	Nicotinic acid mg	Potential nicotinic acid from tryptophan mgTrp ÷60	Vitamin C mg	Vitamin E mg

No	Food	Vitamin B_6 mg	Vitamin B_{12} µg	Folic acid		Panto-thenic acid mg	Biotin µg	Notes
				Free µg	Total µg			

No	Food	Description and number of samples	Edible matter, proportion of weight purchased	Water g	Sugars g	Starch g	Dietary fibre g	Total nitrogen g
675	**Apples, eating**	Flesh only, no skin or core	0.77	84.3	11.8	0.1	2.0[a]	0.04
676	eating (weighed with skin and core)	Calculated from the previous item	0.77	64.9	9.1	0.1	1.5	0.03
677	**cooking** raw	Flesh only, no skin or core	0.81	85.6	9.2	0.4	2.4	0.05
678	baked without sugar	Flesh only, no skin, cored before cooking	0.70	85.0	9.6	0.4	2.5	0.05
679	baked (weighed with skin)	Calculated from the previous item	0.70	68.0	7.7	0.3	2.0	0.04
680	stewed without sugar	Flesh and juice, peeled and cored before cooking	1.00	87.7	7.9	0.3	2.1	0.04
681	stewed with sugar	Flesh and juice, peeled and cored before cooking	1.11	79.3	17.0	0.3	1.9	0.04
682	**Apricots, fresh** raw	Flesh and skin, no stones	0.92	86.6	6.7	0	2.1	0.09
683	raw (weighed with stones)	Calculated from the previous item	0.92	79.6	6.2	0	1.9	0.08
684	stewed without sugar	Fruit and juice, no stones	1.10	87.9	5.6	0	1.7	0.07
685	stewed without sugar (weighed with stones)	Calculated from the previous item	1.10	82.6	5.3	0	1.6	0.07
686	stewed with sugar	Fruit and juice, no stones	1.21	79.4	15.6	0	1.6	0.06
687	stewed with sugar (weighed with stones)	Calculated from the previous item	1.21	74.6	14.7	0	1.5	0.06

[a] Apple peel contains 3.7g per 100g

No	Food	Energy value		Protein (N × 6.25)	Fat	Carbo-hydrate	mg									
		kcal	kJ	g	g	g	Na	K	Ca	Mg	P	Fe	Cu	Zn	S	Cl
675	**Apples, eating**	46	196	0.3	Tr	11.9	2	120	4	5	8	0.3	0.04	0.1	6	1
676	eating (weighed with skin and core)	35	151	0.2	Tr	9.2	2	92	3	4	6	0.2	0.03	0.1	5	1
677	**cooking** raw	37	159	0.3	Tr	9.6	2	120	4	3	16	0.3	0.09	0.1	3	5
678	baked without sugar	39	165	0.3	Tr	10.0	2	130	4	3	17	0.3	0.09	0.1	3	5
679	baked (weighed with skin)	31	133	0.3	Tr	8.0	2	100	3	2	13	0.2	0.07	0.1	2	4
680	stewed without sugar	32	136	0.3	Tr	8.2	2	100	3	3	14	0.3	0.08	0.1	3	4
681	stewed with sugar	66	282	0.3	Tr	17.3	2	94	3	2	13	0.2	0.07	0.1	2	4
682	**Apricots** fresh raw	28	117	0.6	Tr	6.7	Tr	320	17	12	21	0.4	0.12	0.1	6	Tr
683	raw (weighed with stones)	25	108	0.5	Tr	6.2	Tr	290	16	11	20	0.3	0.11	0.1	6	Tr
684	stewed without sugar	23	98	0.4	Tr	5.7	Tr	270	15	10	18	0.3	0.10	0.1	5	Tr
685	stewed without sugar (weighed with stones)	21	92	0.4	Tr	5.3	Tr	250	14	9	17	0.3	0.09	0.1	5	Tr
686	stewed with sugar	60	256	0.4	Tr	15.6	Tr	240	14	10	17	0.2	0.10	0.1	5	Tr
687	stewed with sugar (weighed with stones)	57	242	0.4	Tr	14.7	Tr	230	13	9	16	0.2	0.09	0.1	5	Tr

No	Food	Retinol µg	Carotene µg	Vitamin D µg	Thiamin mg	Riboflavin mg	Nicotinic acid mg	Potential nicotinic acid from tryptophan mg Trp ÷ 60	Vitamin C mg	Vitamin E mg
675	**Apples, eating**	0	30	0	0.04	0.02	0.1	Tr	3 [a]	0.2
676	eating (weighed with skin and core)	0	23	0	0.03	0.02	0.1	Tr	2	0.2
677	**cooking** raw	0	30	0	0.04	0.02	0.1	Tr	15 [b]	0.2
678	baked without sugar	0	30	0	0.03	0.02	0.1	Tr	14	0.2
679	baked (weighed with skin)	0	20	0	0.03	0.01	0.1	Tr	10	0.1
680	stewed without sugar	0	25	0	0.03	0.02	0.1	Tr	12	0.2
681	stewed with sugar	0	25	0	0.03	0.02	0.1	Tr	11	0.2
682	**Apricots** fresh, raw	0	1500 *(1000–2400)*	0	0.04	0.05	0.6	Tr	7	—
683	raw (weighed with stones)	0	1380	0	0.04	0.05	0.5	Tr	6	—
684	stewed without sugar	0	1260	0	0.03	0.04	0.4	Tr	5	—
685	stewed without sugar (weighed with stones)	0	1180	0	0.03	0.04	0.4	Tr	5	—
686	stewed with sugar	0	1150	0	0.03	0.04	0.4	Tr	5	—
687	stewed with sugar (weighed with stones)	0	1080	0	0.03	0.04	0.4	Tr	5	—

No	Food	Vitamin B_6 mg	Vitamin B_{12} µg	Folic acid Free µg	Folic acid Total µg	Pantothenic acid mg	Biotin µg
675	**Apples, eating**	0.03	0	2	5	0.10	0.3
676	eating (weighed with skin and core)	0.02	0	2	4	0.08	0.2
677	**cooking** raw	0.03	0	2	5	0.10	0.3
678	baked without sugar	0.02	0	Tr	3	0.09	0.3
679	baked (weighed with skin)	0.01	0	Tr	2	0.06	0.2
680	stewed without sugar	0.02	0	Tr	2	0.08	0.3
681	stewed with sugar	0.02	0	Tr	2	0.07	0.2
682	**Apricots** fresh, raw	0.07	0	(4)	(5)	0.30	—
683	raw (weighed with stones)	0.06	0	(4)	(5)	0.28	—
684	stewed without sugar	0.04	0	Tr	(2)	0.23	—
685	stewed without sugar (weighed with stones)	0.04	0	Tr	(2)	0.22	—
686	stewed with sugar	0.04	0	Tr	(2)	0.21	—
687	stewed with sugar (weighed with stones)	0.04	0	Tr	(2)	0.20	—

Notes

[a] Value for Cox's Orange Pippin. The vitamin C content varies according to variety, and there is more in the peel than in the flesh. Some values for different varieties are:

	Vitamin C mg/100g Peeled	Unpeeled
Cox's Orange Pippin	3	5
Granny Smith	2	8
Laxton Superb, Golden Delicious, Newton Wonder	3	10
Worcester Pearmain, Lord Lambourne	10	16
Sturmer Pippin	20	30

[b] Value for peeled Bramley's Seedlings. Unpeeled Bramley's Seedlings contain 20 mg per 100 g

No	Food	Description and number of samples	Edible matter, proportion of weight purchased	Water g	Sugars g	Starch g	Dietary fibre g	Total nitrogen g
688	**Apricots** dried, raw	Whole fruit	1.00	14.7	43.4	0	24.0	0.76
689	stewed without sugar	Fruit and juice	2.70	68.4	16.1	0	8.9	0.28
690	stewed with sugar	Fruit and juice	2.81	76.4	19.9	0	8.5	0.27
691	canned	Fruit and syrup	1.00	67.8	27.7	0	1.3	0.08
692	**Avocado pears**	10 pears, flesh only; Fuerte variety	0.71	68.7[a]	1.8	Tr	2.0	0.67
693	**Bananas** raw	Flesh only, no skin	0.59	70.7	16.2	3.0	3.4	0.18
694	raw (weighed with skin)	Calculated from the previous item	0.59	41.6	9.6	1.8	2.0	0.11
695	**Bilberries** raw	Whole fruit; literature sources	0.98	84.9	14.3	0	—	0.10
696	**Blackberries** raw	Whole fruit	1.00	82.0	6.4	0	7.3	0.20
697	stewed without sugar	Fruit and juice	1.17	84.6	5.5	0	6.3	0.17
698	stewed with sugar	Fruit and juice	1.28	76.5	14.8	0	5.7	0.16
699	**Cherries, eating** raw	Flesh and skin, no stalks or stones	0.87	81.5	11.9	0	1.7	0.09
700	raw (weighed with stones)	Calculated from the previous item	0.87	71.0	10.4	0	1.5	0.08
701	**cooking** raw	Flesh and skin, no stalks or stones	0.84	79.8	11.6	0	1.7	0.09
702	raw (weighed with stones)	Calculated from the previous item	0.84	67.0	9.8	0	1.4	0.08
703	stewed without sugar	Fruit and juice, no stones	1.03	83.4	9.7	0	1.4	0.08
704	stewed without sugar (weighed with stones)	Calculated from the previous item	1.03	72.6	8.4	0	1.2	0.07

[a] The water content varies from 52 to 79 g per 100 g according to season (Pearson 1975)

No	Food	Energy value		Protein (N × 6.25)	Fat	Carbo-hydrate	mg									
		kcal	kJ	g	g	g	Na	K	Ca	Mg	P	Fe	Cu	Zn	S	Cl
688	**Apricots** dried, raw	182	776	4.8	Tr	43.4	56	1880	92	65	120	4.1	0.27	0.2	160	35
689	stewed without sugar	66	288	1.8	Tr	16.1	21	700	34	24	44	1.5	0.10	0.1	59	13
690	stewed with sugar	81	347	1.7	Tr	19.9	20	670	33	23	43	1.5	0.10	0.1	57	12
691	canned	106	452	0.5	Tr	27.7	1	260	12	7	13	0.7	0.05	0.1	1	2
692	**Avocado pears**	223	922	4.2	22.2[a]	1.8	2	400	15	29	31	1.5	0.21	—	19	6
693	**Bananas** raw	79	337	1.1	0.3	19.2	1	350	7	42	28	0.4	0.16	0.2	13	79
694	raw (weighed with skin)	47	202	0.7	0.2	11.4	1	210	4	25	17	0.2	0.09	0.2	8	46
695	**Bilberries** raw	56	240	0.6	Tr	14.3	1	65	10	2	9	0.7	0.11	0.1	—	5
696	**Blackberries** raw	29	125	1.3	Tr	6.4	4	210	63	30	24	0.9	0.12	—	9	22
697	stewed without sugar	25	107	1.1	Tr	5.5	3	180	54	26	21	0.8	0.10	—	8	19
698	stewed with sugar	60	254	1.0	Tr	14.8	3	160	49	23	19	0.7	0.09	—	7	17
699	**Cherries, eating** raw	47	201	0.6	Tr	11.9	3	280	16	10	17	0.4	0.07	0.1	7	Tr
700	raw (weighed with stones)	41	175	0.5	Tr	10.4	2	240	14	8	15	0.3	0.06	0.1	6	Tr
701	**cooking** raw	46	196	0.6	Tr	11.6	4	310	20	12	21	0.3	0.10	0.1	8	Tr
702	raw (weighed with stones)	39	165	0.5	Tr	9.8	3	260	17	10	18	0.3	0.08	0.1	7	Tr
703	stewed without sugar	39	165	0.5	Tr	9.8	3	250	18	10	17	0.3	0.08	0.1	7	Tr
704	stewed without sugar (weighed with stones)	33	141	0.4	Tr	8.4	3	220	16	9	15	0.3	0.07	0.1	6	Tr

[a] The fat content varies from 11 to 39 g per 100 g according to season (Pearson 1975)

No	Food	Retinol µg	Carotene µg	Vitamin D µg	Thiamin mg	Riboflavin mg	Nicotinic acid mg	Potential nicotinic acid from tryptophan mg Trp ÷ 60	Vitamin C mg	Vitamin E mg
688	**Apricots** dried, raw	0	3600 *(2400–4400)*	0	Tr	0.20	3.0	0.8	Tr	—
689	stewed without sugar	0	1330	0	Tr	0.06	1.1	0.3	Tr	—
690	stewed with sugar	0	1280	0	Tr	0.06	1.1	0.3	Tr	—
691	canned	0	1000	0	0.02	0.01	0.3	0.1	2	—
692	**Avocado pears**	0	100	—[a]	0.10	0.10	1.0	0.8	15 *(5–30)*	3.2
693	**Bananas** raw	0	200	0	0.04	0.07	0.6	0.2	10	0.2
694	raw (weighed with skin)	0	120	0	0.02	0.04	0.4	0.1	6	0.1
695	**Bilberries,** raw	0	130	0	0.02	0.02	0.4	0.1	22 *(10–44)*	—
696	**Blackberries** raw	0	100	0	0.03	0.04	0.4	0.2	20	3.5[b]
697	stewed without sugar	0	85	0	0.03	0.03	0.3	0.2	15	3.0[c]
698	stewed with sugar	0	80	0	0.02	0.03	0.3	0.2	14	2.7[c]
699	**Cherries, eating** raw	0	120	0	0.05	0.07	0.3	0.1	5	0.1
700	raw (weighed with stones)	0	100	0	0.04	0.06	0.3	0.1	4	0.1
701	**cooking** raw	0	120	0	0.05	0.07	0.3	0.1	5	0.1
702	raw (weighed with stones)	0	100	0	0.04	0.06	0.3	0.1	4	0.1
703	stewed without sugar	0	100	0	0.03	0.06	0.3	0.1	3	0.1
704	stewed without sugar (weighed with stones)	0	85	0	0.03	0.05	0.3	0.1	3	0.1

No	Food	Vitamin B_6 mg	Vitamin B_{12} µg	Folic acid Free µg	Folic acid Total µg	Pantothenic acid mg	Biotin µg
688	**Apricots** dried, raw	0.17	0	10	14	0.70	—
689	stewed without sugar	0.05	0	Tr	2	0.23	—
690	stewed with sugar	0.05	0	Tr	2	0.23	—
691	canned	0.05	0	4	(5)	0.10	—
692	**Avocado pears**	0.42	0	55	66	1.07	3.2
693	**Bananas** raw	0.51	0	14	22	0.26	—
694	raw (weighed with skin)	0.30	0	8	13	0.15	—
695	**Bilberries** raw	0.06	0	2	6	0.16	—
696	**Blackberries** raw	0.05	0	—	—	0.25	0.4
697	stewed without sugar	0.03	0	—	—	0.19	0.3
698	stewed with sugar	0.03	0	—	—	0.18	0.3
699	**Cherries, eating** raw	0.05	0	6	8	0.26	0.4
700	raw (weighed with stones)	0.04	0	5	7	0.23	0.3
701	**cooking** raw	0.05	0	6	8	0.26	0.4
702	raw (weighed with stones)	0.04	0	5	7	0.22	0.3
703	stewed without sugar	0.02	0	Tr	3	0.20	0.3
704	stewed without sugar (weighed with stones)	0.02	0	Tr	3	0.17	0.3

Notes

[a] Avocado pears have been reported to contain vitamin D (Zanobini, Firenzuoli and Bianchi 1974).

[b] Value for wild blackberries, which also contain 4.7 mg γ-tocopherol and 4.5 mg δ-tocopherol. Cultivated blackberries contain 0.6 mg α-tocopherol, 1.1 mg γ-tocopherol and 1.0 mg δ-tocopherol per 100 g (Booth and Bradford 1963a).

[c] Value for wild blackberries.

No	Food	Description and number of samples	Edible matter, proportion of weight purchased	Water g	Sugars g	Starch g	Dietary fibre g	Total nitrogen g
705	**Cherries, cooking** stewed with sugar	Fruit and juice, no stones	1.13	72.8	19.7	0	1.2	0.07
706	stewed with sugar (weighed with stones)	Calculated from the previous item	1.13	64.8	17.5	0	1.1	0.06
707	**Cranberries** raw	Whole fruit	1.00	87.0	3.5	0	4.2	0.06
708	**Currants, black** raw	Whole fruit, no stalks	0.98	77.4	6.6	0	8.7	0.15
709	stewed without sugar	Fruit and juice	1.15	80.7	5.6	0	7.4	0.13
710	stewed with sugar	Fruit and juice	1.26	72.9	15.0	0	6.8	0.12
711	**red** raw	Whole fruit, no stalks	0.97	82.8	4.4	0	8.2	0.18
712	stewed without sugar	Fruit and juice	1.14	85.3	3.8	0	7.0	0.15
713	stewed with sugar	Fruit and juice	1.25	88.2	13.3	0	6.4	0.14
714	**white** raw	Whole fruit, no stalks	0.96	83.3	5.6	0	6.8	0.20
715	stewed without sugar	Fruit and juice	1.13	85.7	4.8	0	5.8	0.17
716	stewed with sugar	Fruit and juice	1.24	77.5	14.2	0	5.3	0.16
717	**dried**	Whole fruit	1.00	22.0	63.1	0	6.5	0.27
718	**Damsons** raw	Flesh and skin, no stalks or stones	0.90	77.5	9.6	0	4.1	0.08
719	raw (weighed with stones)	Calculated from the previous item	0.90	69.8	8.6	0	3.7	0.07
720	stewed without sugar	Fruit and juice, no stones	1.08	80.7	8.1	0	3.5	0.07
721	stewed without sugar (weighed with stones)	Calculated from the previous item	1.08	74.2	7.4	0	3.2	0.06

No	Food	Energy value kcal	Energy value kJ	Protein (N × 6.25) g	Fat g	Carbo-hydrate g	Na mg	K mg	Ca mg	Mg mg	P mg	Fe mg	Cu mg	Zn mg	S mg	Cl mg
705	**Cherries, cooking** stewed with sugar	77	328	0.4	Tr	20.1	2	230	15	9	16	0.2	0.07	0.1	6	Tr
706	stewed with sugar (weighed with stones)	67	287	0.4	Tr	17.5	2	200	13	8	14	0.2	0.06	0.1	5	Tr
707	**Cranberries** raw	15	63	0.4	Tr	3.5	2	120	15	8	11	1.1	0.14	—	11	Tr
708	**Currants, black** raw	28	121	0.9	Tr	6.6	3	370	60	17	43	1.3	0.14	—	33	15
709	stewed without sugar	24	103	0.8	Tr	5.6	3	320	51	16	37	1.1	0.12	—	28	13
710	stewed with sugar	59	254	0.8	Tr	15.0	2	290	47	13	34	1.0	0.11	—	26	12
711	**red** raw	21	89	1.1	Tr	4.4	2	280	36	13	30	1.2	0.12	—	29	14
712	stewed without sugar	18	76	0.9	Tr	3.8	2	240	31	11	26	1.0	0.10	—	25	12
713	stewed with sugar	53	228	0.9	Tr	13.3	2	220	28	10	23	0.9	0.09	—	23	11
714	**white** raw	26	112	1.3	Tr	5.6	2	290	22	13	28	0.9	0.14	—	24	11
715	stewed without sugar	22	96	1.1	Tr	4.8	2	250	19	11	24	0.8	0.12	—	21	9
716	stewed with sugar	57	244	1.0	Tr	14.2	2	230	17	10	22	0.7	0.11	—	19	9
717	**dried**	243	1039	1.7	Tr	63.1	20	710	95	36	40	1.8	0.48	(0.1)	31	16
718	**Damsons** raw	38	162	0.5	Tr	9.6	2	290	24	11	16	0.4	0.08	(0.1)	6	Tr
719	raw (weighed with stones)	34	144	0.4	Tr	8.6	2	260	21	10	15	0.4	0.07	(0.1)	6	Tr
720	stewed without sugar	32	136	0.4	Tr	8.1	2	240	20	10	14	0.3	0.07	(0.1)	5	Tr
721	stewed without sugar (weighed with stones)	29	125	0.4	Tr	7.4	2	220	18	9	13	0.3	0.06	(0.1)	5	Tr

No	Food	Retinol µg	Carotene µg	Vitamin D µg	Thiamin mg	Riboflavin mg	Nicotinic acid mg	Potential nicotinic acid from tryptophan mg Trp ÷ 60	Vitamin C mg	Vitamin E mg
705	**Cherries, cooking** stewed with sugar	0	90	0	0.03	0.06	0.2	0.1	3	0.1
706	stewed with sugar (weighed with stones)	0	80	0	0.03	0.05	0.2	0.1	3	0.1
707	**Cranberries** raw	0	20	0	0.03	0.02	0.1	0.1	12	—
708	**Currants, black** raw	0	200	0	0.03	0.06	0.3	0.1	200 *(150–230)*	1.0
709	stewed without sugar	0	170	0	0.03	0.05	0.3	0.1	150	0.9
710	stewed with sugar	0	160	0	0.02	0.05	0.2	0.1	140[a]	0.8
711	**red** raw	0	70	0	0.04	(0.06)	0.1	0.2	40	0.1
712	stewed without sugar	0	60	0	0.03	(0.05)	0.1	0.1	31	0.1
713	stewed with sugar	0	55	0	0.03	(0.05)	0.1	0.1	28	0.1
714	**white** raw	0	Tr	0	(0.04)	(0.06)	(0.1)	0.2	(40)	(0.1)
715	stewed without sugar	0	Tr	0	(0.03)	(0.05)	(0.1)	0.2	(31)	(0.1)
716	stewed with sugar	0	Tr	0	(0.03)	(0.05)	(0.1)	0.2	(28)	(0.1)
717	**dried**	0	30	0	0.03	(0.08)	(0.5)	0.1	0	—
718	**Damsons** raw	0	(220)	0	0.10	0.03	0.3	0.1	(3)	0.7
719	raw (weighed with stones)	0	(200)	0	0.09	0.03	0.3	0.1	(3)	0.6
720	stewed without sugar	0	(180)	0	0.08	0.03	0.3	0.1	(3)	0.5
721	stewed without sugar (weighed with stones)	0	(170)	0	0.07	0.03	0.3	0.1	(3)	0.5

No	Food	Vitamin B_6 mg	Vitamin B_{12} µg	Folic acid Free µg	Folic acid Total µg	Pantothenic acid mg	Biotin µg	Notes
705	**Cherries, cooking** stewed with sugar	0.02	0	Tr	3	0.13	0.2	[a] Canned blackcurrants contain 100mg per 100g.
706	stewed with sugar (weighed with stones)	0.02	0	Tr	3	0.12	0.2	
707	**Cranberries** raw	0.04	0	1	2	0.22	—	
708	**Currants, black** raw	0.08	0	—	—	0.40	2.4	
709	stewed without sugar	0.06	0	—	—	0.31	2.1	
710	stewed with sugar	0.05	0	—	—	0.28	1.9	
711	**red** raw	0.05	0	—	—	0.06	2.6	
712	stewed without sugar	0.03	0	—	—	0.05	2.2	
713	stewed with sugar	0.03	0	—	—	0.05	2.0	
714	**white** raw	(0.05)	0	—	—	(0.06)	(2.6)	
715	stewed without sugar	(0.03)	0	—	—	(0.05)	(2.2)	
716	stewed with sugar	(0.03)	0	—	—	(0.05)	(2.0)	
717	**dried**	(0.30)	0	4	11	(0.10)	—	
718	**Damsons** raw	(0.05)	0	(1)	(3)	0.27	0.1	
719	raw (weighed with stones)	(0.05)	0	(1)	(3)	0.24	0.1	
720	stewed without sugar	(0.03)	0	Tr	(1)	0.21	0.1	
721	stewed without sugar (weighed with stones)	(0.03)	0	Tr	(1)	0.19	0.1	

No	Food	Description and number of samples	Edible matter, proportion of weight purchased	Water g	Sugars g	Starch g	Dietary fibre g	Total nitrogen g
722	**Damsons** stewed with sugar	Fruit and juice, no stones	1.19	72.0	17.8	0	3.1	0.05
723	stewed with sugar (weighed with stones)	Calculated from the previous item	1.19	67.0	16.6	0	2.9	0.05
724	**Dates** dried	Flesh and skin, no stones	0.86	14.6	63.9	0	8.7	0.32
725	dried (weighed with stones)	Calculated from the previous item	0.86	12.6	54.9	0	7.5	0.28
726	**Figs, green** raw	Whole fruit, no stalks	0.98	84.6	9.5	0	2.5	0.21
727	**dried** raw	Whole fruit	1.00	16.8	52.9	0	18.5	0.57
728	stewed without sugar	Fruit and juice	1.80	53.8	29.4	0	10.3	0.32
729	stewed with sugar	Fruit and juice	1.91	50.7	34.3	0	9.7	0.30
730	**Fruit pie filling** canned	10 cans, blackcurrant, blackberry and apple, gooseberry, apple, cherry	1.00	72.6	23.2	1.9	(1.8)	0.04
731	**Fruit salad** canned	Fruit and syrup	1.00	71.1	25.0	0	1.1	0.04
732	**Gooseberries, green** raw	Flesh, skin and pips, no 'tops' or 'tails'	0.99	89.9	3.4	0	3.2	0.18
733	stewed without sugar	Fruit and juice	1.16	91.4	2.9	0	2.7	0.15
734	stewed with sugar	Fruit and juice	1.27	82.7	12.5	0	2.5	0.14
735	**ripe** raw	Flesh, skin and pips, no 'tops' or 'tails'	0.99	83.7	9.2	0	3.5	0.09
736	**Grapes, black** raw	Flesh only, no skin, pips or stalks	0.81	80.7	15.5	0	0.4	0.09
737	raw (whole grapes weighed)	Calculated from the previous item	0.81	65.2	13.0	0	0.3	0.08

No	Food	Energy value		Protein (N × 6.25)	Fat	Carbo-hydrate	mg									
		kcal	kJ	g	g	g	Na	K	Ca	Mg	P	Fe	Cu	Zn	S	Cl
722	**Damsons** stewed with sugar	69	293	0.3	Tr	18.0	2	220	17	9	13	0.3	0.05	(0.1)	5	Tr
723	stewed with sugar (weighed with stones)	63	271	0.3	Tr	16.6	2	200	16	8	12	0.3	0.05	(0.1)	5	Tr
724	**Dates** dried	248	1056	2.0	Tr	63.9	5	750	68	59	64	1.6	0.21	0.3	51	290
725	dried (weighed with stones)	213	909	1.7	Tr	54.9	4	650	58	50	55	1.4	0.18	0.3	44	250
726	**Figs, green** raw	41	174	1.3	Tr	9.5	2	270	34	20	32	0.4	0.06	0.3	13	18
727	**dried** raw	213	908	3.6	Tr	52.9	87	1010	280	92	92	4.2	0.24	0.9	81	170
728	stewed without sugar	118	504	2.0	Tr	29.4	48	560	160	51	51	2.3	0.13	0.5	45	94
729	stewed with sugar	136	581	1.9	Tr	34.3	46	530	140	48	48	2.2	0.13	0.5	42	89
730	**Fruit pie filling** canned	95	407	0.3	Tr	25.1	30	79	18	4	9	0.5	(0.03)	0.1	—	45
731	**Fruit salad** canned	95	405	0.3	Tr	25.0	2	120	8	8	10	1.0	0.03	—	2	3
732	**Gooseberries, green** raw	17	73	1.1	Tr	3.4	2	210	28	7	34	0.3	0.13	0.1	16	7
733	stewed without sugar	14	62	0.9	Tr	2.9	2	180	24	6	29	0.3	0.11	0.1	14	6
734	stewed with sugar	50	215	0.9	Tr	12.5	2	160	22	5	27	0.2	0.10	0.1	13	5
735	**ripe** raw	37	157	0.6	Tr	9.2	1	170	19	9	19	0.6	0.15	0.1	14	11
736	**Grapes, black** raw	61	258	0.6	Tr	15.5	2	320	4	4	16	0.3	0.08	0.1	7	Tr
737	raw (whole grapes weighed)	51	217	0.5	Tr	13.0	1	270	4	3	14	0.3	0.07	0.1	6	Tr

No	Food	Retinol µg	Carotene µg	Vitamin D µg	Thiamin mg	Riboflavin mg	Nicotinic acid mg	Potential nicotinic acid from tryptophan mg Trp ÷ 60	Vitamin C mg	Vitamin E mg
722	**Damsons** stewed with sugar	0	(170)	0	0.06	0.02	0.2	0.1	(2)	0.5
723	stewed with sugar (weighed with stones)	0	(160)	0	0.06	0.02	0.2	0.1	(2)	0.5
724	**Dates** dried	0	50	0	0.07	0.04	2.0	0.9	0	—
725	dried (weighed with stones)	0	43	0	0.06	0.03	1.7	0.8	0	—
726	**Figs, green** raw	0	(500)	0	0.06	0.05	0.4	0.2	2	—
727	**dried** raw	0	50	0	0.10	0.08	1.7	0.5	0	—
728	stewed without sugar	0	30	0	0.05	0.04	0.9	0.3	0	—
729	stewed with sugar	0	30	0	0.05	0.04	0.9	0.3	0	—
730	**Fruit pie filling** canned	0	—	0	Tr	0.01	0.1	Tr	(Tr)	—
731	**Fruit salad** canned[a]	0	300	0	0.02	0.01	0.3	Tr	3	—
732	**Gooseberries, green** raw	0	180	0	(0.04)	0.03	0.3	0.2	40 *(25–50)*	0.4
733	stewed without sugar	0	150	0	0.03	0.03	0.3	0.2	31	0.3
734	stewed with sugar	0	140	0	0.03	0.02	0.2	0.1	28[b]	0.3
735	**ripe** raw	0	180	0	(0.04)	0.03	0.3	0.1	40 *(25–50)*	0.4
736	**Grapes, black** raw	0	(Tr)	0	0.04	0.02	0.3	Tr	4	—
737	raw (whole grapes weighed)	0	(Tr)	0	0.03	0.02	0.2	Tr	3	—

No	Food	Vitamin B_6 mg	Vitamin B_{12} µg	Folic acid Free µg	Folic acid Total µg	Pantothenic acid mg	Biotin µg
722	**Damsons** stewed with sugar	0.03	0	Tr	(1)	0.18	0.1
723	stewed with sugar (weighed with stones)	0.03	0	Tr	(1)	0.17	0.1
724	**Dates** dried	0.15	0	14	21	0.80	—
725	dried (weighed with stones)	0.13	0	12	18	0.69	—
726	**Figs, green** raw	0.11	0	—	—	0.30	—
727	**dried** raw	0.18	0	3	9	0.44	—
728	stewed without sugar	0.08	0	Tr	2	0.22	—
729	stewed with sugar	0.07	0	Tr	2	0.21	—
730	**Fruit pie filling** canned	(Tr)	0	1	1	(Tr)	(Tr)
731	**Fruit salad** canned [a]	0.01	0	(1)	(4)	0.04	0.1
732	**Gooseberries, green** raw	0.02	0	—	—	0.15	0.5
733	stewed without sugar	0.02	0	—	—	0.12	0.4
734	stewed with sugar	0.02	0	—	—	0.11	0.4
735	**ripe** raw	0.02	0	—	—	0.30	0.1
736	**Grapes, black** raw	0.10	0	3	6	0.05	0.3
737	raw (whole grapes weighed)	0.08	0	2	5	0.04	0.2

Notes

[a] Calculated assuming that canned fruit salad contains canned fruit in the following proportions:

	%
Apricots or peaches	35
Pears	35
Cherries	10
Grapes	10
Pineapple	10

[b] Canned gooseberries contain 24 mg per 100 g.

No	Food	Description and number of samples	Edible matter, proportion of weight purchased	Water g	Sugars g	Starch g	Dietary fibre g	Total nitrogen g
738	**Grapes, white** raw	Flesh and skin, no pips or stalks	0.95	79.3	16.1	0	0.9	0.10
739	raw (whole grapes weighed)	Calculated from the previous item	0.95	75.5	15.3	0	0.9	0.10
740	**Grapefruit** raw	Flesh only, no skin, pith or pips	0.48	90.7	5.3	0	0.6	0.10
741	raw (whole fruit weighed)	Calculated from the previous item	0.48	43.5	2.5	0	0.3	0.05
742	canned	10 cans, fruit and syrup	1.00	81.8	15.5	0	0.4	0.08
743	**Greengages** raw	Flesh and skin, no stalks or stones	0.95	78.2	11.8	0	2.6	0.12
744	raw (weighed with stones)	Calculated from the previous item	0.95	74.4	11.2	0	2.5	0.11
745	stewed without sugar	Fruit and juice, no stones	1.13	81.4	10.0	0	2.2	0.09
746	stewed without sugar (weighed with stones)	Calculated from the previous item	1.13	78.1	9.6	0	2.1	0.09
747	stewed with sugar	Fruit and juice, no stones	1.23	72.8	19.2	0	2.1	0.09
748	stewed with sugar (weighed with stones)	Calculated from the previous item	1.23	70.6	18.6	0	2.0	0.09
749	**Guavas** canned	10 cans, fruit and syrup	1.00	77.6	15.7	Tr	3.6	0.06
750	**Lemons** whole	Whole fruit including skin, no pips	0.99	85.2	3.2	0	5.2	0.12
751	juice, fresh	Strained juice from fresh lemons	0.36	91.3	1.6	0	0	0.05
752	**Loganberries** raw	Whole fruit	1.00	85.0	3.4	0	6.2	0.17
753	stewed without sugar	Fruit and juice	1.08	86.1	3.1	0	5.7	0.16

No	Food	Energy value		Protein (N × 6.25)	Fat	Carbo-hydrate	mg									
		kcal	kJ	g	g	g	Na	K	Ca	Mg	P	Fe	Cu	Zn	S	Cl
738	**Grapes, white** raw	63	268	0.6	Tr	16.1	2	250	19	7	22	0.3	0.10	0.1	9	Tr
739	raw (whole grapes weighed)	60	255	0.6	Tr	15.3	2	240	18	6	21	0.3	0.10	0.1	9	Tr
740	**Grapefruit** raw	22	95	0.6	Tr	5.3	1	230	17	10	16	0.3	0.06	0.1	5	1
741	raw (whole fruit weighed)	11	45	0.3	Tr	2.5	1	110	8	5	8	0.1	0.03	0.1	3	1
742	canned	60	257	0.5	Tr	15.5	10	79	17	7	13	0.7	(0.03)	0.4	—	(5)
743	**Greengages** raw	47	202	0.8	Tr	11.8	1	310	17	8	23	0.4	0.08	(0.1)	3	1
744	raw (weighed with stones)	45	191	0.7	Tr	11.2	1	290	16	7	22	0.4	0.08	(0.1)	3	1
745	stewed without sugar	40	170	0.6	Tr	10.0	1	260	15	6	20	0.3	0.07	(0.1)	3	1
746	stewed without sugar (weighed with stones)	38	164	0.6	Tr	9.6	1	250	14	6	19	0.3	0.07	(0.1)	3	1
747	stewed with sugar	75	321	0.6	Tr	19.4	1	240	13	5	18	0.3	0.06	(0.1)	2	1
748	stewed with sugar (weighed with stones)	72	308	0.6	Tr	18.6	1	230	13	5	17	0.3	0.06	(0.1)	2	1
749	**Guavas** canned	60	258	0.4	Tr	15.7	7	120	8	6	11	0.5	0.10	0.4	—	10
750	**Lemons** whole	15	65	0.8	Tr	3.2	6	160	110	12	21	0.4	0.26	0.1	12	5
751	juice, fresh	7	31	0.3	Tr	1.6	2	140	8	7	10	0.1	0.13	Tr	2	3
752	**Loganberries** raw	17	73	1.1	Tr	3.4	3	260	35	25	24	1.4	0.14	—	18	16
753	stewed without sugar	16	67	1.0	Tr	3.1	3	240	32	23	22	1.3	0.13	—	17	15

No	Food	Retinol µg	Carotene µg	Vitamin D µg	Thiamin mg	Riboflavin mg	Nicotinic acid mg	Potential nicotinic acid from tryptophan mg Trp ÷ 60	Vitamin C mg	Vitamin E mg
738	**Grapes, white** raw	0	Tr	0	0.04	0.02	0.3	Tr	4	—
739	raw (whole grapes weighed)	0	Tr	0	0.04	0.02	0.3	Tr	4	—
740	**Grapefruit** raw	0	Tr	0	0.05	0.02	0.2	0.1	40 *(35–45)*	0.3
741	raw (whole fruit weighed)	0	Tr	0	0.02	0.01	0.1	Tr	19	0.1
742	canned	0	Tr	0	0.04	0.01	0.2	0.1	30	Tr
743	**Greengages** raw	0	—	0	(0.05)	(0.03)	(0.4)	0.1	(3)	(0.7)
744	raw (weighed with stones)	0	—	0	(0.05)	(0.03)	(0.4)	0.1	(3)	(0.7)
745	stewed without sugar	0	—	0	(0.04)	(0.03)	(0.3)	0.1	(3)	(0.6)
746	stewed without sugar (weighed with stones)	0	—	0	(0.04)	(0.03)	(0.3)	0.1	(3)	(0.6)
747	stewed with sugar	0	—	0	(0.04)	(0.02)	(0.3)	0.1	(2)	(0.5)
748	stewed with sugar (weighed with stones)	0	—	0	(0.04)	(0.02)	(0.3)	0.1	(2)	(0.5)
749	**Guavas** canned	0	(100)	0	(0.04)	(0.03)	(0.9)	0.1	180[a]	—
750	**Lemons** whole	0	Tr	0	0.05	0.04	0.2	0.1	80	—
751	juice, fresh	0	Tr	0	0.02	0.01	0.1	Tr	50[b] *(40–60)*	—
752	**Loganberries** raw	0	(80)	0	(0.02)	0.03	(0.4)	0.2	35	(0.3)
753	stewed without sugar	0	(75)	0	(0.02)	0.03	(0.4)	0.2	29	(0.3)

No	Food	Vitamin B_6 mg	Vitamin B_{12} µg	Folic acid		Panto-thenic acid mg	Biotin µg	Notes
				Free µg	Total µg			
738	**Grapes, white** raw	0.10	0	3	6	0.05	0.3	[a] Raw guavas contain about 200mg per 100g; the value may range from 20 to 600mg per 100g.
739	raw (whole grapes weighed)	0.10	0	3	6	0.05	0.3	[b] Limes contain 25mg vitamin C per 100g.
740	**Grapefruit** raw	0.03	0	9	12	0.28	(1.0)	
741	raw (whole fruit weighed)	0.01	0	4	6	0.13	(0.5)	
742	canned	0.02	0	3	4	0.12	1.0	
743	**Greengages** raw	(0.05)	0	(1)	(3)	(0.20)	(Tr)	
744	raw (weighed with stones)	(0.05)	0	(1)	(3)	(0.20)	(Tr)	
745	stewed without sugar	(0.03)	0	(Tr)	(1)	(0.16)	(Tr)	
746	stewed without sugar (weighed with stones)	(0.03)	0	(Tr)	(1)	(0.15)	(Tr)	
747	stewed with sugar	(0.03)	0	(Tr)	(1)	(0.14)	(Tr)	
748	stewed with sugar (weighed with stones)	(0.03)	0	(Tr)	(1)	(0.14)	(Tr)	
749	**Guavas** canned	—	0	—	—	—	—	
750	**Lemons** whole	0.11	0	—	—	0.23	0.5	
751	juice, fresh	0.05	0	7	7	0.10	0.3	
752	**Loganberries** raw	(0.06)	0	—	—	(0.24)	—	
753	stewed without sugar	(0.05)	0	—	—	(0.20)	—	

No	Food	Description and number of samples	Edible matter, proportion of weight purchased	Water g	Sugars g	Starch g	Dietary fibre g	Total nitrogen g
754	**Loganberries** stewed with sugar	Fruit and juice	1.19	77.3	13.4	0	5.2	0.14
755	canned	Fruit and juice	1.00	66.3	26.2	0	3.3	0.10
756	**Lychees** raw	Flesh only; literature sources	0.60	82.0	16.0	0	(0.5)	0.14
757	canned	10 cans ; fruit and syrup	1.00	79.3	17.7	0	0.4	0.06
758	**Mandarin oranges** canned	10 cans; fruit and syrup	1.00	84.3	14.2	0	0.3	0.10
759	**Mangoes** raw	Flesh only; literature sources	0.66	83.0	15.3	Tr	(1.5)	0.08
760	canned	10 cans; fruit and syrup	1.00	74.8	20.2	0.1	1.0	0.05
761	**Medlars** raw	Flesh only, no skin or stones	0.81	74.5	10.6	0	10.2	0.08
762	**Melons, Canteloupe** raw	Flesh only, no skin or seeds	0.59	93.6	5.3	0	1.0	0.16
763	raw (weighed with skin)	Calculated from the previous item	0.59	58.6	3.3	0	0.6	0.10
764	**yellow, Honeydew** raw	Flesh only, no skin or seeds	0.59	94.2	5.0	0	0.9	0.10
765	raw (weighed with skin)	Calculated from the previous item	0.59	59.0	3.1	0	0.6	0.06
766	**watermelon** raw	Flesh only; literature sources	0.50	94.0	5.3	0	—	0.06
767	raw (weighed with skin)	Calculted from the previous item	0.50	47.0	2.7	0	—	0.03
768	**Mulberries** raw	Whole fruit	1.00	85.0	8.1	0	1.7	0.21
769	**Nectarines** raw	Flesh and skin, no stones	0.92	80.2	12.4	0	2.4	0.15
770	raw (weighed with stones)	Calculated from the previous item	0.92	74.0	11.4	0	2.2	0.14

No	Food	Energy value kcal	Energy value kJ	Protein (N× 6.25) g	Fat g	Carbo-hydrate g	Na mg	K mg	Ca mg	Mg mg	P mg	Fe mg	Cu mg	Zn mg	S mg	Cl mg
754	**Loganberries** stewed with sugar	54	230	0.9	Tr	13.4	3	220	29	21	20	1.2	0.12	—	15	13
755	canned	101	429	0.6	Tr	26.2	1	97	18	11	23	1.4	0.04	—	3	5
756	**Lychees** raw	64	271	0.9	Tr	16.0	3	170	8	10	35	0.5	—	—	19	3
757	canned	68	290	0.4	Tr	17.7	2	75	4	6	12	0.7	0.11	0.2	—	(5)
758	**Mandarin oranges** canned	56	237	0.6	Tr	14.2	9	88	18	9	12	0.4	0.05	0.4	—	2
759	**Mangoes** raw	59	253	0.5	Tr	15.3	7	190	10	18	13	0.5	0.12	—	—	—
760	canned	77	330	0.3	Tr	20.3	3	100	10	7	10	0.4	0.09	0.3	—	(5)
761	**Medlars** raw	42	178	0.5	Tr	10.6	6	250	30	11	28	0.5	0.17	—	17	3
762	**Melons, Canteloupe** raw	24	102	1.0	Tr	5.3	14	320	19	20	30	0.8	0.04	0.1	12	44
753	raw (weighed with skin)	15	63	0.6	Tr	3.3	9	200	12	13	19	0.5	0.03	0.1	7	27
764	**yellow, Honeydew** raw	21	90	0.6	Tr	5.0	20	220	14	13	9	0.2	0.04	0.1	6	45
765	raw (weighed with skin)	13	56	0.4	Tr	3.1	12	140	9	8	5	0.2	0.03	0.1	4	28
766	**watermelon** raw	21	92	0.4	Tr	5.3	4	120	5	11	8	0.3	0.03	(0.1)	—	—
767	raw (weighed with skin)	11	47	0.2	Tr	2.7	2	60	3	6	4	0.2	0.02	(Tr)	—	—
768	**Mulberries** raw	36	152	1.3	Tr	8.1	2	260	36	15	48	1.6	0.06	—	9	4
769	**Nectarines** raw	50	214	0.9	Tr	12.4	9	270	4	13	24	0.5	0.06	0.1	10	5
770	raw (weighed with stones)	46	198	0.9	Tr	11.4	8	250	4	12	22	0.4	0.06	0.1	9	4

No	Food	Retinol µg	Carotene µg	Vitamin D µg	Thiamin mg	Riboflavin mg	Nicotinic acid mg	Potential nicotinic acid from tryptophan mgTrp ÷60	Vitamin C mg	Vitamin E mg
754	**Loganberries** stewed with sugar	0	(70)	0	(0.02)	0.03	(0.3)	0.1	26	(0.3)
755	canned	0	(70)	0	(0.01)	0.02	(0.3)	0.1	25	—
756	**Lychees** raw	0	(Tr)	0	0.04	0.04	0.3	0.1	40	—
757	canned	0	(Tr)	0	(0.03)	(0.03)	(0.2)	0.1	8	—
758	**Mandarin oranges** canned	0	50	0	0.07	0.02	0.2	0.1	14	(Tr)
759	**Mangoes** raw	0	1200[a]	0	0.03	0.04	0.3	0.1	30 *(10–180)*	—
760	canned	0	(1200)	0	(0.02)	(0.03)	(0.2)	Tr	10	—
761	**Medlars** raw	0	—	0	—	—	—	0.1	2	—
762	**Melons, Canteloupe** raw	0	2000[b]	0	0.05	0.03	0.5	Tr	25	0.1
763	raw (weighed with skin)	0	1180[b]	0	0.03	0.02	0.3	Tr	15	0.1
764	**yellow, Honeydew** raw	0	100[c]	0	0.05	0.03	0.5	Tr	25	0.1
765	raw (weighed with skin)	0	60[c]	0	0.03	0.02	0.3	Tr	15	0.1
766	**watermelon** raw	0	20	0	0.02	0.02	0.2	0.1	5	(0.1)
767	raw (weighed with skin)	0	10	0	0.01	0.01	0.1	Tr	3	(Tr)
768	**Mulberries** raw	0	Tr	0	0.05	(0.04)	(0.4)	0.2	10	—
769	**Nectarines** raw	0	(500)	0	(0.02)	(0.05)	(1.0)	0.1	(8)	—
770	raw (weighed with stones)	0	(460)	0	(0.02)	(0.05)	(0.9)	0.1	(7)	—

No	Food	Vitamin B_6 mg	Vitamin B_{12} µg	Folic acid Free µg	Folic acid Total µg	Pantothenic acid mg	Biotin µg
754	**Loganberries** stewed with sugar	(0.05)	0	—	—	(0.18)	—
755	canned	(0.04)	0	—	—	(0.17)	—
756	**Lychees** raw	—	0	—	—	—	—
757	canned	—	0	—	—	—	—
758	**Mandarin oranges** canned	0.03	0	5	8	(0.15)	(0.8)
759	**Mangoes** raw	—	0	—	—	0.16	—
760	canned	—	0	—	—	—	—
761	**Medlars** raw	—	0	—	—	—	—
762	**Melons, Canteloupe** raw	0.07	0	30	30	0.23	—
763	raw (weighed with skin)	0.04	0	18	18	0.14	—
764	**yellow, Honeydew** raw	0.07	0	(30)	(30)	0.23	—
765	raw (weighed with skin)	0.04	0	(18)	(18)	0.14	—
766	**watermelon** raw	(0.07)	0	2	3	1.55	—
767	raw (weighed with skin)	(0.04)	0	1	2	0.78	—
768	**Mulberries** raw	(0.05)	0	—	—	(0.25)	(0.4)
769	**Nectarines** raw	(0.02)	0	5	5	(0.15)	—
770	raw (weighed with stones)	(0.02)	0	5	5	(0.14)	—

Notes

[a] Value for ripe, orange coloured mangoes. The carotene content varies according to the colour, and unripe, green mangoes contain about one-tenth of this amount.

[b] Value for orange coloured flesh.

[c] Value for green coloured flesh.

No	Food	Description and number of samples	Edible matter, proportion of weight purchased	Water g	Sugars g	Starch g	Dietary fibre g	Total nitrogen g
771	**Olives** in brine	Bottled in brine; flesh and skin, no stones	0.80	76.5	Tr	0	4.4	0.14
772	in brine (weighed with stones)	Calculated from the previous item	0.80	61.1	Tr	0	3.5	0.11
773	**Oranges** raw	Flesh only, no peel or pips	0.75	86.1	8.5	0	2.0	0.13
774	raw (weighed with peel and pips)	Calculated from the previous item	0.75	64.8	6.4	0	1.5	0.10
775	juice, fresh	Strained juice from fresh oranges	0.46	87.7	9.4	0	0	0.10
776	**Passion fruit** raw	Granadilla; flesh and seeds, no skin	0.42	73.3	6.2	0	15.9	0.44
777	raw (weighed with skin)	Calculated from the previous item	0.42	30.8	2.6	0	6.7	0.18
778	**Paw paw** canned	Papaya; 10 cans, fruit and juice	1.00	80.4	17.0	0	0.5	0.03
779	**Peaches** fresh, raw	Flesh and skin, no stones	0.87	86.2	9.1	0	1.4	0.10
780	raw (weighed with stones)	Calculated from the previous item	0.87	75.1	7.9	0	1.2	0.09
781	dried, raw	Whole fruit	1.00	15.5	53.0	0	14.3	0.55
782	stewed without sugar	Fruit and juice	2.70	68.7	19.6	0	5.3	0.20
783	stewed with sugar	Fruit and juice	2.81	66.0	23.3	0	5.1	0.20
784	canned	Fruit and syrup	1.00	74.3	22.9	0	1.0	0.06
785	**Pears, eating**	Flesh only, no skin or core	0.72	83.2	10.6	0	2.3	0.04
786	eating (weighed with skin and core)	Calculated from the previous item	0.72	59.9	7.6	0	1.7	0.03

No	Food	Energy value kcal	Energy value kJ	Protein (N × 6.25) g	Fat g	Carbo-hydrate g	Na mg	K mg	Ca mg	Mg mg	P mg	Fe mg	Cu mg	Zn mg	S mg	Cl mg
771	**Olives** in brine	103	422	0.9	11.0	Tr	2250	91	61	22	17	1.0	0.23	—	36	3750
772	in brine (weighed with stones)	82	338	0.7	8.8	Tr	1800	73	49	18	13	0.8	0.18	—	29	3000
773	**Oranges** raw	35	150	0.8	Tr	8.5	3	200	41	13	24	0.3	0.07	0.2	9	3
774	raw (weighed with peel and pips)	26	113	0.6	Tr	6.4	2	150	31	10	18	0.3	0.05	0.2	7	2
775	juice, fresh	38	161	0.6	Tr	9.4	2	180	12	12	22	0.3	0.05	0.2	5	1
776	**Passion fruit** raw	34	147	2.8	Tr	6.2	28	350	16	39	54	1.1	0.12	—	19	37
777	raw (weighed with skin)	14	60	1.1	Tr	2.6	12	150	7	16	23	0.5	0.05	—	8	15
778	**Paw paw** canned	65	275	0.2	Tr	17.0	8	110	23	8	6	0.4	0.10	0.3	—	40
779	**Peaches** fresh, raw	37	156	0.6	Tr	9.1	3	260	5	8	19	0.4	0.05	0.1	6	Tr
780	raw (weighed with stones)	32	137	0.6	Tr	7.9	2	230	4	7	16	0.3	0.04	0.1	5	Tr
781	dried, raw	212	906	3.4	Tr	53.0	6	1100	36	54	120	6.8	0.63	—	240	11
782	stewed without sugar	79	336	1.3	Tr	19.6	2	410	13	20	44	2.5	0.23	—	89	4
783	stewed with sugar	93	395	1.3	Tr	23.3	2	390	13	19	43	2.4	0.22	—	85	4
784	canned	87	373	0.4	Tr	22.9	1	150	4	6	10	0.4	0.06	—	1	4
785	**Pears, eating**	41	175	0.3	Tr	10.6	2	130	8	7	10	0.2	0.15	0.1	5	Tr
786	eating (weighed with skin and core)	29	125	0.2	Tr	7.6	1	94	6	5	7	0.1	0.11	0.1	4	Tr

No	Food	Retinol µg	Carotene µg	Vitamin D µg	Thiamin mg	Riboflavin mg	Nicotinic acid mg	Potential nicotinic acid from tryptophan mg Trp ÷ 60	Vitamin C mg	Vitamin E mg
771	**Olives** in brine	0	180[a]	0	Tr	Tr	(Tr)	0.1	0	—
772	in brine (weighed with stones)	0	140	0	Tr	Tr	(Tr)	0.1	0	—
773	**Oranges** raw	0	50	0	0.10	0.03	0.2	0.1	50 *(40–60)*	0.2
774	raw (weighed with peel and pips)	0	38	0	0.08	0.02	0.2	0.1	38	0.2
775	juice, fresh	0	50	0	0.08	0.02	0.2	0.1	50[b] *(40–60)*	Tr
776	**Passion fruit** raw	0	10	0	Tr	0.10	1.5	0.4	20	—
777	raw (weighed with skin)	0	4	0	Tr	0.04	0.6	0.2	8	—
778	**Paw paw** canned	0	(500)	0	0.02	0.02	0.2	Tr	15	—
779	**Peaches** fresh, raw	0	500 *(250–1000)*	0	0.02	0.05	1.0	Tr	8	—
780	raw (weighed with stones)	0	440	0	0.02	0.04	0.9	Tr	7	—
781	dried, raw	0	2000 *(1200–2600)*	0	Tr	0.19	5.3	0.3	Tr	—
782	stewed without sugar	0	740	0	Tr	0.06	2.0	0.1	Tr	—
783	stewed with sugar	0	710	0	Tr	0.06	1.9	0.1	Tr	—
784	canned	0	250	0	0.01	0.02	0.6	Tr	4	—
785	**Pears, eating**	0	10	0	0.03	0.03	0.2	Tr	3	Tr
786	eating (weighed with skin and core)	0	7	0	0.02	0.02	0.1	Tr	2	Tr

No	Food	Vitamin B_6 mg	Vitamin B_{12} µg	Folic acid Free µg	Folic acid Total µg	Pantothenic acid mg	Biotin µg
771	**Olives** in brine	0.02	0	—	—	0.02	(Tr)
772	in brine (weighed with stones)	0.02	0	—	—	0.02	(Tr)
773	**Oranges** raw	0.06	0	30	37	0.25	1.0
774	raw (weighed with peel and pips)	0.05	0	23	28	0.18	0.8
775	juice, fresh	0.04	0	30[b]	37[b]	0.19	0.8
776	**Passion fruit** raw	—	0	—	—	—	—
777	raw (weighed with skin)	—	0	—	—	—	—
778	**Paw paw** canned	—	0	—	—	(0.20)	—
779	**Peaches** fresh, raw	0.02	0	2	3	0.15	(0.2)
780	raw (weighed with stones)	0.02	0	2	3	0.13	(0.2)
781	dried, raw	0.10	0	(10)	(14)	(0.30)	—
782	stewed without sugar	0.03	0	Tr	(2)	(0.10)	—
783	stewed with sugar	0.03	0	Tr	(2)	(0.10)	—
784	canned	0.02	0	(2)	(3)	0.05	0.2
785	**Pears, eating**	0.02	0	4	11	0.07	0.1
786	eating (weighed with skin and core)	0.01	0	3	8	0.05	0.1

Notes

[a] Value for green olives. Ripe, black olives contain 40µg per 100g.

[b] These values also apply to frozen reconstituted orange juice.

No	Food	Description and number of samples	Edible matter, proportion of weight purchased	Water g	Sugars g	Starch g	Dietary fibre g	Total nitrogen g
787	**Pears, cooking** raw	Flesh only, no skin or core	0.77	83.0	9.3	Tr	2.9	0.04
788	stewed without sugar	Flesh and juice, peeled and cored before cooking	0.96	85.5	7.9	Tr	2.5	0.03
789	stewed with sugar	Flesh and juice, peeled and cored before cooking	1.07	77.3	17.1	Tr	2.3	0.03
790	canned	Fruit and syrup	1.00	76.2	20.0	0	1.7	0.06
791	**Pineapple** fresh	Flesh only, no skin or core	0.53	84.3	11.6	0	1.2	0.08
792	canned	Fruit and syrup	1.00	77.1	20.2	0	0.9	0.04
793	**Plums, Victoria dessert** raw	Flesh and skin, no stalks or stones	0.94	84.1	9.6	0	2.1	0.09
794	raw (weighed with stones)	Calculated from the previous item	0.94	79.1	9.0	0	2.0	0.08
795	**cooking** raw	Flesh and skin, no stalks or stones	0.91	85.1	6.2	0	2.5	0.09
796	raw (weighed with stones)	Calculated from the previous item	0.91	77.5	5.6	0	2.3	0.08
797	stewed without sugar	Fruit and juice, no stones	1.09	86.3	5.2	0	2.2	0.08
798	stewed without sugar (weighed with stones)	Calculated from the previous item	1.09	80.3	4.8	0	2.0	0.07
799	stewed with sugar	Fruit and juice, no stones	1.20	77.7	15.1	0	1.9	0.06
800	stewed with sugar (weighed with stones)	Calculated from the previous item	1.20	73.0	14.2	0	1.8	0.06
801	**Pomegranate** juice	Juice from fresh fruit	0.56	85.4	11.6	0	0	0.03

No	Food	Energy value		Protein (N × 6.25)	Fat	Carbo-hydrate	mg									
		kcal	kJ	g	g	g	Na	K	Ca	Mg	P	Fe	Cu	Zn	S	Cl
787	**Pears, cooking**	36	154	0.3	Tr	9.3	3	100	7	4	15	0.2	0.11	0.1	3	2
788	stewed without sugar	30	130	0.2	Tr	7.9	3	85	6	3	13	0.2	0.09	0.1	3	2
789	stewed with sugar	65	277	0.2	Tr	17.1	2	78	5	3	12	0.2	0.09	0.1	2	2
790	canned	77	327	0.4	Tr	20.0	1	90	5	6	5	0.3	0.04	—	1	3
791	**Pineapple** fresh	46	194	0.5	Tr	11.6	2	250	12	17	8	0.4	0.08	0.1	3	29
792	canned	77	328	0.3	Tr	20.2	1	94	13	8	5	0.4	0.05	—	3	4
793	**Plums, Victoria dessert** raw	38	164	0.6	Tr	9.6	2	190	11	7	16	0.4	0.10	Tr	4	Tr
794	raw (weighed with stones)	36	153	0.5	Tr	9.0	2	180	10	7	15	0.3	0.09	Tr	3	Tr
795	**cooking** raw	26	109	0.6	Tr	6.2	2	200	14	8	15	0.3	0.09	Tr	5	Tr
796	raw (weighed with stones)	23	98	0.5	Tr	5.6	2	180	13	7	13	0.3	0.08	Tr	4	Tr
797	stewed without sugar	22	92	0.5	Tr	5.2	2	160	12	6	12	0.3	0.08	Tr	3	Tr
798	stewed without sugar (weighed with stones)	20	84	0.4	Tr	4.8	2	150	11	6	11	0.3	0.07	Tr	3	Tr
799	stewed with sugar	59	252	0.4	Tr	15.3	2	150	11	5	11	0.2	0.06	Tr	3	Tr
800	stewed with sugar (weighed with stones)	55	234	0.4	Tr	14.2	2	140	10	5	10	0.2	0.06	Tr	3	Tr
801	**Pomegranate** juice	44	189	0.2	Tr	11.6	1	200	3	3	8	0.2	0.07	—	4	53

No	Food	Retinol µg	Carotene µg	Vitamin D µg	Thiamin mg	Riboflavin mg	Nicotinic acid mg	Potential nicotinic acid from tryptophan mgTrp ÷ 60	Vitamin C mg	Vitamin E mg
787	**Pears, cooking** raw	0	10	0	0.03	0.03	0.2	Tr	3	Tr
788	stewed without sugar	0	9	0	0.03	0.03	0.2	Tr	3	Tr
789	stewed with sugar	0	8	0	0.02	0.02	0.2	Tr	2	Tr
790	canned	0	10	0	0.01	0.01	0.2	0.1	1	Tr
791	**Pineapple** fresh	0	60	0	0.08	0.02	0.2	0.1	25 *(20–40)*	—
792	canned	0	40	0	0.05	0.02	0.2	Tr	12	—
793	**Plums, Victoria dessert** raw	0	220	0	0.05	0.03	0.5	0.1	3	(0.7)
794	raw (weighed with stones)	0	210	0	0.05	0.03	0.5	0.1	3	(0.7)
795	**cooking** raw	0	220	0	0.05	0.03	0.5	0.1	3	(0.7)
796	raw (weighed with stones)	0	200	0	0.05	0.03	0.5	0.1	3	(0.6)
797	stewed without sugar	0	180	0	0.04	0.03	0.5	0.1	3	(0.5)
798	stewed without sugar (weighed with stones)	0	170	0	0.04	0.03	0.5	0.1	3	(0.5)
799	stewed with sugar	0	170	0	0.04	0.02	0.4	0.1	2	(0.4)
800	stewed with sugar (weighed with stones)	0	160	0	0.04	0.02	0.4	0.1	2	(0.4)
801	**Pomegranate** juice	0	0	0	0.02	0.03	0.2	Tr	8	—

No	Food	Vitamin B_6 mg	Vitamin B_{12} µg	Folic acid Free µg	Folic acid Total µg	Pantothenic acid mg	Biotin µg	Notes
787	**Pears, cooking** raw	0.02	0	4	11	0.07	0.1	
788	stewed without sugar	0.02	0	Tr	5	0.05	0.1	
789	stewed with sugar	0.02	0	Tr	5	0.05	0.1	
790	canned	0.01	0	Tr	(5)	0.02	Tr	
791	**Pineapple** fresh	0.09	0	9	11	0.16	Tr	
792	canned	0.07	0	2	—	0.10	Tr	
793	**Plums, Victoria dessert** raw	0.05	0	1	3	0.15	Tr	
794	raw (weighed with stones)	0.05	0	1	3	0.14	Tr	
795	**cooking** raw	0.05	0	1	3	0.15	Tr	
796	raw (weighed with stones)	0.05	0	1	3	0.14	Tr	
797	stewed without sugar	0.03	0	Tr	1	0.12	Tr	
798	stewed without sugar (weighed with stones)	0.03	0	Tr	1	0.11	Tr	
799	stewed with sugar	0.03	0	Tr	1	0.11	Tr	
800	stewed with sugar (weighed with stones)	0.03	0	Tr	1	0.10	Tr	
801	**Pomegranate** juice	—	0	—	—	—	—	

No	Food	Description and number of samples	Edible matter, proportion of weight purchased	Water g	Sugars g	Starch g	Dietary fibre g	Total nitrogen g
802	**Prunes** dried, raw	Flesh and skin, no stones	0.83	23.3	40.3	0	16.1	0.39
803	raw (weighed with stones)	Calculated from the previous item	0.83	19.3	33.5	0	13.4	0.32
804	stewed without sugar	Fruit and juice, no stones	1.65	60.5	20.4	0	8.1	0.20
805	stewed without sugar (weighed with stones)	Calculated from the previous item	1.65	55.1	18.6	0	7.4	0.18
806	stewed with sugar	Fruit and juice, no stones	1.76	57.1	26.5	0	7.7	0.19
807	stewed with sugar (weighed with stones)	Calculated from the previous item	1.76	52.0	24.1	0	7.0	0.17
808	**Quinces** raw	Flesh only, no skin or core	0.69	84.2	6.3	Tr	6.4	0.05
809	**Raisins** dried	Flesh and skin, no stones	0.92	21.5	64.4	0	6.8	0.17
810	**Raspberries** raw	Whole fruit	1.00	83.2	5.6	0	7.4	0.14
811	stewed without sugar	Fruit and juice	0.95	82.2	5.9	0	7.8	0.15
812	stewed with sugar	Fruit and juice	1.05	72.6	17.3	0	7.0	0.13
813	canned	Fruit and syrup	1.00	74.0	22.5	0	(5.0)	0.10
814	**Rhubarb** raw	Stems only	0.67	94.2	1.0	0	2.6	0.10
815	stewed without sugar	Stems and juice	0.78	94.6	0.9	0	2.4	0.09
816	stewed with sugar	Stems and juice	0.94	85.0	11.4	0	2.2	0.08
817	**Strawberries** raw	Flesh and pips, no stalks	0.97	88.9	6.2	0	2.2	0.10
818	canned	10 cans; fruit and syrup	1.00	79.4	21.1	0	1.0	0.07
819	**Sultanas** dried	Whole fruit	1.00	18.3	64.7	0	7.0	0.28
820	**Tangerines** raw	Flesh only, no peel or pips	0.70	86.7	8.0	0	1.9	0.14
821	raw (weighed with peel and pips)	Calculated from the previous item	0.70	60.6	5.6	0	1.3	0.10

No	Food	Energy value kcal	Energy value kJ	Protein (N × 6.25) g	Fat g	Carbo-hydrate g	Na mg	K mg	Ca mg	Mg mg	P mg	Fe mg	Cu mg	Zn mg	S mg	Cl mg
802	**Prunes** dried, raw	161	686	2.4	Tr	40.3	12	860	38	27	83	2.9	0.16	—	19	3
803	raw (weighed with stones)	134	570	2.0	Tr	33.5	10	720	31	22	69	2.4	0.13	—	15	2
804	stewed without sugar	82	349	1.3	Tr	20.4	7	440	19	13	42	1.4	0.08	—	9	1
805	stewed without sugar (weighed with stones)	74	316	1.1	Tr	18.6	6	400	17	12	38	1.3	0.07	—	8	1
806	stewed with sugar	104	444	1.2	Tr	26.5	5	420	18	13	40	1.4	0.08	—	9	1
807	stewed with sugar (weighed with stones)	95	404	1.1	Tr	24.1	5	380	16	12	36	1.3	0.07	—	8	1
808	**Quinces** raw	25	106	0.3	Tr	6.3	3	200	14	6	19	0.3	0.13	—	5	2
809	**Raisins** dried	246	1049	1.1	Tr	64.4	52	860	61	42	33	1.6	0.24	0.1	23	9
810	**Raspberries** raw	25	105	0.9	Tr	5.6	3	220	41	22	29	1.2	0.21	—	17	22
811	stewed without sugar	26	110	0.9	Tr	5.9	3	230	43	23	31	1.3	0.22	—	18	23
812	stewed with sugar	68	290	0.8	Tr	17.3	3	210	39	21	28	1.1	0.20	—	16	21
813	canned	87	370	0.6	Tr	22.5	4	100	14	11	14	1.7	0.10	—	—	5
814	**Rhubarb** raw	6	26	0.6	Tr	1.0	2	430	100	14	21	0.4	0.13	—	8	87
815	stewed without sugar	6	25	0.6	Tr	0.9	2	400	93	13	19	0.4	0.12	—	7	81
816	stewed with sugar	45	191	0.5	Tr	11.4	2	360	84	12	18	0.3	0.11	—	7	73
817	**Strawberries** raw	26	109	0.6	Tr	6.2	2	160	22	12	23	0.7	0.13	0.1	13	18
818	canned	81	344	0.4	Tr	21.1	7	97	14	7	15	0.9	(0.03)	0.2	—	(5)
819	**Sultanas** dried	250	1066	1.8	Tr	64.7	53	860	52	35	95	1.8	0.35	(0.1)	[illegible]4	16
820	**Tangerines** raw	34	143	0.9	Tr	8.0	2	160	42	11	17	0.3	0.09	0.1	10	2
821	raw (weighed with peel and pips)	23	100	0.6	Tr	5.6	2	110	29	8	12	0.2	0.06	0.1	7	2

No	Food	Retinol µg	Carotene µg	Vitamin D µg	Thiamin mg	Riboflavin mg	Nicotinic acid mg	Potential nicotinic acid from tryptophan mg Trp ÷ 60	Vitamin C mg	Vitamin E mg
802	**Prunes** dried, raw	0	1000	0	0.10	0.20	1.5	0.4	Tr	—
803	raw (weighed with stones)	0	830	0	0.08	0.17	1.2	0.3	Tr	—
804	stewed without sugar	0	510	0	0.04	0.09	0.8	0.2	Tr	—
805	stewed without sugar (weighed with stones)	0	460	0	0.04	0.08	0.7	0.2	Tr	—
806	stewed with sugar	0	470	0	0.04	0.09	0.7	0.2	Tr	—
807	stewed with sugar (weighed with stones)	0	430	0	0.04	0.08	0.6	0.2	Tr	—
808	**Quinces** raw	0	Tr	0	0.02	0.02	0.2	Tr	15	—
809	**Raisins** dried	0	30	0	0.10	0.08	0.5	0.1	0	—
810	**Raspberries** raw	0	80	0	0.02	0.03	0.4	0.1	25 [a] *(14–35)*	0.3 [b]
811	stewed without sugar	0	85	0	0.02	0.03	0.4	0.1	23	0.3 [b]
812	stewed with sugar	0	75	0	0.02	0.03	0.4	0.1	22	0.3 [b]
813	canned	0	(75)	0	0.01	0.03	0.3	0.1	7	—
814	**Rhubarb** raw	0	60	0	0.01	0.03	0.3	0.1	10	0.2
815	stewed without sugar	0	55	0	Tr	0.03	0.3	0.1	8	0.2
816	stewed with sugar	0	50	0	Tr	0.03	0.3	0.1	7 [c]	0.2
817	**Strawberries** raw	0	30	0	0.02	0.03	0.4	0.1	60 [d] *(40–90)*	0.2
818	canned	0	(Tr)	0	0.01	(0.02)	(0.3)	0.1	21	—
819	**Sultanas** dried	0	30	0	0.10	(0.08)	(0.5)	0.1	0	0.7
820	**Tangerines** raw	0	100	0	0.07	0.02	0.2	0.1	30	—
821	raw (weighed with peel and pips)	0	70	0	0.05	0.01	0.1	0.1	21	—

No	Food	Vitamin B_6 mg	Vitamin B_{12} µg	Folic acid Free µg	Folic acid Total µg	Panto-thenic acid mg	Biotin µg
802	**Prunes** dried, raw	0.24	0	1	4	0.46	(Tr)
803	raw (weighed with stones)	0.20	0	1	3	0.38	(Tr)
804	stewed without sugar	0.10	0	Tr	Tr	0.21	(Tr)
805	stewed without sugar (weighed with stones)	0.09	0	Tr	Tr	0.19	(Tr)
806	stewed with sugar	0.10	0	Tr	Tr	0.20	(Tr)
807	stewed with sugar (weighed with stones)	0.09	0	Tr	Tr	0.18	(Tr)
808	**Quinces** raw	—	0	—	—	—	—
809	**Raisins** dried	0.30	0	4	4	0.10	—
810	**Raspberries** raw	0.06	0	—	—	0.24	1.9
811	stewed without sugar	0.05	0	—	—	0.23	2.0
812	stewed with sugar	0.05	0	—	—	0.21	1.8
813	canned	0.04	0	—	—	0.17	—
814	**Rhubarb** raw	0.03	0	8	8	0.08	—
815	stewed without sugar	0.02	0	1	4	0.06	—
816	stewed with sugar	0.02	0	1	4	0.05	—
817	**Strawberries** raw	0.06	0	15	20	0.34	1.1
818	canned	0.03	0	8	20	0.21	(1.0)
819	**Sultanas** dried	(0.30)	0	(4)	(4)	(0.10)	—
820	**Tangerines** raw	0.07	0	19	21	0.20	—
821	raw (weighed with peel and pips)	0.05	0	13	15	0.14	—

Notes

[a] Frozen raspberries contain 20mg per 100g.

[b] Also contains 1.5mg γ-tocopherol and 2.7mg δ-tocopherol per 100g.

[c] Canned rhubarb contains 1mg per 100g.

[d] Frozen strawberries contain 50mg per 100g.

No	Food	Description and number of samples	Edible matter, proportion of weight purchased	Water g	Sugars g	Starch g	Dietary fibre g	Total nitrogen g

No	Food	Energy value		Protein (N × 6.25)	Fat	Carbo-hydrate	mg									
		kcal	kJ	g	g	g	Na	K	Ca	Mg	P	Fe	Cu	Zn	S	Cl

No	Food	Retinol µg	Carotene µg	Vitamin D µg	Thiamin mg	Riboflavin mg	Nicotinic acid mg	Potential nicotinic acid from tryptophan mg Trp ÷ 60	Vitamin C mg	Vitamin E mg

No	Food	Vitamin B_6 mg	Vitamin B_{12} µg	Folic acid		Pantothenic acid mg	Biotin µg	Notes
				Free µg	Total µg			

No	Food	Description and number of samples	Edible matter, proportion of weight purchased	Water g	Sugars g	Starch g	Dietary fibre g	Total nitrogen g
822	**Almonds**	Kernel only, no shell	0.37	4.7	4.3	0	14.3	3.27
823	(weighed with shells)	Calculated from the previous item	0.37	1.7	1.6	0	5.3	1.21
824	**Barcelona nuts**	Kernel only, no shell	0.62	5.7	3.4	1.8	10.3	2.06
825	(weighed with shells)	Calculated from the previous item	0.62	3.5	2.1	1.1	6.4	1.28
826	**Brazil nuts**	Kernel only, no shell	0.45	8.5	1.7	2.4	9.0	2.21
827	(weighed with shells)	Calculated from the previous item	0.45	3.8	0.8	1.1	4.1	0.99
828	**Chestnuts**	Kernel only, no shell	0.83	51.7	7.0	29.6	6.8	0.37
829	(weighed with shells)	Calculated from the previous item	0.83	42.8	5.8	24.6	5.7	0.31
830	**Cob** or **hazel nuts**	Kernel only, no shell	0.36	41.1	4.7	2.1	6.1	1.44
831	(weighed with shells)	Calculated from the previous item	0.36	14.8	1.7	0.8	2.2	0.52
832	**Coconut** fresh	Kernel only, no shell	0.70	42.0	3.7	0	13.6	0.61
833	milk	Drained fluid from fresh coconut	0.15	92.2	4.9	0	(Tr)	0.06
834	desiccated	As purchased	1.00	2.3	6.4	0	23.5	1.05
835	**Peanuts** fresh	Kernel only, no shell	0.69	4.5	3.1	5.5	8.1	4.50
836	(weighed with shells)	Calculated from the previous item	0.69	3.1	2.1	3.8	5.6	3.10
837	roasted and salted	As purchased	1.00	4.5	3.1	5.5	8.1	4.50
838	**Peanut butter** smooth	10 samples, 3 brands	1.00	1.1	6.7	6.4	7.6	4.17
839	**Walnuts**	Kernel only, no shell	0.64	23.5	3.2	1.8	5.2	2.00
840	(weighed with shells)	Calculated from the previous item	0.64	15.0	2.0	1.2	3.3	1.28

No	Food	Energy value		Protein (see p7)	Fat	Carbo-hydrate	mg									
		kcal	kJ	g	g	g	Na	K	Ca	Mg	P	Fe	Cu	Zn	S	Cl
822	**Almonds**	565	2336	16.9	53.5	4.3	6	860	250	260	440	4.2	0.14	3.1	150	2
823	(weighed with shells)	210	865	6.3	19.8	1.6	2	320	92	95	160	1.6	0.05	1.1	54	1
824	**Barcelona nuts**	639	2637	10.9	64.0	5.2	3	940	170	200	300	3.0	0.96	—	180	34
825	(weighed with shells)	396	1632	6.8	39.6	3.2	2	580	110	130	190	1.8	0.60	—	110	21
826	**Brazil nuts**	619	2545	12.0	61.5	4.1	2	760	180	410	590	2.8	1.10	4.2	290	61
827	(weighed with shells)	277	1142	5.4	27.6	1.8	1	340	79	190	270	1.3	0.50	1.9	130	27
828	**Chestnuts**	170	720	2.0	2.7	36.6	11	500	46	33	74	0.9	0.23	—	29	15
829	(weighed with shells)	140	595	1.6	2.2	30.4	9	410	38	27	61	0.7	0.19	—	24	12
830	**Cob** or **hazel nuts**	380	1570	7.6	36.0	6.8	1	350	44	56	230	1.1	0.21	2.4	75	6
831	(weighed with shells)	137	567	2.8	13.0	2.4	1	120	16	20	82	0.4	0.08	0.9	27	2
832	**Coconut** fresh	351	1446	3.2	36.0	3.7	17	440	13	52	94	2.1	0.32	0.5	44	110
833	milk	21	91	0.3	(0.2)	4.9	110	310	29	30	37	0.1	0.04	—	24	180
834	desiccated	604	2492	5.6	62.0	6.4	28	750	22	90	160	3.6	0.55	—	76	200
835	**Peanuts** fresh	570	2364	24.3	49.0	8.6	6	680	61	180	370	2.0	0.27	3.0	380	7
836	(weighed with shells)	394	1631	16.8	33.8	5.9	4	470	42	130	250	1.4	0.19	2.1	260	5
837	roasted and salted	570	2364	24.3	49.0	8.6	440	680	61	180	370	2.0	0.27	3.0	380	660
838	**Peanut butter** smooth	623	2581	22.6	53.7	13.1	350	700	37	180	330	2.1	0.70	3.0	—	500
839	**Walnuts**	525	2166	10.6	51.5	5.0	3	690	61	130	510	2.4	0.31	3.0	100	23
840	(weighed with shells)	336	1388	6.8	33.0	3.2	2	440	39	84	330	1.5	0.20	1.9	67	15

No	Food	Retinol µg	Carotene µg	Vitamin D µg	Thiamin mg	Riboflavin mg	Nicotinic acid mg	Potential nicotinic acid from tryptophan mgTrp ÷ 60	Vitamin C mg	Vitamin E mg
822	**Almonds**	0	0	0	0.24 [a]	0.92	2.0	2.7	Tr	20.0 [b]
823	(weighed with shells)	0	0	0	0.09	0.34	0.7	1.0	Tr	7.4
824	**Barcelona nuts**	0	0	0	0.11	—	—	3.1	Tr	—
825	(weighed with shells)	0	0	0	0.07	—	—	1.9	Tr	—
826	**Brazil nuts**	0	0	0	1.00	0.12	1.6	2.6	Tr	6.5 [b]
827	(weighed with shells)	0	0	0	0.45	0.05	0.7	1.2	Tr	2.9
828	**Chestnuts**	0	0	0	0.20	0.22	0.2	0.4	Tr	0.5 [b]
829	(weighed with shells)	0	0	0	0.17	0.18	0.2	0.4	Tr	0.4
830	**Cob** or **hazel nuts**	0	0	0	0.40	—	0.9	2.2	Tr	21.0 [b]
831	(weighed with shells)	0	0	0	0.14	—	0.3	0.8	Tr	7.6
832	**Coconut**, fresh	0	0	0	0.03	0.02	0.3	0.7	2	0.7 [b]
833	milk	0	0	0	Tr	Tr	0.1	0.1	2	Tr
834	desiccated	0	0	0	0.06	0.04	0.6	1.2	0	—
835	**Peanuts** fresh	0	0	0	0.90	0.10	16	5.3	Tr	8.1 [b]
836	(weighed with shells)	0	0	0	0.62	0.07	11	3.6	Tr	5.6
837	roasted and salted	0	0	0	0.23	0.10	16	5.3	Tr	(8.1) [b]
838	**Peanut butter** smooth	0	0	0	0.17	0.10	15	4.9	Tr	4.7 [b]
839	**Walnuts**	0	0	0	0.30	0.13	1.0	2.0	Tr [c]	0.8 [b]
840	(weighed with shells)	0	0	0	0.19	0.08	0.6	1.3	Tr	0.5

No	Food	Vitamin B_6 mg	Vitamin B_{12} µg	Folic acid Free µg	Folic acid Total µg	Panto-thenic acid mg	Biotin µg
822	**Almonds**	0.10	0	33	96	0.47[a]	0.4
823	(weighed with shells)	0.04	0	12	36	0.17	0.1
824	**Barcelona nuts**	—	0	—	—	—	—
825	(weighed with shells)	—	0	—	—	—	—
826	**Brazil nuts**	0.17	0	—	—	0.23	—
827	(weighed with shells)	0.08	0	—	—	0.10	—
828	**Chestnuts**	0.33	0	—	—	0.47	1.3
829	(weighed with shells)	0.27	0	—	—	0.39	1.1
830	**Cob** or **hazel nuts**	0.55	0	23	72	1.15	—
831	(weighed with shells)	0.20	0	8	26	0.41	—
832	**Coconut** fresh	0.04	0	9	26	0.20	—
833	milk	0.03	0	—	—	0.05	—
834	desiccated	—	0	—	—	—	—
835	**Peanuts** fresh	(0.50)	0	28	110	2.7	—
836	(weighed with shells)	(0.35)	0	19	76	1.9	—
837	**Peanuts** roasted and salted	0.40	0	—	—	2.1	—
838	**Peanut butter** smooth	0.50	0	16	53	(2.1)	—
839	**Walnuts**	0.73	0	48	66	0.90	2.0
840	(weighed with shells)	0.47	0	31	42	0.58	1.3

Notes

[a] The thiamin content is reduced to 0.05 mg per 100 g on roasting, and the pantothenic acid to 0.25 mg.

[b] Most nuts contain only γ-tocopherol in addition to α-tocopherol. Values for γ-tocopherol are:

	γ-tocopherol mg/100 g
Almonds	3.0
Brazil nuts	11.0
Chestnuts	7.0
Cob nuts	1.5
Coconut, fresh	0.3
Peanuts	8.8
Peanut butter	2.9
Walnuts	18.0

[c] Value for ripe walnuts. Unripe walnuts contain 1300–3000 mg per 100 g.

No	Food	Description and number of samples	Water g	Sugars g	Starch and dextrins g	Dietary fibre g	Total nitrogen g
	Sugars						
841	**Glucose** liquid, BP	1 sample	20.4	40.2	44.5	0	Tr
842	**Sugar** Demerara	5 samples	Tr	104.5[a]	0	0	0.08
843	white	Granulated and loaf	Tr	105.0[b]	0	0	Tr
844	**Syrup** golden	3 samples of the same brand	20.0	79.0	0	0	0.05
845	**Treacle** black	3 samples	28.5	67.2	0	0	0.19
	Preserves						
846	**Cherries** glacé	3 samples	—	55.8	0	—	0.10
847	**Honey** comb	2 samples	20.2	74.4	0	—	0.09
848	in jars	2 samples	23.0	76.4	0	—	0.06
849	**Jam** fruit with edible seeds	Blackberry, blackcurrant, gooseberry, raspberry, strawberry; 2 samples of each, different brands	29.8	69.0	0	1.1	0.10
850	stone fruit	Apricot, damson, greengage, plum; 2 samples of each, different brands	29.6	69.3	0	1.0	0.06
851	**Lemon curd** starch base	10 jars, 4 brands	30.1	40.4	22.3	0.2	0.09
852	home made	Recipe p 344	42.1	41.3	0	0	0.53
853	**Marmalade**	4 brands	28.0	69.5	0	0.7	0.01
854	**Marzipan** almond paste	Recipe p 344	10.5	49.2	0	6.4	1.62
855	**Mincemeat**	10 samples of the same brand	27.5	62.1	Tr	3.3	0.10

[a] 99.3g per 100g expressed as sucrose [b] 99.9g per 100g expressed as sucrose

No	Food	Energy value kcal	Energy value kJ	Protein (N × 6.25) g	Fat g	Carbo-hydrate g	Na mg	K mg	Ca mg	Mg mg	P mg	Fe mg	Cu mg	Zn mg	S mg	Cl mg
	Sugars															
841	**Glucose** liquid, BP	318	1355	Tr	0	84.7	150	3	8	2	11	0.5	0.09	—	—	190
842	**Sugar** Demerara	394	1681	0.5	0	104.5 [a]	6	89	53	15	20	0.9	0.06	—	14	35
843	white	394	1680	Tr	0	105.0 [b]	Tr	2	2	Tr	Tr	Tr	0.02	—	Tr	Tr
844	**Syrup** golden	298	1269	0.3	0	79.0	270	240	26	10	20	1.5	0.09	—	54	42
845	**Treacle** black	257	1096	1.2	0	67.2	96	1470	500	140	31	9.2	0.43	—	69	820
	Preserves															
846	**Cherries** glacé	212	903	0.6	0	55.8	65	18	44	8	18	2.9	1.28	—	21	71
847	**Honey** comb	281	1201	0.6	4.6 [c]	74.4	7	35	8	2	32	0.2	0.04	—	1	26
848	in jars	288	1229	0.4	Tr	76.4	11	51	5	2	17	0.4	0.05	—	1	18
849	**Jam** fruit with edible seeds	261	1114	0.6	0	69.0	16	110	24	10	18	1.5	0.23	—	7	9
850	stone fruit	261	1116	0.4	0	69.3	12	100	12	5	18	1.0	0.12	—	3	4
851	**Lemon curd** starch base	283	1202	0.6	5.1	62.7	65	11	9	2	15	0.5	(0.03)	1.3	—	150
852	home made	290	1216	3.3	13.5	41.3	150	66	18	5	62	0.6	0.06	0.4	48	220
853	**Marmalade**	261	1114	0.1	0	69.5	18	44	35	4	13	0.6	0.12	—	2	7
854	**Marzipan** almond paste	443	1856	8.7	24.9	49.2	13	400	120	120	220	2.0	0.08	1.5	81	13
855	**Mincemeat**	235	1163	0.6	4.3	62.1	140	190	30	10	17	1.5	0.20	0.2	—	200

[a] 99.3g per 100g expressed as sucrose [b] 99.9g per 100g expressed as sucrose
[c] Waxy material, probably not available as fat; disregarded in calculating energy values

No	Food	Retinol µg	Carotene µg	Vitamin D µg	Thiamin mg	Riboflavin mg	Nicotinic acid mg	Potential nicotinic acid from tryptophan mgTrp ÷ 60	Vitamin C mg	Vitamin E mg
	Sugars									
841	**Glucose** liquid, BP	0	0	0	0	0	0	0	0	0
842	**Sugar** Demerara	0	0	0	Tr	Tr	Tr	Tr	0	0
843	white	0	0	0	0	0	0	0	0	0
844	**Syrup** golden	0	0	0	Tr	Tr	Tr	Tr	0	0
845	**Treacle** black	0	0	0	Tr	Tr	Tr	Tr	0	0
	Preserves									
846	**Cherries** glacé	0	—	0	Tr	Tr	Tr	Tr	Tr	Tr
847	**Honey** comb	0	0	0	Tr	0.05	0.2	Tr	Tr	—
848	in jars	0	0	0	Tr	0.05	0.2	Tr	Tr	—
849	**Jam** fruit with edible seeds	0	Tr	0	Tr	Tr	Tr	Tr	10[a]	Tr
850	stone fruit	0	Tr	0	Tr	Tr	Tr	Tr	Tr	Tr
851	**Lemon curd** starch base	(10)	0	(0.10)	Tr	(0.02)	Tr	0.1	(Tr)	—
852	home made	130	60	0.55	0.02	0.12	Tr	1.0	8	0.4
853	**Marmalade**	0	50	0	Tr	Tr	Tr	Tr	10	Tr
854	**Marzipan** almond paste	10	0	0.13	0.12	0.45	0.9	1.5	2	9.1
855	**Mincemeat**	0	(10)	0	(0.03)	(0.02)	(0.2)	0.1	Tr	—

No	Food	Vitamin B_6 mg	Vitamin B_{12} µg	Folic acid Free µg	Folic acid Total µg	Panto-thenic acid mg	Biotin µg
	Sugars						
841	**Glucose** liquid, BP	0	0	0	0	0	0
842	**Sugar** Demerara	Tr	0	Tr	Tr	Tr	Tr
843	white	0	0	0	0	0	0
844	**Syrup** golden	Tr	0	Tr	Tr	Tr	Tr
845	**Treacle** black	Tr	0	Tr	Tr	Tr	Tr
	Preserves						
846	**Cherries** glacé	Tr	0	Tr	Tr	Tr	Tr
847	**Honey** comb	—	0	—	—	—	—
848	in jars	—	0	—	—	—	—
849	**Jam** fruit with edible seeds	Tr	0	Tr	Tr	Tr	Tr
850	stone fruit	Tr	0	Tr	Tr	Tr	Tr
851	**Lemon curd** starch base	Tr	Tr	Tr	Tr	(0.10)	1
852	home made	0.03	Tr	4	4	0.49	7
853	**Marmalade**	Tr	0	5	5	Tr	Tr
854	**Marzipan** almond paste	0.06	Tr	17	45	0.35	2
855	**Mincemeat**	(0.10)	0	(Tr)	(Tr)	0.03	(Tr)

Notes

[a] Blackcurrant jam contains 24mg per 100g.

No	Food	Description and number of samples	Water g	Sugars g	Starch and dextrins g	Dietary fibre g	Total nitrogen g
856	**Boiled sweets**	6 samples	—	86.9	0.4	0	0.01
	Chocolate						
857	milk	10 samples of the same brand	2.2	56.5	2.9	—	1.35
858	plain	10 samples of the same brand	0.6	59.5	5.3	—	0.75
859	fancy and filled	8 samples of different brands, mixed, milk and plain	5.7	65.8	7.5	—	0.66
860	Bounty Bar	8 samples	7.6	53.7	4.6	—	0.77
861	Mars Bar	8 samples	6.9	65.8	0.7	—	0.84
862	**Fruit gums**	8 samples of the same brand	12.0	42.6	2.2	—	0.16
863	**Liquorice allsorts**	6 samples	6.6	67.2	6.9	—	0.63
864	**Pastilles**	6 samples of different brands	10.2	61.9	—	—	0.84
865	**Peppermints**	Several samples of 6 different brands	0.2	102.2	0	0	0.08
866	**Toffees** mixed	8 samples of different brands	4.8	70.1	1.0	—	0.34

No	Food	Energy value		Protein (N × 6.25)	Fat	Carbo-hydrate	mg									
		kcal	kJ	g	g	g	Na	K	Ca	Mg	P	Fe	Cu	Zn	S	Cl
856	**Boiled sweets**	327	1397	Tr	Tr	87.3	25	8	5	2	12	0.4	0.09	—	—	68
	Chocolate															
857	milk	529	2214	8.4	30.3	59.4	120	420	220	55	240	1.6	0.30	0.2	—	270
858	plain	525	2197	4.7	29.2	64.8	11	300	38	100	140	2.4	0.70	0.2	—	100
859	fancy and filled	460	1938	4.1	18.8	73.3	60	240	92	51	120	1.8	0.45	—	—	180
860	Bounty Bar	473	1980	4.8	26.1	58.3	180	320	110	43	140	1.3	0.47	—	—	400
861	Mars Bar	441	1853	5.3	18.9	66.5	150	250	160	35	150	1.1	0.31	—	—	300
862	**Fruit gums**	172	734	1.0	0	44.8	64	360	360	110	4	4.2	1.43	—	—	160
863	**Liquorice allsorts**	313	1333	3.9	2.2	74.1	75	220	63	38	29	8.1	0.39	—	—	120
864	**Pastilles**	253	1079	5.2	0	61.9	77	40	40	12	Tr	1.4	0.32	—	—	120
865	**Peppermints**	392	1670	0.5	0.7	102.2	9	Tr	7	3	Tr	0.2	0.04	—	—	22
866	**Toffees** mixed	430	1810	2.1	17.2	71.1	320	210	95	25	64	1.5	0.40	—	—	480

No	Food	Retinol µg	Carotene µg	Vitamin D µg	Thiamin mg	Riboflavin mg	Nicotinic acid mg	Potential nicotinic acid from tryptophan mg Trp ÷ 60	Vitamin C mg	Vitamin E mg
856	**Boiled sweets**	0	0	0	0	0	0	0	0	0
	Chocolate									
857	milk	Tr	(40)	Tr	0.10	0.23	0.2	1.4	0	0.5 [a]
858	plain	0	(40)	0	0.07	0.08	0.4	0.8	0	0.5 [b]
859	fancy and filled	0	(40)	Tr	(0.10)	(0.10)	(0.3)	0.7	0	—
860	Bounty Bar	0	(40)	Tr	(0.04)	(0.10)	(0.3)	0.8	0	—
861	Mars Bar	0	(40)	Tr	(0.05)	(0.20)	(0.3)	0.9	0	—
862	**Fruit gums**	0	0	0	0	0	0	0	0	0
863	**Liquorice allsorts**	0	0	0	0	0	0	0.7	0	0
864	**Pastilles**	0	0	0	0	0	0	0	0	0
865	**Peppermints**	0	0	0	0	0	0	0	0	0
866	**Toffees** mixed	0	—	0	0	0	0	0.4	0	—

No	Food	Vitamin B_6 mg	Vitamin B_{12} µg	Folic acid Free µg	Folic acid Total µg	Panto-thenic acid mg	Biotin µg	Notes
856	**Boiled sweets**	0	0	0	0	0	0	
	Chocolate							
857	milk	(0.02)	Tr	(9)	(10)	(0.6)	(3)	
858	plain	(0.02)	0	(9)	(10)	(0.6)	(3)	
859	fancy and filled	(0.02)	Tr	(9)	(10)	(0.6)	(3)	
860	Bounty Bar	(0.02)	Tr	—	—	(0.6)	(3)	
861	Mars Bar	(0.02)	Tr	—	—	(0.6)	(3)	
862	**Fruit gums**	0	0	0	0	0	0	
863	**Liquorice allsorts**	0	0	0	0	0	0	
864	**Pastilles**	0	0	0	0	0	0	
865	**Peppermints**	0	0	0	0	0	0	
866	**Toffees** mixed	0	0	0	0	0	0	

Notes

[a] Also contains 2.4 mg γ-tocopherol per 100 g.

[b] Also contains 3.5 mg γ-tocopherol per 100 g.

No	Food	Description and number of samples	Water g	Sugars g	Starch and dextrins g	Dietary fibre g	Total nitrogen g
867	**Bournvita**	6 samples	1.5	52.0	27.0 [a]	—	1.39
868	**Cocoa powder**	10 samples, 2 brands	3.4	Tr	11.5	— [b]	3.70 [c]
869	**Coffee and chicory essence**	7 bottles of the same brand	36.9	53.8	2.2	— [b]	0.33 [d]
870	**Coffee** ground, roasted	5 samples	4.1	Tr	28.5	— [b]	2.04 [e]
871	infusion, 5 minutes	60g coffee from mixed sample; boiled in percolater with 900ml water and strained	—	Tr	0.3	—	0.04
872	instant	10 jars, 2 brands	3.4	6.5	4.5 [a]	— [b]	3.26 [f]
873	**Drinking chocolate**	10 tins, 3 brands	2.1	73.8	3.6	— [b]	1.04 [g]
874	**Horlicks malted milk**	Mixed sample	2.5	49.4	23.5 [a]	—	2.21
875	**Ovaltine**	Mixed sample	2.3	73.0	8.2 [a]	—	1.57
876	**Tea** Indian	5 samples	9.3	(3.0)	Tr	—	4.08 [h]
877	Indian, infusion	10g from mixed sample; infused with 1000ml boiling water 2–10 minutes and strained	—	Tr	0	—	Tr

[a] Dextrins only
[b] Complex polysaccharides, which are probably unavailable, are present in these foods
[c] Includes 0.74g purine nitrogen
[d] Includes 0.08g purine nitrogen
[e] Includes 0.38g purine nitrogen
[f] Includes 0.93g purine nitrogen
[g] Includes 0.16g purine nitrogen
[h] Includes 0.95g purine nitrogen

No	Food	Energy value		Protein (N × 6.25)	Fat	Carbo- hydrate	mg									
		kcal	kJ	g	g	g	Na	K	Ca	Mg	P	Fe	Cu	Zn	S	Cl
867	**Bournvita**	377	1601	8.7	5.1	79.0	460	380	93	110	350	1.9	0.5	1.1	—	—
868	**Cocoa powder**	312	1301	18.5[a]	21.7	11.5	950	1500	130	520	660	10.5	3.9	6.9	—	460
869	**Coffee and chicory essence**	218	931	1.6[a]	0.2	56.0	65	750	30	39	90	0.7	0.6	—	—	85
870	**Coffee** ground, roasted	287	1203	10.4[a]	15.4	28.5	74	2020	130	240	160	4.1	0.82	—	110	24
871	infusion, 5 minutes	2	8	0.2	Tr	0.3	Tr	66	2	6	2	Tr	Tr	—	—	Tr
872	instant	100	424	14.6[a]	0	11.0	41	4000	160	390	350	4.4	0.05	0.5	—	50
873	**Drinking chocolate**	366	1554	5.5[a]	6.0	77.4	250	410	33	150	190	2.4	1.1	1.9	—	130
874	**Horlicks malted milk**	396	1679	13.8	7.5	72.9	350	750	230	46	300	1.8	0.8	—	68	610
875	**Ovaltine**	378	1606	9.8	3.8	81.2	150	850	36	150	400	2.6	1.2	—	—	—
876	**Tea** Indian	108	455	19.6[a]	(2.0)	(3.0)	45	2160	430	250	630	15.2	1.6	(3.0)	180	52
877	Indian, infusion	<1	2	0.1	Tr	Tr	Tr	17	Tr	1	1	Tr	Tr	Tr	—	Tr

[a] (Total N − purine N) × 6.25

No	Food	Retinol µg	Carotene µg	Vitamin D µg	Thiamin mg	Riboflavin mg	Nicotinic acid mg	Potential nicotinic acid from tryptophan mg Trp ÷ 60	Vitamin C mg	Vitamin E mg
867	**Bournvita**	0	0	0	—	—	—	1.9	0	—
868	**Cocoa powder**	0	(40)	0	0.16	0.06	1.7	5.6	0	0.4 [a]
869	**Coffee and chicory essence**	0	—	0	0	0.03	2.8	—	0	—
870	**Coffee** ground, roasted	0	—	0	—	0.20	10 [b]	—	0	—
871	infusion, 5 minutes	0	—	0	—	0.01	0.7	Tr	0	—
872	instant	0	—	0	0	0.11	22 [c]	2.9	0	—
873	**Drinking chocolate**	0	—	0	0.06	0.04	0.5	1.6	0	0.1 [d]
874	**Horlicks malted milk**	465	—	1.55	0.84	1.06	11.2	2.9	0	—
875	**Ovaltine**	—	—	(30.6)	(1.76)	—	—	2.1	—	—
876	**Tea** Indian	0	Tr	0	0.14	1.2	7.5	—	Tr	—
877	Indian, infusion	0	0	0	Tr	0.01	0.1	0	0	—

No	Food	Vitamin B_6 mg	Vitamin B_{12} µg	Folic acid Free µg	Folic acid Total µg	Panto-thenic acid mg	Biotin µg
867	**Bournvita**	—	0	—	—	—	—
868	**Cocoa powder**	0.07	0	31	38	—	—
869	**Coffee and chicory essence**	—	0	—	—	—	—
870	**Coffee** ground, roasted	—	0	—	—	—	—
871	infusion, 5 minutes	—	0	—	—	—	—
872	instant	0.03	0	—	—	0.4	—
873	**Drinking chocolate**	0.02	0	9	10	—	—
874	**Horlicks malted milk**	—	—	—	—	—	—
875	**Ovaltine**	—	—	—	—	—	—
876	**Tea** Indian	—	0	—	—	1.3	—
877	Indian, infusion	—	0	—	—	Tr	—

Notes

[a] Also contains 2.8 mg γ-tocopherol per 100 g.

[b] Increases during roasting of coffee beans; a dark roasted variety may contain 3 or 4 times as much.

[c] Can be as high as 39 mg per 100 g. Decaffeinated instant coffee contains about the same amount.

[d] Also contains 0.8 mg γ-tocopherol per 100 g.

No	Food	Description and number of samples	Specific gravity	Water g	Sugars g	Starch and dextrins g	Dietary fibre g	Total nitrogen g
878	**Coca-cola**	8 cans and 5 bottles	1.039	89.8	10.5	Tr	0	Tr
879	**Grapefruit juice** canned							
	unsweetened	10 cans, 7 brands	—	89.8	7.9	Tr	0	0.05
880	sweetened	5 cans of different brands	—	87.3	9.7	Tr	0	0.08
881	**Lemonade** bottled	7 bottles of the same brand	1.015	94.6	5.6	0	0	Tr
882	**Lime juice cordial** undiluted	6 bottles of the same brand	1.102	70.5	24.8	Tr	0	0.01
883	**Lucozade**	Mixed sample	1.074	81.7	9.0	9.0[a]	0	Tr
884	**Orange drink** undiluted	Mixed sample	1.116	71.2	28.5	0	0	Tr
885	**Orange juice** canned, unsweetened	9 cans, 6 brands	—	88.7	8.5	Tr	0	0.07
886	sweetened	5 cans, 4 brands	—	85.8	12.8	Tr	0	0.11
887	**Pineapple juice** canned	6 cans of different brands	1.054	86.1	13.4	Tr	0	0.05
888	**Ribena** undiluted	Mixed sample	1.283	39.8	60.9	0	0	0.02
889	**Rosehip syrup** undiluted	9 bottles, 4 brands	—	32.5	61.8	0.1	0	—
890	**Tomato juice** canned	10 cans, 6 brands	—	93.3	3.2	0.2	—	0.12

[a] Dextrins only

No	Food	Energy value		Protein (N × 6.25)	Fat	Carbo-hydrate	mg									
		kcal	kJ	g	g	g	Na	K	Ca	Mg	P	Fe	Cu	Zn	S	Cl
878	**Coca-cola**	39	168	Tr	0	10.5	8	1	4	1	15	Tr	(0.03)	Tr	—	(10)
879	**Grapefruit juice** canned															
	unsweetened	31	132	0.3	Tr	7.9	3	110	9	8	12	0.3	(0.03)	0.4	—	(10)
880	sweetened	38	164	0.5	Tr	9.7	2	110	9	9	12	0.3	(0.03)	0.3	—	(10)
881	**Lemonade** bottled	21	90	Tr	0	5.6	7	1	5	Tr	Tr	Tr	0.01	—	—	Tr
882	**Lime juice cordial** undiluted	112	479	0.1	0	29.8	8	49	9	4	5	0.3	0.07	—	—	4
883	**Lucozade**	68	288	Tr	0	18.0	29	1	5	1	4	0.1	0.04	—	—	35
884	**Orange drink** undiluted	107	456	Tr	0	28.5	21	17	8	3	2	0.1	0.01	—	—	4
885	**Orange juice** canned															
	unsweetened	33	143	0.4	Tr	8.5	4	130	9	9	15	0.5	(0.03)	0.3	—	(10)
886	sweetened	51	217	0.7	Tr	12.8	3	120	9	8	14	0.3	(0.03)	0.3	—	(10)
887	**Pineapple juice** canned	53	225	0.4	0.1	13.4	1	140	12	12	10	0.7	0.09	—	—	38
888	**Ribena** undiluted	229	976	0.1	0	60.9	20	86	9	5	10	0.5	0.02	—	—	7
889	**Rosehip syrup** undiluted	232	990	(Tr)	0	61.9	280	26	—	—	—	—	—	—	—	—
890	**Tomato juice** canned	16	66	0.7	Tr	3.4	230	260	10	10	20	0.5	0.05	0.4	—	370

No	Food	Retinol µg	Carotene µg	Vitamin D µg	Thiamin mg	Riboflavin mg	Nicotinic acid mg	Potential nicotinic acid from tryptophan mg Trp ÷ 60	Vitamin C mg	Vitamin E mg
878	**Coca-cola**	0	0	0	0	0	0	0	0	0
879	**Grapefruit juice** canned									
	unsweetened	0	Tr	0	(0.04)	(0.01)	(0.2)	0.1	28	Tr
880	sweetened	0	Tr	0	(0.04)	(0.01)	(0.2)	0.1	29	Tr
881	**Lemonade** bottled	0	Tr	0	Tr	Tr	Tr	Tr	Tr[a]	Tr
882	**Lime juice cordial** undiluted	0	Tr	0	Tr	Tr	Tr	Tr	Tr	Tr
883	**Lucozade**	0	0	0	Tr	Tr	Tr	Tr	3	0
884	**Orange drink** undiluted	0	—	0	Tr	Tr	Tr	Tr	Tr[b]	Tr
885	**Orange juice** canned, unsweetened	0	(50)	0	(0.07)	(0.02)	(0.2)	0.1	35	Tr
886	sweetened	0	(50)	0	(0.07)	(0.02)	(0.2)	0.1	31	Tr
887	**Pineapple juice** canned	0	(40)	0	0.05	0.02	0.2	0.1	8	—
888	**Ribena** undiluted	0	—	0	—	—	—	Tr	210	—
889	**Rosehip syrup** undiluted	0	—	0	0	Tr	Tr	Tr	295	Tr
890	**Tomato juice** canned	0	(500)	0	(0.06)	(0.03)	(0.7)	0.1	20	(0.2)

No	Food	Vitamin B_6 mg	Vitamin B_{12} µg	Folic acid Free µg	Folic acid Total µg	Panto-thenic acid mg	Biotin µg
878	**Coca-cola**	0	0	0	0	0	0
879	**Grapefruit juice** canned unsweetened	(0.01)	0	4	6	(0.12)	(1)
880	sweetened	(0.01)	0	4	6	(0.12)	(1)
881	**Lemonade** bottled	Tr	0	Tr	Tr	Tr	Tr
882	**Lime juice cordial** undiluted	Tr	0	Tr	Tr	Tr	Tr
883	**Lucozade**	Tr	0	Tr	Tr	Tr	Tr
884	**Orange drink** undiluted	Tr	0	Tr	Tr	Tr	Tr
885	**Orange juice** canned unsweetened	(0.04)	0	7	7	(0.15)	(1)
886	sweetened	(0.04)	0	7	7	(0.15)	(1)
887	**Pineapple juice** canned	0.10	0	2	—	0.10	—
888	**Ribena** undiluted	—	0	—	—	—	—
889	**Rosehip syrup** undiluted	Tr	0	Tr	Tr	Tr	Tr
890	**Tomato juice** canned	(0.11)	0	4	13	(0.20)	(1)

Notes

[a] Vitamin C may be added to some brands, and the content may range from 5 to 15 mg per 100 g.

[b] Vitamin C is added to some brands, and the content may range from 20 to 60 mg per 100 g. There is a loss on storage of opened bottles, particularly when exposed to light.

No	Food	Description and number of samples	Specific gravity	Alcohol g	Solids g	Sugars g	Total nitrogen g
	Beers						
891	**Brown ale** bottled	6 samples from different brewers	1.008	2.2	4.2	3.0	0.04
892	**Canned beer** bitter	6 samples	1.008	3.1	3.3	2.3	0.04
893	**Draught** bitter	5 samples from different brewers	1.004	3.1	3.3	2.3	0.04
894	mild	5 samples from different brewers	1.001	2.6	2.5	1.6	0.03
895	**Keg** bitter	6 samples from different brewers	1.009	3.0	3.6	2.3	0.04
896	**Lager** bottled	6 samples	1.005	3.2	2.4	1.5	0.03
897	**Pale ale** bottled	6 samples from diffeent brewers	1.003	3.3	3.3	2.0	0.05
898	**Stout** bottled	4 samples from different brewers	1.014	2.9	5.8	4.2	0.05
899	**Stout** extra	6 samples of the same brand	1.002	4.3	3.6	2.1	0.05
900	**Strong ale**	6 samples from different brewers, barley wine type	1.018	6.6	8.0	6.1	0.11
	Ciders						
901	**Cider** dry	3 samples of different brands	1.007	3.8	3.7	2.6	Tr
902	sweet	3 samples of different brands	1.012	3.7	5.1	4.3	Tr
903	vintage	3 samples of the same brand	1.017	10.5	8.9	7.3	Tr
	Wines						
904	**Red wine**	3 samples, Beaujolais, Burgundy, claret	0.998	9.5	2.3	0.3	0.03
905	**Rosé** medium	5 samples from different vintners	1.003	8.7	4.0	2.5	0.01
906	**White wine** dry	5 samples from different vintners	0.995	9.1	1.8	0.6	0.02
907	medium	1 sample, Graves	1.005	8.8	5.4	3.4	0.02
908	sweet	1 sample, Sauternes	1.016	10.2	9.2	5.9	0.03
909	sparkling	1 sample, Champagne	0.995	9.9	3.3	1.4	0.04

No	Food	Energy value		Protein (N × 6.25)	Fat	Carbo-hydrate	mg									
		kcal	kJ	g	g	g	Na	K	Ca	Mg	P	Fe	Cu	Zn	S	Cl
	Beers															
891	**Brown ale** bottled	28	117	0.3	Tr	3.0	16	33	7	6	11	0.03	0.07	—	—	37
892	**Canned beer** bitter	32	132	0.3	Tr	2.3	9	37	8	7	11	0.01	Tr	Tr	—	—
893	**Draught** bitter	32	132	0.3	Tr	2.3	12	38	11	9	13	0.01	0.08	—	—	32
894	mild	25	104	0.2	Tr	1.6	11	33	10	8	12	0.02	0.05	—	—	34
895	**Keg** bitter	31	129	0.3	Tr	2.3	8	35	8	7	9	0.01	0.01	0.02	—	30
896	**Lager** bottled	29	120	0.2	Tr	1.5	4	34	4	6	12	Tr	Tr	—	—	19
897	**Pale ale** bottled	32	133	0.3	Tr	2.0	10	49	9	10	15	0.02	0.04	—	—	31
898	**Stout** bottled	37	156	0.3	Tr	4.2	23	45	8	8	17	0.05	0.08	—	—	48
899	**Stout** extra	39	163	0.3	Tr	2.1	4	86	5	9	28	0.02	0.03	—	—	24
900	**Strong ale**	72	301	0.7	Tr	6.1	15	110	14	20	40	0.03	0.08	—	—	57
	Ciders															
901	**Cider** dry	36	152	Tr	0	2.6	7	72	8	3	3	0.49	0.04	—	—	6
902	sweet	42	176	Tr	0	4.3	7	72	8	3	3	0.49	0.04	—	—	6
903	vintage	101	421	Tr	0	7.3	2	97	5	4	9	0.31	0.02	—	—	5
	Wines															
904	**Red wine**	68	284	0.2	0	0.3	10	130	7	11	14	0.90	0.12	—	—	18
905	**Rosé** medium	71	294	0.1	0	2.5	4	75	12	7	6	0.95	0.02	0.04	—	7
906	**White wine** dry	66	275	0.1	0	0.6	4	61	9	8	6	0.50	0.01	0.01	—	10
907	medium	75	311	0.1	0	3.4	21	88	14	9	8	1.21	0.01	—	—	4
908	sweet	94	394	0.2	0	5.9	13	110	14	11	13	0.58	0.05	—	—	7
909	sparkling	76	315	0.3	0	1.4	4	57	3	6	7	0.50	0.01	—	—	7

No	Food	Retinol µg	Carotene µg	Vitamin D µg	Thiamin mg	Riboflavin mg	Nicotinic acid mg	Potential nicotinic acid from tryptophan mg Trp ÷ 60	Vitamin C mg	Vitamin E mg
	Beers									
891	**Brown ale** bottled	0	Tr	0	Tr	0.02	0.26	0.13	0	—
892	**Canned beer** bitter	0	Tr	0	Tr	(0.03)	(0.30)	0.13	0	—
893	**Draught** bitter	0	Tr	0	Tr	0.04	0.47	0.13	0	—
894	mild	0	Tr	0	Tr	(0.03)	(0.30)	0.10	0	—
895	**Keg** bitter	0	Tr	0	Tr	0.03	0.32	0.13	0	—
896	**Lager** bottled	0	Tr	0	Tr	0.02	0.33	0.21	0	—
897	**Pale ale** bottled	0	Tr	0	Tr	0.02	0.35	0.17	0	—
898	**Stout** bottled	0	Tr	0	Tr	0.03	0.26	0.17	0	—
899	**Stout** extra	0	Tr	0	Tr	0.04	0.51	0.17	0	—
900	**Strong ale**	0	Tr	0	Tr	0.06	0.83	0.37	0	—
	Ciders									
901	**Cider** dry	0	Tr	0	Tr	Tr	0.01	Tr	0	—
902	sweet	0	Tr	0	Tr	Tr	0.01	Tr	0	—
903	vintage	0	Tr	0	Tr	Tr	(0.01)	Tr	0	—
	Wines									
904	**Red wine**	0	Tr	0	Tr	0.02	0.09	Tr	0	—
905	**Rosé** medium	0	Tr	0	Tr	0.01	0.07	Tr	0	—
906	**White wine** dry	0	Tr	0	Tr	0.01	0.06	Tr	0	—
907	medium	0	Tr	0	Tr	0.01	0.08	Tr	0	—
908	sweet	0	Tr	0	Tr	0.01	0.08	Tr	0	—
909	sparkling	0	Tr	0	Tr	0.01	0.07	Tr	0	—

No	Food	Vitamin B_6 mg	Vitamin B_{12} µg	Folic acid Free µg	Folic acid Total µg	Pantothenic acid mg	Biotin µg	Notes
	Beers							
891	**Brown ale** bottled	0.012	0.11	2.3	4.0	(0.10)	(0.5)	
892	**Canned beer** bitter	(0.020)	(0.15)	(3.5)	(4.0)	(0.10)	(0.5)	
893	**Draught** bitter	0.023	0.17	4.1	8.8	(0.10)	(0.5)	
894	mild	(0.020)	(0.15)	(4.0)	(4.5)	(0.10)	(0.5)	
895	**Keg** bitter	0.019	0.15	3.9	4.6	(0.10)	(0.5)	
896	**Lager** bottled	0.021	0.14	4.3	4.3	(0.10)	(0.5)	
897	**Pale ale** bottled	0.014	0.14	3.5	4.1	(0.10)	(0.5)	
898	**Stout** bottled	0.014	0.11	3.8	4.4	(0.10)	(0.5)	
899	**Stout** extra	0.016	0.17	5.5	5.9	—	—	
900	**Strong ale**	0.042	0.37	5.9	8.8	—	—	
	Ciders							
901	**Cider** dry	0.005	—	—	—	0.04	0.6	
902	sweet	0.005	—	—	—	0.03	0.6	
903	vintage	(0.005)	—	—	—	(0.03)	(0.6)	
	Wines							
904	**Red wine**	0.015	Tr	0.2	0.2	(0.04)	—	
905	**Rosé** medium	0.023	Tr	0.2	0.2	(0.04)	—	
906	**White wine** dry	0.020	Tr	0.2	0.2	(0.03)	—	
907	medium	0.014	Tr	0.2	0.2	(0.03)	—	
908	sweet	0.012	Tr	0.1	0.1	(0.03)	—	
909	sparkling	0.017	Tr	0.1	0.1	(0.03)	—	

No	Food	Description and number of samples	Specific gravity	Alcohol g	Solids g	Sugars g	Total nitrogen g
	Wines, liqueur (fortified)						
910	**Port**	2 samples	1.026	15.9	13.0	12.0	0.02
911	**Sherry** dry	1 sample	0.988	15.7	3.3	1.4	0.03
912	medium	5 samples from different importers	0.998	14.8	4.7	3.6	0.02
913	sweet	1 sample	1.009	15.6	9.6	6.9	0.05
	Vermouths						
914	**Vermouth** dry	5 samples of different brands	1.005	13.9	6.6	5.5	0.01
915	sweet	5 samples of different brands	1.046	13.0	16.4	15.9	Tr
	Liqueurs						
916	**Advocaat**	4 samples of different brands	1.093	12.8	39.6	28.4	0.75
917	**Cherry brandy**	6 samples of different brands	1.093	19.0	33.3	32.6	Tr
918	**Curaçao**	4 samples of different brands	1.052	29.3	27.8	28.3	Tr
	Spirits						
919	**70% proof**	Mean of brandy, gin, rum, whisky	0.950	31.7	Tr	Tr	Tr

No	Food	Energy value		Protein (N × 6.25)	Fat	Carbo-hydrate	mg									
		kcal	kJ	g	g	g	Na	K	Ca	Mg	P	Fe	Cu	Zn	S	Cl
	Wines, liqueur (fortified)															
910	**Port**	157	655	0.1	0	12.0	4	97	4	11	12	0.40	0.10	—	—	8
911	**Sherry** dry	116	481	0.2	0	1.4	10	57	7	13	11	0.39	0.03	—	—	12
912	medium	118	489	0.1	0	3.6	6	89	9	8	7	0.53	0.10	0.27	—	7
913	sweet	136	568	0.3	0	6.9	13	110	7	11	10	0.37	0.11	—	—	14
	Vermouths															
914	**Vermouth** dry	118	493	0.1	0	5.5	17	40	7	5	7	0.34	0.06	0.04	—	9
915	sweet	151	631	Tr	0	15.9	28	30	6	4	6	0.36	0.04	0.03	—	16
	Liqueurs															
916	**Advocaat**	272	1139	4.7	6.3	28.4	—	—	—	—	—	—	—	—	—	—
917	**Cherry brandy**	255	1073	Tr	0	32.6	—	—	—	—	—	—	—	—	—	—
918	**Curaçao**	311	1303	Tr	0	28.3	—	—	—	—	—	—	—	—	—	—
	Spirits															
919	**70% proof**	222	919	Tr	0	Tr	Tr	Tr	Tr	Tr	Tr	Tr	Tr	Tr	Tr	Tr

No	Food	Retinol µg	Carotene µg	Vitamin D µg	Thiamin mg	Riboflavin mg	Nicotinic acid mg	Potential nicotinic acid from tryptophan mg Trp ÷ 60	Vitamin C mg	Vitamin E mg
	Wines, liqueur (fortified)									
910	**Port**	0	Tr	0	Tr	0.01	0.06	Tr	0	0
911	**Sherry** dry	0	Tr	0	Tr	0.01	0.10	Tr	0	0
912	medium	0	Tr	0	Tr	0.01	0.08	Tr	0	0
913	sweet	0	Tr	0	Tr	0.01	0.07	Tr	0	0
	Vermouths									
914	**Vermouth** dry	0	0	0	Tr	Tr	0.04	0	0	0
915	sweet	0	0	0	Tr	Tr	0.04	0	0	0
	Liqueurs									
916	**Advocaat**	—	—	Tr	—	—	—	1.4	0	—
917	**Cherry brandy**	0	Tr	0	Tr	Tr	Tr	0	0	0
918	**Curaçao**	0	0	0	Tr	Tr	Tr	0	0	0
	Spirits									
919	**70% proof**	0	0	0	0	0	0	0	0	0

No	Food	Vitamin B_6 mg	Vitamin B_{12} µg	Folic acid		Panto-thenic acid mg	Biotin µg	Notes
				Free µg	Total µg			
	Wines, liqueur (fortified)							
910	**Port**	0.010	Tr	0.1	0.1	—	—	
911	**Sherry** dry	0.008	Tr	0.1	0.1	—	—	
912	medium	0.009	Tr	0.1	0.1	—	—	
913	sweet	0.008	Tr	0.1	0.1	—	—	
	Vermouths							
914	**Vermouth** dry	0.008	Tr	Tr	Tr	—	—	
915	sweet	0.004	Tr	Tr	Tr	—	—	
	Liqueurs							
916	**Advocaat**	—	—	—	—	—	—	
917	**Cherry brandy**	Tr	Tr	Tr	Tr	—	—	
918	**Curacao**	Tr	Tr	Tr	Tr	—	—	
	Spirits							
919	**70% proof**	0	0	0	0	0	0	

No	Food	Description and number of samples	Water g	Sugars g	Starch and dextrins g	Dietary fibre g	Total nitrogen g
920	**Bread sauce**	Recipe p 344	76.0	4.4	8.4	0.5	0.70
921	**Brown sauce** bottled	6 bottles of different brands	64.0	23.1	2.1	—	0.18
922	**Cheese sauce**	Recipe p 344	66.5	4.2	4.8	0.2	1.31
923	**Chutney** apple	Recipe p 345	45.0	50.1	0.4	1.8	0.12
924	tomato	Recipe p 345	58.2	39.5	0.2	1.9	0.18
925	**French dressing**	Recipe p 345	23.5	0.2	0	0	0.02
926	**Mayonnaise**	Recipe p 345	28.0	0.1	0	0	0.29
927	**Onion sauce**	Recipe p 345	81.5	4.2	3.7	0.7	0.46
928	**Piccalilli**	12 jars, 5 brands	86.4	2.6	3.4	1.9	0.18
929	**Pickle** sweet	10 jars, 3 brands, including Branston, Pan Yan	58.9	32.6	1.8	1.7	0.09
930	**Salad cream**	6 bottles of the same brand	52.7	13.4	1.7	—	0.30
931	**Tomato ketchup**	6 bottles of different brands	64.8	22.9	1.1	—	0.34
932	**Tomato purée**	Purée or paste in tubes; calculated values	65.7	11.4	0	—	0.97
933	**Tomato sauce**	Recipe p 346	81.5	3.6	4.5	1.9	0.39
934	**White sauce** savoury	Recipe p 346	73.2	5.1	5.9	0.3	0.68
935	sweet	Recipe p 346	67.6	13.5	5.5	0.3	0.63

No	Food	Energy value		Protein	Fat	Carbo-hydrate	mg									
		kcal	kJ	g	g	g	Na	K	Ca	Mg	P	Fe	Cu	Zn	S	Cl
920	**Bread sauce**	110	463	4.3	5.0	12.8	490	140	120	17	100	0.4	0.05	0.5	40	780
921	**Brown sauce** bottled	99	422	1.1	Tr	25.2	980	390	43	29	36	3.1	0.33	—	—	1550
922	**Cheese sauce**	198	825	8.3	14.6	9.0	450	150	260	18	190	0.3	0.03	1.1	—	730
923	**Chutney** apple	193	824	0.7	0.1	50.5	180	200	26	18	30	0.9	0.10	0.1	29	280
924	tomato	154	658	1.1	0.1	39.7	130	310	30	20	39	1.1	0.14	0.2	32	230
925	**French dressing**	658	2706	0.1	73.0	0.2	960	22	5	13	8	0.1	0.01	—	6	1480
926	**Mayonnaise**	718	2952	1.8	78.9	0.1	360	24	16	7	59	0.7	0.03	0.4	21	570
927	**Onion sauce**	99	413	2.9	6.4	7.9	440	130	96	14	75	0.3	0.05	0.3	—	690
928	**Piccalilli**	33	141	1.1	0.7	6.0	1200	55	24	10	23	0.9	0.10	0.2	—	1700
929	**Pickle** sweet	134	572	0.6	0.3	34.4	1700	110	19	10	11	2.0	0.10	1.4	—	2600
930	**Salad cream**	311	1288	1.9	27.4	15.1	840	80	34	21	90	0.8	0.08	—	—	1300
931	**Tomato ketchup**	98	420	2.1	Tr	24.0	1120	590	25	19	43	1.2	0.40	—	—	1810
932	**Tomato purée**	67	286	6.1	Tr	11.4	20[a]	1540	51	66	130	5.1	0.63	1.7	—	290[a]
933	**Tomato sauce**	86	359	2.4	5.1	8.1	340	320	28	14	37	0.7	0.12	0.4	—	560
934	**White sauce** savoury	151	630	4.3	10.3	11.0	410	160	140	16	110	0.3	0.04	0.4	—	650
935	sweet	172	722	3.9	9.5	19.0	110	150	130	13	100	0.2	0.04	0.4	—	180

[a] Values for a brand without added salt. If salt is added at a level of 1 per cent, the purée contains about 420 mg sodium and 890 mg chloride per 100 g

No	Food	Retinol µg	Carotene µg	Vitamin D µg	Thiamin mg	Riboflavin mg	Nicotinic acid mg	Potential nicotinic acid from tryptophan mg Trp ÷ 60	Vitamin C mg	Vitamin E mg
920	**Bread sauce**	40	15	0.16	0.05	0.12	0.3	1.0	Tr	0.2
921	**Brown sauce** bottled	0	—	0	—	—	—	0.2	—	—
922	**Cheese sauce**	140	50	0.56	0.05	0.23	0.2	1.9	Tr	0.6
923	**Chutney** apple	0	10	0	0.02	0.03	0.1	0.1	4	0.1
924	tomato	0	360	0	0.04	0.05	0.5	0.1	8	0.8
925	**French dressing**	0	0	0	0	0	0	0	0	3.9
926	**Mayonnaise**	80	Tr	1.0	0.06	0.11	Tr	1.0	0	4.9
927	**Onion sauce**	60	10	0.39	0.04	0.12	0.2	0.7	2	1.0
928	**Piccalilli**	0	—	0	0.16	0.01	0.2	0.2	Tr	—
929	**Pickle** sweet	0	—	0	0.03	0.01	0.2	0.1	—	—
930	**Salad cream**	—	—	—	—	—	—	0.4	0	—
931	**Tomato ketchup**	0	—	0	—	—	—	0.3	—	—
932	**Tomato purée**	0	2860	0	0.34	0.17	4.0	0.8	(100)	6.9
933	**Tomato sauce**	30	1230	0.27	0.08	0.05	1.0	0.4	10	1.4
934	**White sauce** savoury	100	20	0.63	0.06	0.16	0.2	1.0	Tr	0.7
935	sweet	90	20	0.58	0.05	0.15	0.2	0.9	Tr	0.6

No	Food	Vitamin B_6 mg	Vitamin B_{12} µg	Folic acid		Panto-thenic acid mg	Biotin µg	Notes
				Free µg	Total µg			
920	**Bread sauce**	0.03	Tr	3	5	0.3	2	
921	**Brown sauce** bottled	—	0	Tr	Tr	—	—	
922	**Cheese sauce**	0.05	Tr	3	5	0.3	2	
923	**Chutney** apple	0.05	0	3	4	0.1	Tr	
924	tomato	0.09	0	7	11	0.2	1	
925	**French dressing**	0	0	0	0	0	0	
926	**Mayonnaise**	0.10	1	14	(14)	1.0	12	
927	**Onion sauce**	0.05	Tr	2	4	0.3	1	
928	**Piccalilli**	—	0	—	—	—	Tr	
929	**Pickle** sweet	—	0	—	—	—	Tr	
930	**Salad cream**	—	—	—	—	—	—	
931	**Tomato ketchup**	—	0	—	—	—	—	
932	**Tomato purée**	0.63	0	(63)	(140)	1.1	8	
933	**Tomato sauce**	0.11	0	2	15	0.3	2	
934	**White sauce** savoury	0.04	Tr	3	4	0.3	2	
935	sweet	0.04	Tr	3	4	0.3	2	

No	Food	Description and number of samples	Water g	Sugars g	Starch and dextrins g	Dietary fibre g	Total nitrogen g
937	**Bone and vegetable broth** [a]	Mean of 6 samples, analysed as served in hospital	90.3	1.0	0.1	—	0.59
	Soup						
938	**Chicken, cream of** canned, ready to serve	10 cans, 3 brands	87.9	1.1	3.4	—	0.27
939	condensed	7 cans of the same brand	82.2	1.4	4.6	—	0.41
940	condensed, as served	Diluted with an equal volume of water	91.1	0.7	2.3	—	0.20
941	**Chicken noodle** dried	10 packets, 5 brands	4.8	10.2	50.7	—	2.20
942	dried, as served	Calculated from 35 g soup powder to 570 ml water	94.2	0.6	3.1	—	0.13
943	**Lentil**	Recipe p 346	77.8	2.0	9.9	2.2	0.71
944	**Minestrone** dried	10 packets, 3 brands	3.9	15.0	32.6	6.6	1.62
945	dried, as served	Calculated from 45 g soup powder to 570 ml water	92.6	1.2	2.5	0.5	0.12
946	**Mushroom, cream of** canned, ready to serve	10 cans, 3 brands	89.2	0.8	3.1	—	0.17
947	**Oxtail** canned, ready to serve	10 cans, 3 brands	88.5	0.9	4.2	—	0.38
948	dried	10 packets, 5 brands	3.0	9.2	41.8	3.8	2.81
949	dried, as served	Calculated from 45 g soup powder to 570 ml water	92.5	0.7	3.2	0.3	0.22
950	**Tomato, cream of** canned, ready to serve	10 cans, 3 brands	84.2	2.6	3.3	—	0.13
951	condensed	7 cans, 2 brands	70.6	11.2	3.4	—	0.27
952	condensed, as served	Diluted with an equal volume of water	85.3	5.6	1.7	—	0.14
953	**dried**	10 packets, 4 brands	2.8	36.1	28.9	3.3	1.05
954	dried, as served	Calculated from 58 g soup powder to 570 ml water	90.6	3.5	2.8	0.3	0.10
955	**Vegetable** canned, ready to serve	10 cans, 4 brands	86.4	2.5	4.2	—	0.24

[a] See McCance, Sheldon and Widdowson (1934)

No	Food	Energy value		Protein (N × 6.25)	Fat	Carbo-hydrate	mg									
		kcal	kJ	g	g	g	Na	K	Ca	Mg	P	Fe	Cu	Zn	S	Cl
937	**Bone and vegetable broth**	60	251	3.7	4.6	1.1	74	64	17	3	10	0.3	0.04	—	—	75
	Soup															
938	**Chicken, cream of** canned, ready to serve	58	242	1.7	3.8	4.5	460	41	27	5	27	0.4	0.02	0.3	—	700
939	condensed	98	407	2.6	7.2	6.0	710	(62)	(41)	(7)	(41)	(0.5)	(0.03)	(0.5)	—	1070
940	condensed, as served	49	203	1.3	3.6	3.0	350	(31)	(20)	(4)	(20)	(0.3)	(0.02)	(0.3)	—	530
941	**Chicken noodle** dried	329	1394	13.8	5.0	60.9	6120	270	45	44	160	2.7	0.26	1.2	—	9030
942	dried, as served	20	84	0.8	0.3	3.7	370	16	3	3	10	0.2	0.02	0.1	—	550
943	**Lentil**	99	402	4.4	3.7	11.9	190	160	40	16	59	1.2	0.11	0.5	—	290
944	**Minestrone** dried	298	1259	10.1	8.8	47.6	5600	800	120	48	160	2.8	0.29	1.0	—	7670
945	dried, as served	23	99	0.8	0.7	3.7	430	62	9	7	12	0.2	0.02	0.1	—	590
946	**Mushroom, cream of** canned, ready to serve	53	222	1.1	3.8	3.9	470	55	30	4	30	0.3	0.04	0.3	—	750
947	**Oxtail** canned, ready to serve	44	185	2.4	1.7	5.1	440	93	40	6	37	1.0	0.04	0.4	—	660
948	dried	356	1504	17.6	10.5	51.0	5250	700	140	44	260	4.3	0.25	2.4	—	7670
949	dried, as served	27	116	1.4	0.8	3.9	400	54	11	3	20	0.3	0.02	0.2	—	590
950	**Tomato, cream of** canned, ready to serve	55	230	0.8	3.3	5.9	460	190	17	8	20	0.4	0.06	0.2	—	740
951	condensed	123	514	1.7	6.8	14.6	830	(360)	(32)	(15)	(38)	(0.7)	(0.11)	0.3	—	1320
952	condensed, as served	62	258	0.9	3.4	7.3	410	(180)	(16)	(8)	(19)	(0.3)	(0.06)	0.2	—	660
953	**dried**	321	1359	6.6	5.6	65.0	4040	920	140	39	130	1.8	0.33	0.8	—	6620
954	dried, as served	31	130	0.6	0.5	6.3	390	89	14	4	13	0.2	0.03	0.1	—	640
955	**Vegetable** canned, ready to serve	37	159	1.5	0.7	6.7	500	140	17	10	27	0.6	0.06	0.3	—	750

No	Food	Retinol µg	Carotene µg	Vitamin D µg	Thiamin mg	Riboflavin mg	Nicotinic acid mg	Potential nicotinic acid from tryptophan mgTrp ÷ 60	Vitamin C mg	Vitamin E mg
937	**Bone and vegetable broth**	0	—	0	—	—	—	0.8	0	—
	Soup									
938	**Chicken, cream of** canned, ready to serve	0	0	0	0.01	0.03	0.2	0.3	0	—
939	condensed	0	0	0	(0.02)	0.04	0.6	0.5	0	—
940	condensed, as served	0	0	0	(0.01)	0.02	0.3	0.2	0	—
941	**Chicken noodle** dried	0	0	0	0.23	0.08	2.2	2.6	0	—
942	dried, as served	0	0	0	0.01	Tr	0.1	0.2	0	—
943	**Lentil**	40	430	0.28	(0.07)	(0.05)	(0.3)	0.8	Tr	—
944	**Minestrone** dried	0	—	0	0.21	0.15	3.1	1.9	0	—
945	dried, as served	0	—	0	0.02	0.01	0.2	0.1	0	—
946	**Mushroom, cream of** canned ready to serve	0	0	0	Tr	0.05	0.3	0.2	0	—
947	**Oxtail** canned, ready to serve	0	0	0	0.02	0.03	0.7	0.5	0	—
948	**dried**	0	0	0	10.4 [a]	0.30	3.5	3.8	0	—
949	dried, as served	0	0	0	0.8 [a]	0.02	0.3	0.3	0	—
950	**Tomato, cream of** canned, ready to serve	0	210	0	0.03	0.02	0.5	0.1	(Tr)	—
951	condensed	0	(400)	0	(0.06)	0.05	1.0	0.2	(Tr)	—
952	condensed, as served	0	(200)	0	(0.03)	0.03	0.5	0.1	(Tr)	—
953	**dried**	0	—	0	0.23	0.18	1.9	0.9	(Tr)	—
954	dried, as served	0	—	0	0.02	0.02	0.2	0.1	(Tr)	—
955	**Vegetable** canned, ready to serve	0	Tr [b]	0	0.03	0.02	0.4	0.2	(Tr)	—

No	Food	Vitamin B_6 mg	Vitamin B_{12} µg	Folic acid Free µg	Folic acid Total µg	Panto-thenic acid mg	Biotin µg
937	**Bone and vegetable broth**	—	0	—	—	—	—
	Soup						
938	**Chicken, cream of** canned, ready to serve	0.01	0	—	—	—	—
939	condensed	—	0	—	—	—	—
940	condensed, as served	—	0	—	—	—	—
941	**Chicken noodle** dried	—	0	—	—	—	—
942	dried, as served	—	0	—	—	—	—
943	**Lentil**	(0.07)	0	—	—	—	—
944	**Minestrone** dried	—	0	—	—	—	—
945	dried, as served	—	0	—	—	—	—
946	**Mushroom, cream of** canned, ready to serve	0.01	0	(2)	—	—	—
947	**Oxtail** canned, ready to serve	0.03	0	(2)	—	—	—
948	dried	—	0	—	—	—	—
949	dried, as served	—	0	—	—	—	—
950	**Tomato, cream of** canned, ready to serve	0.06	0	6	12	—	—
951	condensed	—	0	—	—	—	—
952	condensed, as served	—	0	—	—	—	—
953	**dried**	—	0	21	52	—	—
954	dried, as served	—	0	—	—	—	—
955	**Vegetable** canned, ready to serve	0.05	0	2	10	—	—

Notes

[a] This remarkably high content is derived from the flavouring agent.

[b] 36µg α-carotene per 100g was found, but no β-carotene.

No	Food	Description and number of samples	Water g	Sugars g	Starch and dextrins g	Dietary fibre g	Total nitrogen g
956	**Baking powder**	6 samples of the same brand	6.3	Tr	37.8	—	0.91
957	**Bovril**	9 jars	38.7	0	2.9	0	6.25[a]
958	**Curry powder**	2 samples	—[b]	—	—	—	1.52
959	**Gelatin**	Literature sources	13.0	0	0	0	15.20
960	**Ginger** ground	3 samples	—[b]	—	—	—	1.19
961	**Marmite**	7 jars	25.4	0	1.8	—	6.62[c]
962	**Oxo cubes**	10 samples	9.1	—	12.0	0	6.29
963	**Mustard** powder	2 brands	—[b]	—	—	—	4.62
964	**Pepper**	3 samples	—[b]	—	—	—	1.40
965	**Salt** block	2 samples	0.2	0	0	0	0
966	table	2 samples	Tr	0	0	0	0
967	**Vinegar**[d]	4 samples	—	0.6	0	0	0.07
968	**Yeast** bakers', compressed	Literature sources	70.0	Tr	1.1	6.9	2.02[e]
969	dried	Literature sources	5.0	Tr	(3.5)	(21.9)	6.32[e]

[a] Includes 0.17g purine nitrogen

[b] The loss of weight at 100°C cannot be used to determine the amount of water present, since these substances contain volatile essential oils

[c] Includes 0.27g purine nitrogen

[d] Contains 4.8ml acetic acid per cent

[e] Purine nitrogen forms about 10 per cent of the total nitrogen

No	Food	Energy value		Protein	Fat	Carbo-hydrate	mg									
		kcal	kJ	g	g	g	Na	K	Ca	Mg	P	Fe	Cu	Zn	S	Cl
956	**Baking powder**	163	693	5.2	Tr	37.8	11 800 [a]	49	11300 [a]	9	8430 [a]	Tr	Tr	—	—	29
957	**Bovril**	174	737	39.1 [b]	0.7	2.9	4800	1200	40	61	590	14.0	0.45	1.8	—	6800
958	**Curry powder**	233	979	9.5	10.8	26.1	450	1830	640	280	270	75.0 [c]	1.04	—	86	470
959	**Gelatin**	338	1435	84.4	Tr	0	—	—	—	—	—	—	—	—	—	—
960	**Ginger** ground	258	1101	7.4	3.3 [d]	60.0	34	910	97	130	140	17.2	0.45	(6.8)	150	40
961	**Marmite**	179	759	41.4 [b]	0.7	1.8	4500	2600	95	180	1700	3.7	0.30	2.1	—	6600
962	**Oxo cubes**	229	969	38.3 [b]	3.4	12.0	10 300	730	180	59	360	24.5	0.71	—	—	16 000
963	**Mustard** powder	452	1884	28.9	28.7	20.7	5	940	330	260	180	10.9	0.20	(6.5)	1280	62
964	**Pepper**	308	1312	8.8	6.5 [d]	68.0	7	42	130	45	130	10.2	1.13	(1.8)	99	60
965	**Salt** block	0	0	0	0	0	38 700	Tr	230	140	Tr	0.3	0.39	—	400	59 600
966	table	0	0	0	0	0	38 850	Tr	29	290	8	0.2	0.10	—	23	59 900
967	**Vinegar**	4	16	0.4	0	0.6	20	89	15	22	32	0.5	0.04	—	19	47
968	**Yeast** bakers', compressed	53	226	11.4 [b]	0.4	1.1	16	610	25	59	390	5.0	(1.6)	(2.6)	—	—
969	dried	169	717	35.6 [b]	1.5	3.5	(50)	(2000)	80	230	(1290)	20.0	5.0	8.0	—	—

[a] The sodium, calcium and phosphorus content will depend on the brand

[b] (Total N — purine N) × 6.25

[c] This high value has been confirmed by a recent analysis giving 95 mg per 100g

[d] By Soxhlet extraction. The figure for fat obtained by von Lieberman's method is 0.4g per 100g in ginger and 2.0g per 100g in pepper, and these have been used for calculating energy values

No	Food	Retinol µg	Carotene µg	Vitamin D µg	Thiamin mg	Riboflavin mg	Nicotinic acid mg	Potential nicotinic acid from tryptophan mg Trp ÷ 60	Vitamin C mg	Vitamin E mg
956	**Baking powder**	0	0	0	Tr	Tr	Tr	1	0	Tr
957	**Bovril**	0	0	0	9.1	7.4	82	3	0	—
958	**Curry powder**	0	—	0	—	—	—	—	0	—
959	**Gelatin**	0	0	0	Tr	Tr	Tr	0	0	0
960	**Ginger** ground	0	—	0	—	—	—	—	0	—
961	**Marmite**	0	0	0	3.1	11	58	9	0	—
962	**Oxo cubes**	0	0	0	—	—	—	—	0	—
963	**Mustard** powder	0	—	0	—	—	—	—	0	—
964	**Pepper**	0	—	0	—	—	—	—	0	—
965	**Salt**, block	0	0	0	0	0	0	0	0	0
966	table	0	0	0	0	0	0	0	0	0
967	**Vinegar**	0	0	0	0	0	0	0	0	0
968	**Yeast** bakers', compressed	0	Tr	0	0.71	1.7	11	2	Tr	Tr
969	dried	0	Tr	0	2.33[a]	4.0	36	7	Tr	Tr

No	Food	Vitamin B_6 mg	Vitamin B_{12} µg	Folic acid Free µg	Folic acid Total µg	Panto-thenic acid mg	Biotin µg	Notes
956	**Baking powder**	Tr	0	Tr	Tr	Tr	Tr	[a] Value for bakers' yeast. Brewers' yeast contains 15.6 mg per 100 g.
957	**Bovril**	0.53	8.3	750	1040	—	—	
958	**Curry powder**	—	0	—	—	—	—	
959	**Gelatin**	Tr	0	Tr	Tr	Tr	Tr	
960	**Ginger** ground	—	0	—	—	—	—	
961	**Marmite**	1.3	0.5	83	1010	—	—	
962	**Oxo cubes**	—	—	—	—	—	—	
963	**Mustard** powder	—	0	—	—	—	—	
964	**Pepper**	—	0	—	—	—	—	
965	**Salt** block	0	0	0	0	0	0	
966	table	0	0	0	0	0	0	
967	**Vinegar**	0	0	0	0	0	0	
968	**Yeast** bakers', compressed	0.6	Tr	(42)	(1250)	3.5	(60)	
969	dried	2.0	Tr	130	4000	11.0	200	

No	Food	Description and number of samples	Water g	Sugars g	Starch and dextrins g	Dietary fibre g	Total nitrogen g

No	Food	Energy value		Protein	Fat	Carbo-hydrate	mg									
		kcal	kJ	g	g	g	Na	K	Ca	Mg	P	Fe	Cu	Zn	S	Cl

No	Food	Retinol µg	Carotene µg	Vitamin D µg	Thiamin mg	Riboflavin mg	Nicotinic acid mg	Potential nicotinic acid from tryptophan mg Trp ÷ 60	Vitamin C mg	Vitamin E mg

No	Food	Vitamin B_6 mg	Vitamin B_{12} µg	Folic acid		Panto-thenic acid mg	Biotin µg	Notes
				Free µg	Total µg			

The tables

continued

Section 2
Amino acid composition

mg per g nitrogen

Isoleucine	Ile	Arginine	Arg
Leucine	Leu	Histidine	His
Lysine	Lys	Alanine	Ala
Methionine	Met	Aspartic acid	Asp
Cystine	Cys	Glutamic acid	Glu
Phenylalanine	Phe	Glycine	Gly
Tyrosine	Tyr	Proline	Pro
Threonine	Thr	Serine	Ser
Tryptophan	Trp		
Valine	Val		

Asterisk indicates new analytical data

Amino acids (mg per g nitrogen)

No	Food		Ile	Leu	Lys	Met	Cys	Phe	Tyr	Thr	Trp	Val	Arg	His	Ala	Asp	Glu	Gly	Pro	Ser
	Cereals																			
2002	**Barley** pearl		220	420	160	100	140	320	190	210	100	310	300	130	260	350	1470	240	680	250
2005	**Bran** wheat		210	410	270	100	170	260	200	220	80	310	490	190	350	510	1280	400	390	300
2006	**Cornflour** maize		230	780	170	120	100	310	240	230	40	300	260	170	470	390	1180	230	560	310
	Flour																			
2009	wholemeal		210	420	150	100	160	280	190	170	70	280	290	130	230	310	1710	250	660	330
2010	brown		210	420	140	100	160	280	190	170	70	270	260	130	190	270	2020	200	770	350
2011	white		240	440	120	100	160	300	160	170	70	270	220	130	190	270	2060	200	790	350
2017	**Oatmeal**		240	450	230	110	170	310	210	210	80	320	390	130	280	480	1310	290	320	290
2019	**Rice** milled		240	510	230	130	100	300	250	210	80	360	470	150	360	600	1200	270	290	290
2021	**Rye** whole		220	390	210	90	120	280	120	210	70	300	290	140	270	450	1510	270	590	240
2024	**Soya** flour		280	490	400	80	100	310	200	240	80	300	450	160	270	730	1170	260	340	320
	Milk and eggs																			
2123	**Milk** cows'		350	640	510	180	60	340	280	310	90	460	250	190	240	530	1440	140	590	370
2138	**Milk** human	*	320	580	430	90	120	230	180	270	140	410	230	150	250	540	1070	150	580	260
2161	**Yogurt**	*	390	720	430	130	50	370	310	280	80	430	180	210	240	430	1150	160	480	360
	Eggs																			
2165	whole	*	350	520	390	200	110	320	250	320	110	470	380	150	340	670	750	190	240	490
2166	white		350	510	360	220	110	360	250	300	110	490	340	140	360	680	760	200	250	460
2167	yolk		360	530	450	160	100	250	250	350	110	430	450	160	310	660	680	170	220	540

Amino acids (mg per g nitrogen)

No	Food		Ile	Leu	Lys	Met	Cys	Phe	Tyr	Thr	Trp	Val	Arg	His	Ala	Asp	Glu	Gly	Pro	Ser
	Meat																			
2210	**Bacon** lean	*	300	460	530	160	80	270	220	270	70	320	400	200	380	560	1020	330	320	260
2237	**Beef** lean	*	320	500	570	170	80	280	240	290	80	330	420	230	400	600	1080	350	320	280
2266	**Lamb** lean	*	290	450	610	160	80	240	220	290	80	300	380	200	360	570	1050	310	290	270
2296	**Pork** lean	*	280	440	600	170	80	240	230	270	70	300	370	270	340	560	1000	330	300	260
	Poultry and game																			
2314	**Chicken**	*	290	470	560	150	80	280	220	260	70	300	390	190	360	570	1030	310	260	250
2327	**Duck**	*	310	490	550	170	80	280	230	280	80	320	420	160	380	580	1040	320	310	260
2340	**Turkey**	*	310	480	560	180	70	280	210	260	70	320	390	180	360	580	990	310	310	260
2350	**Rabbit**	*	310	480	550	170	80	290	230	270	70	320	400	160	380	600	1030	320	320	260
	Offal																			
2354	**Brain,** calf and lamb	*	270	510	560	130	110	320	240	320	80	360	390	250	340	570	870	290	340	330
2357	**Heart,** lamb and ox	*	330	570	540	140	100	290	210	290	80	350	400	160	390	530	950	380	220	330
2363	**Kidney,** lamb, ox and pig	*	260	490	510	130	90	310	200	270	80	350	350	220	330	550	770	370	350	300
2370	**Liver,** calf, chicken, lamb ox and pig	*	270	490	530	150	90	310	190	270	80	360	330	230	330	540	760	310	330	290
2381	**Oxtail**	*	300	430	560	120	80	250	210	260	80	300	380	250	360	520	980	390	400	250
2384	**Sweetbread**	*	220	400	540	90	80	210	150	230	80	270	370	180	320	430	870	390	320	250
2386	**Tongue,** lamb and ox	*	290	440	660	110	90	220	190	260	80	300	370	250	330	540	980	350	330	260
2391	**Tripe**	*	250	420	500	150	80	240	180	270	80	310	440	180	420	530	940	650	530	310

Amino acids (mg per g nitrogen)

No	Food		Ile	Leu	Lys	Met	Cys	Phe	Tyr	Thr	Trp	Val	Arg	His	Ala	Asp	Glu	Gly	Pro	Ser
	Canned meats																			
2393	**Beef,** corned	*	290	460	580	150	90	260	240	280	90	330	410	180	390	580	1020	420	390	250
2394	**Ham**	*	290	480	560	170	80	250	200	270	60	310	410	230	380	570	1030	390	320	230
2395	**Ham and pork** chopped	*	270	430	530	140	80	230	210	270	70	300	410	200	370	570	1000	440	400	280
2396	**Luncheon meat**	*	230	390	410	120	90	240	170	220	80	290	380	150	390	480	960	540	550	270
2397	**Stewed steak with gravy**	*	270	440	540	140	70	250	180	280	70	290	400	210	370	530	1050	430	390	290
2398	**Tongue**	*	290	460	600	150	90	300	230	270	90	320	410	150	380	540	960	430	390	280
2400	**Veal,** jellied	*	300	450	540	150	80	270	210	270	70	310	430	180	360	550	990	390	340	270
	Offal products																			
2401	**Black pudding**	*	140	610	460	90	80	360	150	250	80	450	300	310	460	590	930	400	440	300
2402	**Faggots**	*	250	440	420	100	90	300	140	220	70	320	400	140	450	480	1270	620	640	290
2403	**Haggis**	*	240	460	420	100	80	270	170	260	70	370	380	170	370	510	1010	440	420	300
2404	**Liver sausage**	*	250	470	490	120	80	270	170	340	70	370	370	210	370	520	980	490	380	310
	Sausages																			
2405	**Frankfurters**	*	290	460	490	140	70	270	210	250	60	320	410	180	400	560	1070	480	490	280
2406	**Polony**	*	240	410	460	120	70	230	140	220	70	300	360	150	350	460	1340	490	550	250
2407	**Salami**	*	300	450	520	130	80	260	190	280	70	310	380	170	370	580	1060	430	370	250
2408	**Sausages** beef	*	260	430	380	120	100	260	160	240	80	310	380	160	400	500	1280	550	550	270
2411	**Sausages** pork	*	260	410	410	130	110	250	160	240	80	300	370	170	370	490	1150	460	510	270
2414	**Saveloy**	*	220	380	460	110	80	210	150	210	70	270	370	160	360	470	1000	530	510	240

Amino acids (mg per g nitrogen)

No	Food		Ile	Leu	Lys	Met	Cys	Phe	Tyr	Thr	Trp	Val	Arg	His	Ala	Asp	Glu	Gly	Pro	Ser
	Meat products																			
2415	**Beefburgers**	*	280	450	490	160	70	270	200	250	70	310	400	200	400	540	1170	420	400	260
2417	**Brawn**	*	200	340	450	110	70	270	160	190	70	260	440	130	450	460	900	780	440	250
2418	**Meat paste**	*	300	450	490	150	80	240	140	400	70	330	390	170	410	580	1050	430	410	270
2419	**White pudding**	*	250	450	310	120	90	290	190	300	70	380	390	190	290	490	1330	370	400	340
	Meat and pastry products																			
2420	**Cornish pastie**	*	230	400	250	110	100	290	170	200	80	280	260	140	250	400	1680	280	570	250
2421	**Pork pie**	*	260	400	330	120	100	310	170	210	80	280	330	160	320	430	1390	390	570	280
2425	**Steak and kidney pie** individual	*	290	510	340	120	110	380	130	250	70	360	380	190	360	450	1690	390	710	290
	Fish																			
2435	**White and fatty fish** all kinds		330	530	610	180	70	260	220	300	70	360	400	180	430	650	950	290	260	310
2516	**Crustacea** all kinds		290	540	490	180	80	250	230	290	70	300	520	120	420	680	980	410	270	320
2530	**Molluscs** all kinds		300	480	500	170	100	260	260	290	80	390	470	150	350	700	880	320	260	320

Amino acids (mg per g nitrogen)

No	Food	Ile	Leu	Lys	Met	Cys	Phe	Tyr	Thr	Trp	Val	Arg	His	Ala	Asp	Glu	Gly	Pro	Ser
	Vegetables																		
2558	**Asparagus**	160	280	280	80	60	160	130	180	70	230	240	100	360	680	1300	210	360	200
2561	**Beans, French**	230	430	340	80	70	270	210	240	90	310	270	150	280	750	670	240	240	330
2564	**broad**	250	440	400	40	50	270	200	210	60	280	560	150	260	700	940	260	250	280
2565	**butter**	310	510	470	90	90	380	200	260	60	320	370	200	290	770	820	260	290	410
2567	**haricot**	260	480	450	70	50	330	160	250	60	290	360	180	260	750	920	240	220	350
2572	**red kidney**	260	480	450	70	50	330	160	250	60	290	360	180	260	750	920	240	220	350
2574	**Beetroot**	150	280	330	120	70	220	220	210	60	150	410	80	140	1130	950	130	160	220
2576	**Broccoli tops**	240	330	320	90	70	230	—	230	70	310	360	110	—	—	—	—	—	—
2578	**Brussels sprouts**	260	340	340	60	40	230	—	270	70	300	390	140	—	—	—	—	—	—
2585	**Cabbage**	190	330	190	60	70	190	120	230	60	260	520	160	320	410	540	300	230	260
2587	**Carrots**	190	280	240	70	70	170	140	180	50	280	280	90	300	730	1210	180	180	200
2591	**Cauliflower**	270	420	350	120	—	210	90	260	90	360	280	120	500	520	480	430	—	—
2594	**Celery**	240	430	130	110	30	280	80	210	70	300	250	90	—	—	—	—	—	—
2597	**Cucumber**	190	260	270	60	—	140	—	160	50	210	470	90	—	—	—	—	—	—
2603	**Lentils**	270	480	450	50	60	330	200	250	60	310	540	170	270	720	1040	260	270	330
2606	**Lettuce**	240	390	240	110	—	320	170	260	50	340	280	100	270	720	640	260	330	210
2609	**Mushrooms**	140	230	280	90	50	130	120	170	60	160	370	80	290	280	440	160	320	170
2613	**Onions**	90	170	280	70	—	170	210	90	90	140	800	60	—	—	930	—	—	—
2620	**Peas**	270	430	470	60	70	290	170	250	60	290	590	140	260	690	1010	250	240	270
2630	**chick**	280	470	430	80	90	360	180	240	50	280	590	170	270	730	990	250	260	320
2633	**red pigeon**	190	390	480	80	70	520	130	180	30	230	300	230	260	600	1170	200	250	260

Amino acids (mg per g nitrogen)

No	Food	Ile	Leu	Lys	Met	Cys	Phe	Tyr	Thr	Trp	Val	Arg	His	Ala	Asp	Glu	Gly	Pro	Ser
	***Vegetables** contd*																		
2639	**Potatoes**	260	380	340	100	80	270	190	240	90	320	310	120	230	1150	800	210	240	260
2657	**Spinach**	300	590	450	110	100	380	310	330	100	380	400	160	400	620	730	320	300	300
2664	**Sweet potatoes**	230	340	210	100	70	240	150	240	110	280	310	80	300	830	540	230	220	260
2666	**Tomatoes**	120	170	180	40	40	110	80	140	50	130	130	90	150	720	—	110	100	160
2669	**Turnips**	160	260	120	70	—	130	90	180	80	160	100	50	290	360	580	160	220	210
2671	**Turnip tops**	210	420	310	90	80	280	170	250	80	270	240	110	330	490	680	270	250	230
2673	**Yam**	220	380	260	90	80	290	200	210	80	270	480	120	270	660	780	220	230	320
	Fruit																		
2675	**Apples**	220	390	370	50	80	160	90	230	60	250	170	120	280	1300	700	240	200	270
2682	**Apricots**	110	180	180	30	—	100	80	130	50	150	80	100	220	1470	370	110	170	180
2692	**Avocado pears**	210	340	310	100	—	220	140	180	70	290	210	110	380	1410	770	250	240	260
2693	**Bananas**	250	320	270	80	170	260	160	190	70	260	360	380	280	660	580	260	260	240
2724	**Dates**	140	270	170	80	130	180	90	170	170	200	210	90	290	370	650	280	360	210
2726	**Figs**	190	270	250	50	100	150	270	200	50	240	140	90	380	1500	600	210	410	310
2736	**Grapes**	50	130	140	210	100	130	110	170	30	170	460	23[illegible]	260	760	1300	190	210	300
2762	**Melons, Canteloupe**	—	—	160	20	—	190	—	—	10	—	—	—	—	—	—	—	—	—
2766	**watermelon**	—	—	—	—	—	150	150	—	—	—	—	—	—	—	—	—	—	—
2773	**Oranges**	180	170	330	90	80	230	130	90	40	240	400	90	390	880	760	640	350	180
2779	**Peaches**	100	220	230	240	70	140	160	210	30	310	130	130	310	710	1100	120	210	260
2785	**Pears**	—	—	—	—	—	150	240	—	—	—	—	—	—	—	—	—	—	—
2791	**Pineapple**	—	—	140	20	—	160	160	—	80	—	—	—	—	—	—	—	—	—
2817	**Strawberries**	140	320	250	10	50	180	210	190	70	180	270	120	320	1400	920	250	200	240

Amino acids (mg per g nitrogen)

No	Food		Ile	Leu	Lys	Met	Cys	Phe	Tyr	Thr	Trp	Val	Arg	His	Ala	Asp	Glu	Gly	Pro	Ser
	Nuts																			
2822	**Almonds**		220	390	140	80	90	300	180	150	50	320	610	140	240	590	1370	330	300	220
2826	**Brazil nuts**		180	430	170	360	130	240	170	160	70	270	830	140	220	460	1160	280	300	270
2830	**Cob** or **hazel nuts**		360	390	180	60	70	230	230	180	90	390	910	120	—	440	1280	590	350	600
2832	**Coconut**		240	420	220	110	100	280	170	210	70	340	820	130	280	550	1170	280	230	300
2835	**Peanuts**		210	400	220	70	80	310	240	160	70	260	700	150	240	710	1140	350	270	300
2838	**Peanut butter**	*	230	420	210	80	80	310	210	160	90	280	750	150	280	760	1240	360	290	360
2839	**Walnuts**		250	450	120	90	110	270	210	190	60	300	790	130	—	—	—	—	—	—
	Confectionery																			
2857	**Chocolate** milk	*	400	680	480	160	80	410	200	300	120	440	270	220	280	550	1390	180	660	440
2858	plain	*	240	400	260	120	120	340	130	260	90	360	400	120	290	620	1160	260	350	330
	Beverages																			
2868	**Cocoa powder**	*	180	290	190	80	100	210	140	190	80	280	340	80	220	480	820	210	280	280
2872	**Coffee** instant	*	110	250	20	90	30	160	100	80	80	170	0	70	180	290	910	230	170	70
	Beers																			
2895	**Keg** bitter	*	90	180	140	60	140	140	160	170	190	180	190	120	250	350	1050	270	890	210
2896	**Lager**	*	100	180	230	80	130	140	150	180	410	180	180	120	270	350	1190	290	880	240

Amino acids (mg per g nitrogen)

No	Food		Ile	Leu	Lys	Met	Cys	Phe	Tyr	Thr	Trp	Val	Arg	His	Ala	Asp	Glu	Gly	Pro	Ser
	Miscellaneous																			
2957	**Bovril**	*	160	290	330	80	20	190	110	220	30	250	350	140	480	440	750	710	530[a]	240
2959	**Gelatin**		90	180	250	50	Tr	130	20	120	0	140	490	40	610	370	630	1510	860[b]	230
2961	**Marmite**	*	290	360	430	80	60	240	110	310	80	370	160	140	410	580	780	310	300	310
2968	**Yeast**		310	450	510	110	60	270	230	300	70	400	320	180	400	620	670	290	260	340

[a] Also contains 340 mg hydroxyproline

[b] Also contains 730 mg hydroxyproline

The tables

continued

Section 3
Fatty acid composition

g fatty acids per 100g total fatty acids

Common names of the most frequently occurring fatty acids

Carbon : Double bonds	Common name
Saturated	
C4 : 0	Butyric
C6 : 0	Caproic
C8 : 0	Caprylic
C10 : 0	Capric
C12 : 0	Lauric
C14 : 0	Myristic
C16 : 0	Palmitic
C18 : 0	Stearic
C20 : 0	Arachidic
C22 : 0	Behenic
C24 : 0	Lignoceric
Mono-unsaturated	
C16 : 1	Palmitoleic
C18 : 1	Oleic
C20 : 1	Eicosenoic
C22 : 1	Erucic
Polyunsaturated	
C18 : 2	Linoleic
C18 : 3	Linolenic
C20 : 4	Arachidonic

Asterisk indicates new analytical data

No	Food		Saturated						Mono-unsaturated				Polyunsaturated			Other
			10:0	12:0	14:0	16:0	18:0	20:0	16:1	18:1	20:1	22:1	18:2	18:3	20:4	
3002	**Barley**		0	Tr	0.4	23.6	0.7	0	0.1	11.8	Tr	0	57.4	6.1	0	
3005	**Bran** wheat		0	Tr	Tr	18.4	1.0	0.8	0.5	15.3	0.4	0	59.4	4.1	0.2	
3008	**Flour** wholemeal, brown and white		0	Tr	Tr	18.4	1.0	0.8	0.5	15.3	0.4	0	59.4	4.1	0.2	
3017	**Oatmeal**		0	Tr	Tr	17.0	1.1	0.5	0.2	38.4	Tr	0	40.7	2.2	Tr	
3019	**Rice**		0	Tr	0.9	24.0	2.5	0.4	0.1	29.6	0	0	41.2	1.1	Tr	
3021	**Rye**		0	Tr	Tr	18.8	0.6	Tr	0.5	13.7	0.8	0	56.8	8.8	0	
3030	**Bread** wholemeal	*	0	Tr	1.7	19.2	4.2	Tr	1.9	16.9	Tr	0	50.2	3.6	Tr	
3033	white	*	0	0.7	1.6	23.0	5.0	Tr	1.1	17.7	Tr	0	45.4	3.3	Tr	
	Biscuits															
3058	**chocolate** full coated [a]	*	1.0	10.7	5.1	25.7	20.1	0.5	0.7	29.4	0.3	0	4.1	0.2	0	
3060	**crispbread** rye	*	0	0.2	0.2	18.2	0.7	0	0.5	14.8	1.6	0	56.2	7.2	0	
3061	wheat, starch reduced [a]	*	0	0.6	0.9	32.4	3.3	0.2	0.3	27.4	0.6	0	32.7	1.6	0	
3063	**digestive** chocolate [a]	*	0	0.6	2.2	31.1	17.7	0.9	1.0	35.2	1.6	1.6	6.6	0.2	0	
3064	**ginger nuts** [a]	*	0	1.1	2.8	35.8	9.2	0.6	1.6	36.6	1.0	0.8	9.1	0.5	0	
3066	**Matzo**	*	0	0.8	0.3	20.3	0.7	0	0.2	10.0	0.8	0	62.6	4.0	0	
3067	**oatcakes** [a]	*	0	0.3	1.9	15.4	4.3	0.6	1.4	38.1	4.2	3.1	27.2	2.4	0	
3068	**sandwich** [a]	*	1.5	14.0	5.9	26.9	8.5	0.4	0.7	31.0	—	0.5	6.0	0	—	8:0 1.4; 20:1–20:5 1.4
3069	**semi-sweet** [a]	*	0	2.2	3.1	35.5	8.8	0.6	1.2	35.6	—	0.5	9.4	0	—	20:1–20:5 1.4
3070	**short-sweet** [a]	*	0	3.0	3.8	36.5	8.2	0.9	1.4	34.0	—	0	8.2	0.5	—	20:1–20:5 2.2
3072	**wafers** filled [a]	*	2.8	28.7	10.5	14.4	5.1	0	0.3	28.5	0	0	3.1	0	0	10:0 3.4

[a] The composition may vary according to type of fat used in the manufacture of the biscuits

No	Food		Saturated							Mono-unsaturated				Polyunsaturated			
			8:0	10:0	12:0	14:0	16:0	18:0	20:0	16:1	18:1	20:1	22:1	18:2	18:3	20:4	Other
	Cakes																
3074	**Fancy iced cakes** [a]	*	2.5	2.1	21.6	9.4	18.1	11.9	Tr	2.6	20.5	1.6	3.1	5.1	0.7	0	
3077	**Fruit cake** plain [a]	*	0	0	0.5	5.2	30.6	10.0	0.8	6.4	29.4	4.0	2.4	8.5	0.7	0	
3079	**Madeira cake** [a]	*	1.7	1.6	10.1	8.3	22.7	9.1	0.7	4.9	22.8	2.8	4.6	9.1	0	0	22:0 0.6
3083	**Spongecake** jam filled [a]	*	0	0	Tr	3.4	25.7	10.1	0.8	4.6	26.0	3.3	7.3	14.0	2.9	Tr	
	Puddings																
3108	**Ice cream** non-dairy [a]	*	0	0	Tr	2.5	43.2	7.6	Tr	0.8	35.0	1.7	1.1	7.1	0	0	

[a] The composition may vary according to type of fat used in the manufacture of these foods

No	Food		Saturated								Mono-unsaturated			Polyunsaturated		Other
			4:0	6:0	8:0	10:0	12:0	14:0	16:0	18:0	14:1	16:1	18:1	18:2	18:3	
3107	**Ice cream** dairy	*	2.9	2.0	1.1	2.8	3.7	10.7	31.8	12.4	0.8	1.8	23.7	1.6	1.0	15:0 1.1 17:0 0.5 17:1 1.0
3123	**Milk, cows'**		3.2 *(2.6–3.9)*	2.0 *(1.5–2.3)*	1.2 *(0.9–1.4)*	2.8 *(2.5–3.2)*	3.5 *(3.1–4.0)*	11.2 *(10.4–12.4)*	26.0 *(24.1–32.0)*	11.2 *(9.2–13.2)*	1.4 *(1.1–1.6)*	2.7 *(2.1–3.1)*	27.8 *(22.0–30.7)*	1.4 *(0.8–1.9)*	1.5 *(0.6–2.5)*	15:0 1.1 15:1 0.7 17:0 1.0 17:1 1.1
3137	**Milk, goats'**		2.1	2.4	3.2	9.1	4.5	11.3	27.0	9.6	Tr	2.4	26.0	2.3	—	
3138	**Milk, human**	*	0	0	Tr	1.4 *(0.5–2.0)*	5.4 *(3.3–8.2)*	7.3 *(5.6–8.5)*	26.5 *(20.2–26.8)*	9.5 *(4.7–10.3)*	Tr	4.0 *(2.4–5.7)*	35.4 *(35.4–46.4)*	7.2 *(4.1–13.0)*	0.8 *(0.4–3.4)*	17:1 0.6 20:1 0.5
3165	**Eggs**		0	0	0	0	0	Tr	28.6	9.3	Tr	4.2	42.9	11.1	Tr	20:4 0.8 22:6 1.2

No	Food	Saturated						Mono-unsaturated				Polyunsaturated				
		12:0	14:0	16:0	18:0	20:0	22:0	16:1	18:1	20:1	22:1	18:2	18:3	20:4 20:5	22:5 22:6	Other
3183	**Compound cooking fat** [a]	0.4	6.2	22.2	8.5	2.6	2.2	6.5	19.5	8.0	8.3	5.0	0.5	6.3	3.0	
3184	**Dripping, beef**	Tr	3.2	26.9	13.0	Tr	0	6.3	42.0	0	0	2.0	1.3	1.0[b]	0	14:1 1.5 15:0 0.6 17:0 1.2 17:1 1.0
3185	**Lard**	Tr	1.6	26.8	15.6	Tr	0	2.5	40.7	0.8	0	8.7	0.8	Tr	0	17:0 0.2
6186	**Low fat spread** [a]	5.8	2.9	13.7	4.8	0.5	0.5	0.6	31.2	1.9	6.2	27.7	2.8	0.5	Tr	
	Margarine [a]															
3188	**hard** animal and vegetable oils	0.3	5.8	19.6	7.8	2.0	2.0	5.8	21.6	8.5	8.8	4.6	0.2	7.2	4.0	17:0 1.0
3189	vegetable oils only	0.1	1.2	28.0	7.4	1.0	0.5	1.2	42.6	1.6	3.5	9.9	0.5	1.2	1.0	17:0 0.3
3190	**soft** animal and vegetable oils	1.0	4.5	15.7	5.3	1.7	2.5	5.5	25.7	6.5	9.4	8.8	0.4	6.7	4.5	17:0 0.9
3191	vegetable oils only	1.5	1.4	23.7	5.2	0.8	0.5	1.4	36.9	1.4	3.8	21.1	2.0	Tr	Tr	
3192	**polyunsaturated** vegetable oils only	2.6	1.4	10.8	8.7	0.5	0.7	0.4	18.8	0.6	0.7	53.7	0.7	0.2	Tr	
3194	**Suet** shredded *	Tr	3.3	27.8	26.6	Tr	0	2.2	34.3	Tr	0	1.3	0	Tr	Tr	15:0 0.5 17:0 1.2 15:1 0.6 17:1 1.6

[a] As these products are made from a mixture of oils and fats, the composition may vary throughout the year depending on raw materials

[b] Only 20:4

No	Food	Saturated							Mono-unsaturated				Poly-unsaturated	
		12:0	14:0	16:0	18:0	20:0	22:0	24:0	16:1	18:1	20:1	22:1	18:2	18:3
	Vegetable oils													
3196	**coconut**[a]	47.7 *(44–51)*	15.8	9.0	2.4	1.0	0	0	0.4	6.6	0	0	1.8	0
3197	**cottonseed**	0.4	0.8	23.0 *(15–28)*	2.4	0.2	Tr	Tr	1.3	21.0 *(13–27)*	Tr	Tr	49.0 *(33–64)*	1.4
3198	**maize, corn**	0	0.6	14.0 *(6–22)*	2.3	0.3	Tr	Tr	0.3	30.0 *(19–50)*	0.2	0.2	50.0 *(34–62)*	1.6
3199	**olive**	0	Tr	12.0 *(7–20)*	2.3	0.4	0	0	1.0	72.0 *(65–85)*	0	0	11.0 *(4–20)*	0.7
3200	**palm**	0.2	1.1	41.5 *(32–51)*	4.3	0.3	0	0	0.3	43.3 *(35–52)*	0	0	8.4 *(5–12)*	0.3
3201	**peanut, groundnut, arachis**	0.1	0.5	10.7 *(6–15)*	2.7	1.2	3.4	1.1	Tr	49.0 *(35–72)*	1.1	Tr	29.0 *(13–45)*	0.8
3202	**rapeseed** high erucic acid	0	Tr	3.5	1.0	0.8	0.2	0.1	0.2	24.1 *(3–45)*	10.0 *(3–15)*	33.0 *(12–61)*	15.5 *(11–29)*	10.5 *(5–16)*
3203	low erucic acid	0	Tr	4.5	1.2	0.8	0.3	0.1	2.4	54.0 *(43–70)*	1.5	2.0 *(0–11)*	23.0 *(15–31)*	10.0 *(6–15)*
3204	**safflowerseed**	0	Tr	8.0	2.5	0.2	Tr	0	0.1	13.0 *(7–42)*	0.1	0	75.0 *(55–81)*	0.5
3205	**soyabean**	0.1	0.2	10.0 *(6–19)*	4.0	0.3	0.1	Tr	0.2	25.0 *(14–35)*	0.2	Tr	52.0 *(40–62)*	7.4 *(4–11)*
3206	**sunflowerseed**	0	0.1	5.8	6.3 *(0.2–12)*	0.6	0.7	0.2	0.1	33.0 *(5–60)*	0.2	Tr	52.0 *(17–78)*	0.3

[a] 8:0 7.5, 10:0 7.1

No	Food		Saturated					Mono-unsaturated				Polyunsaturated						Other	
			14:0	15:0	16:0	17:0	18:0	16:1	17:1	18:1	20:1	18:2	18:3	20:3	20:4	20:5	22:5		
3209	**Bacon**	*	1.6	Tr	27.4	Tr	14.3	3.5	Tr	43.8	0.6	7.2	0.6	0	Tr	Tr	0		
3240	**Beef**		3.2	0.6	26.9	1.2	13.0	6.3	1.0	42.0	Tr	2.0	1.3	Tr	1.0	Tr	Tr	14:1	1.5
3269	**Lamb**		5.4	0.6	24.2	1.0	20.9	1.3	1.0	38.2	Tr	2.5	2.5	0	0	Tr	Tr		
3299	**Pork** [a]		1.6	Tr	27.1	Tr	13.8	3.4	Tr	43.8	0.7	7.4	0.9	0	Tr	Tr	Tr		
3314	**Chicken** [a]	*	1.3	Tr	26.7	Tr	7.1	7.2	Tr	39.8	0.6	13.5	0.7	Tr	0.7		Tr	22:6	1.0
3328	**Duck**	*	0.6	Tr	22.8	Tr	5.5	4.4	Tr	52.8	Tr	12.1	0.6	Tr	Tr	Tr	Tr		
3332	**Grouse**		0.6	0.1	16.7	1.5	5.7	1.9	0.3	10.7	0.1	31.9	30.3	Tr	Tr	Tr	Tr		
3334	**Partridge**		0.9	0.1	21.0	1.8	3.8	7.6	1.8	39.8	0.1	15.5	9.6	Tr	Tr	Tr	Tr		
3336	**Pheasant**		1.0	0.1	28.1	0.1	5.9	11.4	0.1	40.4	0.1	6.1	6.7	Tr	Tr	Tr	Tr		
3340	**Turkey** [a]		1.0	Tr	25.0	0.5	10.0	5.0	Tr	21.5	0.4	20.0	1.0	Tr	5.0	1.5	2.0	22:6	5.0
3350	**Rabbit**	*	2.6	0.7	30.0	0.8	9.2	2.1	Tr	18.7	Tr	20.9	9.9	Tr	1.9	0	1.3		
	Offal																		
3356	**Brain, lamb**	*	0.8	0.5	21.5	0.5	18.1	1.4	1.9 [b]	28.4	3.3	0.4	0	1.5	4.2	0.7	3.4	22:1	0.6
																		22:4	0.8
																		22:6	9.5
3358	**Heart, lamb**	*	3.3	0.5	19.7	1.1	24.2	2.1	2.4	33.1	Tr	7.3	2.7	Tr	2.1		Tr		
3360	**ox**	*	3.3	0.8	27.4	1.3	29.2	1.9	1.8	29.2	Tr	2.5	0.5	Tr	0.7		Tr	15:1	0.6
3364	**Kidney, lamb**	*	2.3	Tr	19.9	1.0	22.2	2.1	1.3 [b]	28.2	Tr	8.1	4.0	0.5	7.1	Tr	Tr		
3366	**ox**	*	2.7	0.7	25.0	1.0	26.9	2.0	1.4 [b]	31.0	Tr	4.8	0.5	Tr	2.6	Tr	Tr	15:1	0.6
3368	**pig**	*	1.0	Tr	24.8	Tr	17.7	2.0	Tr	32.3	0.5	11.7	0.5	0.6	6.7	Tr	Tr		
3371	**Liver, calf**	*	0.8	Tr	16.5	0.6	23.3	1.9	0.7	20.8	Tr	15.0	1.4	2.1	9.0	0.3	4.0	22:6	2.5
3373	**chicken**	*	0.5	0	24.6	Tr	17.0	3.2	Tr	26.3	Tr	14.8	0.6	0.7	5.5	0	0.9	22:6	4.8
3375	**lamb**	*	1.3	0.5	20.4	1.0	18.3	3.5	1.6	29.7	0	5.0	3.8	0.6	5.1	0	3.0	15:1 [c]	0.5
																		22:6	2.4
3377	**ox**	*	0.9	0.8	17.2	0.6	30.1	1.4	1.2	18.2	0	7.4	2.5	4.6	6.4	0.7	5.6	15:1	0.8
																		22:6	1.2
3379	**pig**	*	Tr	Tr	17.7	0.5	23.4	1.7	Tr	19.2	Tr	14.7	0.5	1.3	14.3	0	2.3	22:6	3.8

[a] The fatty acid composition varies according to the diet fed to the animal

[b] Includes branch chain C17

[c] Also contains 3.2 g C18: unidentified polyunsaturated fatty acids

No	Food		Saturated					Mono-unsaturated				Polyunsaturated				Other
			14:0	15:0	16:0	17:0	18:0	16:1	17:1	18:1	20:1	18:2	18:3	20:4	20:5	
	Offal *contd*															
3384	**Sweetbread, lamb**	*	4.9	0.6	21.7	1.1	22.4	2.6	1.3	37.1	Tr	2.1	2.2	1.3	Tr	12:0 0.7
3391	**Tripe, ox**	*	2.8	0.9	23.4	1.3	30.5	2.2	1.5	33.3	0	1.5	0.6	Tr	Tr	15:1 1.0
	Canned meats															
3394	**Ham**	*	1.5	Tr	26.1	Tr	11.6	3.8	Tr	44.4	0.6	9.5	Tr	0.9		
3395	**Ham and pork** chopped	*	1.6	Tr	25.0	Tr	12.5	3.4	Tr	45.6	0.8	9.6	0.5	0.3		
3396	**Luncheon meat**	*	1.9	Tr	26.7	0.3	11.5	3.8	0.4	45.0	1.1	7.8	0.7	0.2		20:3 0.1 22:5 0.3
3399	**Tongue, lamb**	*	2.7	0.5	19.7	1.0	16.9	2.9	1.4	46.2	0	4.0	3.4	Tr		
	Sausages															
3404	**Liver sausage**	*	1.6	Tr	25.4	Tr	12.3	3.9	Tr	44.7	0.8	9.1	0.5	Tr		
3408	**Sausages**, beef	*	3.6	Tr	25.5	0.5	14.8	4.2	0.9	44.0	Tr	3.9	1.1	0		
3411	pork	*	1.9	Tr	25.8	Tr	13.6	3.6	Tr	45.4	0.5	7.7	0.6	0		
	Other products															
3425	**Steak and kidney pie** individual	*	3.1	Tr	25.7	Tr	14.0	3.3	Tr	37.3	1.3	10.0	0.9	Tr		20:0 0.6 22:0 0.5 22:1 1.1

From 3394 onward, one value is printed across the 20:4 and 20:5 columns for each food (bracketed together under both headings).

		Saturated					Mono-unsaturated				Polyunsaturated							
No	Food	14:0	15:0	16:0	17:0	18:0	16:1	18:1	20:1	22:1	18:2	18:3	18:4	20:4	20:5	22:5	22:6	Other
	White fish																	
3438	**Cod** raw	0.9	0.2	21.5	Tr	3.5	2.3	11.0	1.8	0.8	0.5	0.1	0.2	3.9	17.2	1.5	33.4	
3451	**Haddock** raw	1.5	0.5	20.0	1.5	6.1	4.0	14.2	2.6	0.1	2.2	0.4	0.5	3.3	12.0	2.4	24.5	22:4 1.8
3458	**Halibut** raw	2.8	0.3	10.3	0.8	2.4	8.1	17.5	6.1	Tr	1.6	3.5	1.9	8.5[a]	10.3	4.9	15.0	17:1 0.5
																		20:2 0.5
																		22:4 2.3
3461	**Lemon sole** raw	2.7	0.8	12.5	0.8	3.3	8.1	13.2	4.8	1.3	0.5	0.5	1.7	30.3		3.4	11.2	17:1 1.8
3466	**Plaice** raw	3.5	0.5	16.4	0.4	2.6	14.7	17.9	3.7	2.2	0.7	0.7	2.1	16.3		3.6	10.3	16:2 1.1
																		17:1 1.0
																		20:3 0.7
3471	**Saithe** raw	1.4	0.2	9.9	0.2	5.3	3.7	20.5	5.1	6.2	1.1	0.6	0.7	13.9		1.8	27.1	17:1 1.6
3474	**Whiting** raw	2.7	0.3	13.3	0.2	3.9	5.2	20.8	7.9	6.3	1.5	0.5	1.8	12.3		2.1	18.0	17:1 0.8
	Fatty fish																	
3482	**Herring** raw	6.7	0.5	13.7	0.2	1.2	10.0	15.2	13.2	17.4	1.4	1.2	1.8	0.6	7.0	1.1	6.5	
3491	**Mackerel** raw	5.0	0.5	17.6	0.4	3.5	5.9	18.4	7.1	9.9	1.6	1.1	1.9	1.0	7.3	1.6	12.6	16:2 0.7
																		24:1 1.5
3494	**Pilchards** canned in tomato sauce *	8.6	Tr	20.1	Tr	6.4	11.3	12.4	1.1	0.5	1.2	0.6	1.5	0.6	21.5	2.5	4.1	16:2 5.2
																		20:3 0.8
3498	**Salmon** canned *	4.8	Tr	18.8	Tr	3.9	6.0	22.8	8.4	5.4	1.4	0.8	1.7	0.5	8.2	2.7	11.0	16:2 0.6
																		20:3 1.4
3501	**Sardines** canned in oil (fish plus oil) *	3.1	Tr	14.0	Tr	3.5	4.3	49.7	1.4	2.0	7.4	1.0	1.1	0	5.5	0.7	4.3	16:2 0.8
3502	canned in tomato sauce *	5.2	Tr	21.1	Tr	5.0	6.2	22.8	1.4	0.5	6.6	1.7	2.5	0	9.5	1.3	12.3	16:2 1.1
																		20:3 0.8
3503	**Sprats** raw	7.9	0.6	20.8	0.3	2.0	5.8	18.2	8.9	13.9	1.4	1.1	1.9	0.7	6.0	0.7	7.7	24:1 2.1
3508	**Tuna** canned in oil *	Tr	0	17.1	Tr	1.7	Tr	41.0	Tr	0	36.6	0.9	0	Tr	Tr	0	0.8	

[a] Includes 20:3

No	Food	Saturated					Mono-unsaturated				Polyunsaturated							Other
		14:0	15:0	16:0	17:0	18:0	16:1	18:1	20:1	22:1	18:2	18:3	18:4	20:4	20:5	22:5	22:6	
	Cartilaginous fish																	
3510	**Dogfish** raw	1.7	0.5	15.4	0.3	2.2	6.3	20.1	6.7	4.9	1.7	1.1	1.4	11.8 (20:4 + 20:5)		2.9	20.3	17:1 1.1 20:3 1.2
3513	**Skate** raw	1.9	0.3	11.1	0.3	2.7	6.0	13.2	3.6	2.8	1.2	0.5	1.6	13.6 (20:4 + 20:5)		4.2	34.2	17:1 0.8 20:3 0.8
	Crustacea																	
3517	**Crab** raw	1.4	0.7	9.2	1.2	4.3	5.0	15.0	3.5	3.9	3.2	4.6	2.3	0.6	21.5	1.4	10.2	16:2 2.9 20:2 2.4 22:3 1.6 22:4 1.1 24:6 1.0
3526	**Shrimps** raw	2.5	0.6	15.8	0.5	2.4	5.9	18.8	2.6	1.7	1.4	1.3	1.1	1.3	21.6	1.2	15.4	17:1 1.1 22:3 1.6 22:4 1.1
	Molluscs [a]																	
3532	**Mussels** raw	2.5	0.8	12.2	1.3	7.4	6.6	7.1	7.5	5.2	2.0	1.6	5.5	4.5	11.3	2.0	4.6	16:2 1.8 16:4 1.0 19:3 1.3 20:2 1.2 22:2 1.5 22:3 1.0 24:0 1.1

[a] These fish also contain small quantities (less than 1 per cent) of other branch chain and polyunsaturated fatty acids which are not listed here

No	Food																	Other
	***Molluscs** contd*																	
3535	**Oysters** raw	4.6	1.1	20.2	1.6	4.0	3.8	5.8	2.8	3.0	1.9	1.5	4.5	1.6	15.2	1.8	17.2	22:4 1.8
3537	**Scallops** raw	4.3	1.2	17.6	0.8	4.1	3.9	5.1	3.1	2.0	0.8	1.5	3.5	3.8	11.6	0.5	8.2	12:0 5.4 13:0 3.0 16:2 1.9 22:2 1.4 22:3 2.7
	Fish products																	
3550	**Roe, cod** hard, raw	1.3	0.2	21.8	—	1.8	6.4	19.4	1.9	0.7	0.5	0.1	0.2	3.1	16.8	1.4	23.3	

Vegetables and fruit

Fatty acids g per 100 g total fatty acids

		Saturated						Mono-unsaturated			Polyunsaturated			
No	Food	12:0	14:0	16:0	17:0	18:0	20:0	16:1	18:1	20:1	18:2	18:3	20:4	Other
	Vegetables													
3562	**Beans, runner**	0	0	21.8	Tr	4.0	0	0.5	3.6	0	28.2	40.0	0	
3569	**baked** canned in tomato sauce	0	0	19.3	Tr	2.0	0	0	12.1	0	24.9	38.1	0	
3597	**Cucumber**	0.2	0.8	36.8	Tr	3.7	0	0.2	3.2	0	29.0	26.3	0	
3609	**Mushrooms**	0	0.4	27.4	Tr	1.9	0	0.3	1.3	0	13.2	55.4	0	
3620	**Peas** fresh	0	2.4	29.7	Tr	14.6	0	2.6	36.3	0.5	10.9	3.1	0	
3634	**Peppers, green**	0.1	2.9	15.4	Tr	4.2	0.1	0.5	7.3	0	56.3	12.0	0	
3639	**Potatoes**	0	0	17.8	Tr	5.4	0	1.3	1.9	0	56.5	17.2	0	
3657	**Spinach**	0	1.0	12.1	Tr	—	0	1.9	7.3	0	12.5	62.1	0	
3664	**Sweet potatoes**	5.5	0.7	35.0	1.5	3.4	1.2	1.5	6.7	0	34.3	6.9	0	10:0 1.5
3669	**Turnips**	0	0	14.1	Tr	1.4	0	0.8	8.0	0	16.4	57.8	0	17:1 1.7
	Fruit													
3675	**Apples**	0	0	24.5	Tr	3.8	0	0	6.9	0	54.0	10.8	0	
3692	**Avocado pears**	0	Tr	12.3 *(7–22)*	Tr	Tr	Tr	3.5 *(3–11)*	75.1 *(59–81)*	0	8.6 *(7–14)*	0.4	0.1	
3693	**Bananas**	0	Tr	41.9	Tr	3.8	Tr	2.1	14.4	0	16.1	21.7	0	

No	Food		Saturated						Mono-unsaturated			Polyunsaturated		
			10:0	12:0	14:0	16:0	18:0	20:0	16:1	18:1	20:1	18:2	18:3	Other
	Nuts													
3822	**Almonds**		0	0	0.1	6.3	1.7	0.2	0.7	70.9	Tr	19.1	0.5	
3826	**Brazil nuts**		0	0	0.2	15.6	10.9	0	0.4	33.9	Tr	39.0	0	
3828	**Chestnuts**		0	0	0	16.7	1.0	0.5	0.7	38.5	Tr	37.7	4.2	
3830	**Cob** or **hazel nuts**		0	0	0.3	5.2	1.8	0.2	0.3	80.7	0.1	10.7	0.2	
3832	**Coconut**		7.1	47.7	15.8	9.0	2.4	1.0	0.4	6.6	0	1.8	0	6:0 0.7 8:0 7.5
3835	**Peanuts**		Tr	0.1	0.5	10.7	2.7	1.2	Tr	49.0	1.1	29.0	0.8	22:0 3.4 24:0 1.1
3838	**Peanut butter**	*	0	0	0	11.3	5.5	1.4	Tr	51.6	1.4	26.2	0	22:0 2.5
3839	**Walnuts**		0	0	1.1	7.5	2.1	0.7	0.2	16.1	Tr	60.0	11.4	
	Chocolate products													
3857	**Chocolate**, milk	*	0.6	0.9	3.4	28.2	26.8	0.6	0.6	33.1	Tr	2.9	0.8	4:0 0.5 6:0 0.4
3858	plain	*	Tr	0.5	1.6	26.8	32.9	0.5	Tr	33.6	0	3.3	0	
3868	**Cocoa powder**	*	0	Tr	Tr	26.2	34.5	0.9	Tr	34.8	0	3.0	0	
3873	**Drinking chocolate**	*	0	Tr	0.5	26.6	33.5	0.8	Tr	34.7	0	3.2	0	

The tables

continued

Section 4
Subsidiary data:
cholesterol,
phytic acid phosphorus,
iodine,
organic acids

Cholesterol

A number of sterols (Kritchevsky, 1963; Lange, 1950) are found in foods either in the free form or esterified with fatty acids.

Although a range of sterols (the phytosterols) can be isolated from individual plants, most animal tissues and products contain virtually only one—cholesterol. Cholesterol is therefore the characteristic sterol of the animal kingdom and does not appear to occur in plants. Molluscs appear to be an exception in this matter and a number of different sterols have been found in the species examined.

A number of phytosterols are known and of these the most common is β-sitosterol; α-spinasterol occurs in spinach and stigmasterol is also widely distributed. Traces of other sterols are found in many plant oils. The phytosterols do not appear to be absorbed by the mammalian digestive tract and can often be recovered unchanged in faeces. The values in this section are therefore restricted to animal products. Lange (1950) gives some values for the phytosterol content of a range of plant foods.

Sources of values

The figures given are derived from two sources, direct analyses of samples collected for this edition and selected values derived from the literature (see Appendix 6). The values are given as mg cholesterol per 100 g and can be converted to mmol cholesterol by dividing by 386.6.

Asterisk indicates new analytical data

Table 4.1
Cholesterol per 100 g

No	Food	Cholesterol mg per 100 g	No	Food	Cholesterol mg per 100 g
	Cereal products				
	Cakes		4089	**Pastry, choux** raw	110
4074	**Fancy iced cakes**	—	4090	cooked	170
4075	**Fruit cake** rich	50	4091	**flaky** raw	—[a]
4076	rich, iced	40	4092	cooked	—[a]
4077	plain	—	4093	**shortcrust** raw	—[a]
4078	**Gingerbread**	60	4094	cooked	—[a]
4079	**Madeira cake**	—	4095	**Scones**	(5)
4080	**Rock cakes**	40	4096	**Scotch pancakes**	50
4081	**Sponge cake** with fat	130		***Puddings***	
4082	without fat	260	4097	**Apple crumble**	—[a]
4083	jam filled	—	4098	**Bread and butter pudding**	100
	Buns and Pastries				
4084	**Currant buns**	—	4099	**Cheesecake**	95
4085	**Doughnuts**	—	4100	**Christmas pudding**	60
4086	**Eclairs**	90	4101	**Custard** egg	100
4087	**Jam tarts**	—[a]	4102	made with powder	16
4088	**Mince pies**	—[a]	4103	**Custard tart**	60

No	Food		Cholesterol mg per 100g	No	Food		Cholesterol mg per 100g
4104	**Dumpling**		8	4113	**Meringues**		0
4105	**Fruit pie** individual, with pastry top and bottom		—	4114	**Milk pudding**		15
4106	**Fruit pie** with pastry top		—[a]	4115	canned, rice		—
4107	**Ice cream** dairy	*	21	4116	**Pancakes**		65
4108	non-dairy	*	11	4117	**Queen of puddings**		100
4109	**Jelly** packet, cubes		0	4118	**Sponge pudding** steamed		80
4110	made with water		0	4119	**Suet pudding** steamed		4
4111	made with milk		6	4120	**Treacle tart**		—[a]
4112	**Lemon meringue pie**		90	4121	**Trifle**		50
				4122	**Yorkshire pudding**		70

[a] Contains only a trace if made with vegetable fats

Milk, milk products and eggs

No	Food		Cholesterol mg per 100g
	Milk		
4123	**cows'** fresh, whole		14
4126	fresh, whole, Channel Islands		18
4129	sterilized		14
4130	longlife		14
4131	fresh, skimmed		2
4132	condensed, whole, sweetened		34
4133	condensed, skimmed, sweetened		3
4134	evaporated, whole, unsweetened		34
4135	dried, whole	*	120
4136	dried, skimmed		18
4137	**Milk, goats'**		—
4138	**Milk, human** mature	*	16
4139	transitional		—
4140	**Butter** salted		230
4141	**Cream** single		66
4144	double		140
4147	whipping		100
4150	sterilized, canned		73
	Cheese		
4151	Camembert type		72
4152	Cheddar type	*	70
4153	Danish Blue type		88
4154	Edam type		72
4155	Parmesan		90
4156	Stilton		120
4157	cottage cheese	*	13
4158	cream cheese	*	94
4159	processed cheese		88
4160	cheese spread		71
	Yogurt low fat		
4161	natural	*	7
4162	flavoured	*	7
4163	fruit	*	6
4164	hazelnut	*	7
	Eggs		
4165	whole, raw	*	450
4166	white, raw		0
4167	yolk, raw		1260
4168	dried		1780
4169	boiled		450
4170	fried		—[a]
4171	poached		480
4172	omelette		410
4173	scrambled		410
	Egg and cheese dishes		
4174	**Cauliflower cheese**		17
4175	**Cheese pudding**		130
4176	**Cheese soufflé**		180
4177	**Macaroni cheese**		20
4178	**Pizza, cheese and tomato**		20
4179	**Quiche Lorraine**		130
4180	**Scotch egg**		220
4181	**Welsh rarebit**		67

[a] The cholesterol content will depend on the fat used for frying

Fats and oils

No	Food		Cholesterol mg per 100g	No	Food		Cholesterol mg per 100g
4183	**Compound cooking fat**		—[a]	4187	**Margarine**		—[a]
4184	**Dripping, beef**		(60)	4193	**Suet** block		(60)
4185	**Lard**		(70)	4194	shredded	*	74
4186	**Low fat spread**		Tr	4195	**Vegetable oils**		Tr

[a] The cholesterol content will depend on the blend of oils used. If made with vegetable oils only, these foods will contain only a trace of cholesterol

Meat

No	Food		Cholesterol mg per 100g	No	Food		Cholesterol mg per 100g
4209	**Bacon** raw, lean and fat	*	57		*Offal*		
4210	lean only	*	51				
4224	fried, lean and fat	*	80	4354	**Brain, calf** and **lamb**		
4225	lean only	*	87		raw	*	2200
4229	grilled, lean and fat	*	74	4355	**calf** boiled	*	3100
4230	lean only	*	77	4356	**lamb** boiled	*	2200
4236	**Beef** raw, lean and fat	*	65	4358	**Heart, lamb** raw	*	140
4237	lean only	*	59	4359	**sheep** roast		260
4238	cooked, lean and fat	*	82	4360	**ox** raw	*	140
4239	lean only	*	82	4361	stewed	*	230
4265	**Lamb** raw, lean and fat	*	78	4364	**Kidney, lamb** raw	*	400
4266	lean only	*	79	4365	fried	*	610
4267	cooked, lean and fat	*	110	4366	**ox** raw	*	400
4268	lean only	*	110	4367	stewed	*	690
4295	**Pork** raw, lean and fat	*	72	4368	**pig** raw	*	410
4296	lean only	*	69	4369	stewed	*	700
4297	cooked, lean and fat	*	110	4371	**Liver, calf** raw	*	370
4298	lean only	*	110	4372	fried	*	330
4316	**Chicken** raw, light meat	*	69	4373	**chicken** raw	*	380
4317	dark meat	*	110	4374	fried	*	350
4319	boiled, light meat	*	80	4375	**lamb** raw	*	430
4320	dark meat	*	110	4376	fried	*	400
4323	roast, light meat	*	74	4377	**ox** raw	*	270
4324	dark meat	*	120	4378	stewed	*	240
4327	**Duck** raw, meat only	*	110	4379	**pig** raw	*	260
4329	roast, meat only	*	160	4380	stewed	*	290
4342	**Turkey** raw, light meat	*	49	4381	**Oxtail** raw	*	75
4343	dark meat	*	81	4382	stewed	*	110
4346	roast, light meat	*	62	4384	**Sweetbread, lamb**		
4347	dark meat	*	100		raw	*	260
4350	**Rabbit** raw	*	71	4385	fried	*	380

No	Food		Cholesterol mg per 100g
	***Offal** contd*		
4387	**Tongue, lamb** raw	*	180
4388	**sheep** stewed		(270)
4389	**ox** pickled, raw	*	78
4390	boiled		(100)
4391	**Tripe**, dressed	*	95
4392	stewed	*	160
	Meat products and dishes		
	Canned meats		
4393	**Beef, corned**	*	85
4394	**Ham**	*	33
4395	**Ham and pork** chopped	*	60
4396	**Luncheon meat**	*	53
4397	**Stewed steak with gravy**	*	44
4398	**Tongue**	*	110
4400	**Veal,** jellied	*	97
	Offal products		
4401	**Black pudding** fried	*	68
4402	**Faggots**	*	79
4403	**Haggis** boiled	*	91
4404	**Liver sausage**	*	120
	Sausages		
4405	**Frankfurters**	*	46
4406	**Polony**	*	40
4407	**Salami**	*	79
4408	**Sausages, beef** raw	*	40
4409	fried	*	42
4410	grilled	*	42
4411	**pork** raw	*	47
4412	fried	*	53
4413	grilled	*	53
4414	**Saveloy**	*	45
	Other products		
4415	**Beefburgers** frozen, raw	*	59
4416	fried	*	68
4417	**Brawn**	*	52
4418	**Meat paste**	*	68
4419	**White pudding**	*	22
	Meat and pastry products		
4420	**Cornish pastie**	*	49
4421	**Pork pie** individual	*	52
4422	**Sausage roll** flaky pastry		20
4423	short pastry		30
4424	**Steak and kidney pie** pastry top only		125
	Cooked dishes		
4426	**Beef steak pudding**		30
4427	**Beef stew**		30
4428	**Bolognese sauce**		25
4429	**Curried meat**		25
4430	**Hot pot**		25
4431	**Irish stew**		35
4433	**Moussaka**		40
4434	**Shepherd's pie**		25

Fish

No	Food	Cholesterol mg per 100g
	White fish	
4438	**Cod** fresh, raw	50
4440	baked	60
4442	fried	—
4443	grilled	60
4444	poached	60
4446	steamed	60
4448	**Cod, smoked** raw	(50)
4449	poached	(60)
4451	**Haddock, fresh** raw	60
4452	fried	—
4454	steamed	75
4456	**smoked** steamed	(75)
4458	**Halibut** raw	50
4459	steamed	60
4461	**Lemon sole** raw	60
4462	fried	—
4464	steamed	60
4466	**Plaice** raw	70
4467	fried in batter	—
4468	fried in crumbs	—
4469	steamed	90
4471	**Saithe** raw	60
4472	steamed	75
4477	**Whiting** steamed	110

No	Food	Cholesterol mg per 100g
	Fatty fish	
4480	**Eel** raw	—
4481	stewed	—
4482	**Herring** raw	70
4483	fried	(80)
4485	grilled	80
4487	**Bloater** grilled	80
4489	**Kipper** baked	80
4491	**Mackerel** raw	80
4492	fried	(90)
4494	**Pilchards** canned in tomato sauce	(70)
4495	**Salmon** raw	(70)
4496	steamed	(80)
4498	canned	90
4499	smoked	(70)
4500	**Sardines** canned in oil, fish only	100
4501	fish plus oil	80
4502	canned in tomato sauce	(100)
4504	**Sprats** fried	—
4506	**Trout** steamed	(80)
4508	**Tuna** canned in oil	65
	Cartilaginous fish	
4510	**Dogfish**	—
4513	**Skate**	—
	Crustacea	
4517	**Crab** fresh	100
4520	canned	100
4521	**Lobster**	150
4523	**Prawns**	200
4525	**Scampi**	110
4526	**Shrimps**	200
	Molluscs [a]	
4531	**Cockles**	40
4532	**Mussels** raw	100
4535	**Oysters** raw	50
4537	**Scallops** raw	40
4539	**Whelks**	100
4541	**Winkles**	100
	Fish products and dishes	
4543	**Fish cakes** frozen	—
4544	fried	—
4545	**Fish fingers** frozen	(50)
4546	fried	(50)
4547	**Fish paste**	—
4548	**Fish pie**	20
4549	**Kedgeree**	120
4550	**Roe, cod** hard, raw	(500)
4551	fried	(500)
4552	**herring** soft, raw	700
4553	fried	(700)

Products containing eggs

No	Food	Cholesterol mg per 100g
4852	**Lemon curd** home made	150
4854	**Marzipan** home made	35
4926	**Mayonnaise** home made	260

[a] Other sterols are present in these fish, cholesterol forms about 40 per cent of the total sterol in cockles, mussels, oysters and scallops. It forms about 90 per cent of the total sterol in whelks and about 70 per cent in winkles.

Phytic acid phosphorus

Table 4.2*
Phytic acid phosphorus

	Phytic acid phosphorus as per cent of total phosphorus
Cereals and cereal foods	
All-Bran Kellogg's	76
Barley, pearl	66
Biscuits, digestive	61
Bread brown (92%)	55
national wheatmeal (85%)	30
white	15
Hovis	38
Cornflakes	25
Flour English or Manitoba, 100% extraction	70
85% extraction	55
80% extraction	47
white	30
Oatmeal raw	70
Rice polished	61
Rye 100% extraction	72
85% extraction	54
75% extraction	44
60% extraction	31
Ryvita	54
Sago	Tr
Shredded Wheat	80
Soya full fat or low fat flour or grits	31
Tapioca	0
Vita-Wheat	59
Fruit	
Apples	0
Bananas	0
Blackberries	16
Figs dried	13
Prunes dried	0

	Phytic acid phosphorus as per cent of total phosphorus
Nuts	
Almonds	82
Barcelona nuts	83
Brazil nuts	86
Chestnuts	18
Cob nuts	74
Coconuts	81
Peanuts	57
Walnuts	42
Vegetables	
Artichokes, Jerusalem boiled	25
Beans, broad boiled	5
butter, raw	84
haricot, raw	73
Carrots raw	16
Cauliflower boiled	0
Celery raw	0
Lentils raw	51
Mushrooms raw	0
Onions raw	0
Parsnips raw	31
Peas fresh, raw	11
dried, raw	80
split, raw	57
canned	17
Potatoes old, boiled	19
new, boiled	23
Spinach boiled	0
Swedes raw	0
Turnips raw	0
Cocoa and chocolate	
Chocolate, milk	18
Cocoa	15

*Reprinted from the third edition of these tables. For further discussion of phytic acid in foods see McCance and Widdowson 1935 and 1942.

Iodine

Occurrence

Iodine occurs in foods almost entirely as the inorganic iodide. Seafoods are by far the richest sources. Some green vegetables, notably spinach, lettuce and watercress, contain moderate amounts; most other foods contain only small amounts. The occurrence of iodine in foods is extremely variable, depending on the locality where it is produced. The small quantities that are present led to difficulties in earlier methods of analysis, and methods used after about 1932 are held to be more reliable. A commonly used method is that employing the colour reaction of ceric sulphate and arsenious acid (Broadhead, Pearson and Wilson, 1965; Rodgers and Poole 1958); alternative methods using neutron activation analysis have also been described (Bowen 1959; Johansen and Steinnes, 1976).

Amounts in foods

The most comprehensive study of the iodine content of foods is the annotated bibliography of the Chilean Iodine Educational Bureau (1952). Since the 1950's there have been relatively few reports on the iodine content of foods (Vought and London 1964). The Chilean Iodine Educational Bureau closed down in 1971, and there has been no study on foods by the Bureau since the 1952 publication. Some of the more important studies on British and Irish foods are discussed below.

Milk

Iodine crosses the mammary barrier very readily; consequently the iodine in milk closely reflects the iodine in the cow's diet. Iodine is commonly added to the winter feed of cows, and winter milk has a higher iodine content than does summer milk (Alderman and Stranks, 1967; Broadhead *et al.*, 1965).

Fish

The iodine content of fish has been reported by Broadhead *et al.* (1965) and Wayne, Koutras and Alexander (1964). The losses of iodine during the cooking of fish were found to be about 20 per cent in grilling and frying and 60 per cent in boiling (Harrison, McFarlane, Harden and Wayne, 1965).

Other foods

There is one recent report in Great Britain on the iodine content of *eggs* (Wayne *et al.*, 1964). There are very few studies on *meat*, the most useful one being for a number of cooked meats and meat products in Ireland (Mason, O'Donovan and Kilbride, 1945). *Cereals* also have not been studied in this country to any great extent. The iodine content of *vegetables* has been reported in more detail because of their content of goitrogens, particularly in the *Brassica* genus. There are no recent British studies, but values for cabbages, carrots, lettuce, onions, potato and watercress were reported in detail by Orr (1931), and some cooked vegetables by Mason *et al.* (1945).

Goitrogens

There are a number of substances present in plants of the *Brassica* and other Cruciferae and some Leguminosae genera which interfere with the uptake and utilisation of iodine. The goitrogens also pass into milk. The occurrence and mode of action of a number of goitrogenic substances have been described by Gontzea and Sutzescu (1968).

Organic acids

Organic acids are found in many foods and, while for the most part they are minor constituents, in some foods the concentrations of individual acids may be nutritionally significant. The major organic acids most frequently found in foods are given in table 4.3. This list is not exclusive and many hydroxy and other organic acids are found in some foods, usually in low concentrations.

Occurrence

The organic acids are quantitatively more important in fruits and fruit products and in the majority of vegetables the concentrations are quite low. Many manufactured products contain added citric acid and acetic acid is widely used as a preservative in pickling. Lactic acid is found in fermented products such as yogurt and sauerkraut and at low concentrations in cheese.

Citric and *malic* acids are the major organic acids in most fruits and vegetables and the many other organic acids are residual components of the various oxidative and other metabolic pathways in the plant. The concentrations of all these organic acids depend very greatly on the state of maturity of the food and the conditions under which it has been grown. One factor which may be of special importance is the level of illumination the plant has received. The concentrations of organic acids are furthermore dependent on the post-harvest metabolism of the plant and in fruits the changes during the interval between harvest and consumption are often very large (Hulme, 1971).

Tartaric acid is the major organic acid in wines and in the fruit of the tamarind and the beobab tree.

Oxalic acid, often as the calcium salt, is found in many vegetables and some fruits and inclusions of oxalate crystals can frequently be seen in sections of plant tissue.

Absorption and metabolism

The organic acids are usually well absorbed, although oxalic acid may be an exception as it forms insoluble salts with calcium at the pH of the intestinal contents and some oxalate is excreted in this way. The toxicity of oxalate shows, however, that some absorption does occur.

Many of the organic acids in foods are intermediates in metabolic processes common to most living cells and it is reasonable to assume that they are completely metabolised within the body. The extent to which tartaric acid is metabolised is uncertain but it is believed that digestion mainly occurs as a result of fermentation in the large intestine by the microflora and that any absorbed tartrate is excreted unchanged.

Energy value

The organic acids that are normal constituents of metabolic pathways will provide energy to the body and should therefore not be ignored in the calculation of the energy value of foods. The energy conversion factors suggested in table 4.3 are the heats of combustion of the acids as it is reasonable to expect these acids to be completely absorbed and metabolised to carbon dioxide and water.

Amounts in foods

Few detailed studies of the concentrations of organic acids in foods in themselves have been reported, although there is a large body of work on the concentration and metabolism of organic acids in fruits.

Vegetables

Before the development of chromatographic methods, separation of the individual acids was difficult and most estimates were based on total titratable acidity. This procedure may underestimate total organic anions but the errors involved are thought to be small for most foods. Using this technique, Hartmann and Hillig (1934) showed that the total organic acids in vegetables ranged between 0.1 and 0.8 g/100 g, and in most of those examined the concentration did not exceed 0.5 g/100 g. For practical nutritional purposes the organic acids in vegetables can therefore be discounted in energy calculations.

Fruits

In fruits the situation is more complex; Merrill and Watt (1955) grouped the fruits into five classes depending on the range of reported values for total organic acids present. A large number of fruits contained less than 1 g/100 g and only in lemons and limes did the concentrations exceed 3 g/100 g. Actual values reported in the literature are very variable and depend on variety, state of maturity and especially on the extent of post-harvest metabolism. It is therefore very difficult to give meaningful typical values and the method adopted by Merrill and Watt is probably the best practical guide.

This shows that lemons and limes have more than 3 g/100 g; cranberries, currants (black, red or white) and gooseberries between 2 and 3 g/100 g; and apricots, grapefruit, loganberries, nectarines, oranges, plums, pomegranates, raspberries, strawberries and tangerines between 1 and 2 g/100 g.

Beverages

Citric acid is added to many types of beverage, especially those with a fruit character. In most fruit drinks the concentrations are less than 1 g/100 ml in the ready-to-consume drink, but lemon juices may contain up to 6 g/100 ml of mainly citric acid (≡14.8 kcal). Beverages designated as 'low calorie' may have only minor amounts of energy from organic acids (usually less than 1 kcal/100 ml) but some have citric acid concentrations which would provide up to 7 kcal/100 ml.

Vinegar

Vinegar contains about 5 g acetic acid/100 ml and this will provide 17 kcal/100 ml.

Conclusions

For most nutritional purposes the concentrations of organic acids in foods are of academic interest. Oxalate concentrations are of course of great significance in the toxicological properties of foods. Organic acids can, in certain fruits and beverages, make a significant contribution to the energy value and it is hoped that as more values become available it will be possible to include this contribution in the total calculated energy contents given in section 1. For this edition it was considered that the data were too fragmentary to adopt this approach and the contributions from organic acids have not been included.

Table 4.3
Major organic acids in foods[a]

Acid	Energy value[b] (per g) kcal	kJ	Occurrence
Acetic acid	3.49	14.6	Fermented products, vinegar, pickles etc
Citric acid	2.47	10.3	Most fruits and fruit products; low concentrations in vegetables, many processed foods and drinks
Lactic acid	3.62	15.1	Fermented products, yogurts, sauerkraut
Malic acid	2.39	10.0	Many fruits and fruit products—major acid in many fruits, low concentrations in vegetables
Oxalic acid	[c]	[c]	Many vegetables and some fruits
Tartaric acid	[c]	[c]	Some fruits, major acid in wines and a few fruits

[a] Many other hydroxy and other organic acids occur in foods usually at low concentrations (less than 0.1 g/100 g)

[b] Based on heats of combustion

[c] These acids are probably not metabolised to any significant extent to provide energy

Appendix I: Details of analytical procedures

The analytical methods used to obtain the values presented in the tables are described in this appendix. The aim of this is to enable the user to see the principles behind the methods and, by having procedural details or references to the literature, to set up and use similar methods.

Analytical methods for foods, despite many contrary impressions, are in a continuous state of development, modification and appraisal. In most the principle of the method used is more important than the precise technical procedure. Instrumental methods have transformed many determinations that were difficult or time-consuming into simple rapid procedures, and in a few cases have made determinations that were impracticable or virtually impossible to perform on more than the occasional sample routine procedures. Although instrumental analysis is rapidly becoming the rule it is incorrect to assume that these modern procedures are, in themselves, more accurate or precise than the methods they have replaced. In most cases they enable more samples to be analysed in a given time or make fewer demands on the manipulative skills of the analyst. In the course of this revision many comparisons between results obtained by new and older methods have been made. These have shown that in the majority of cases the skill and care of the analysts in the past has compensated for the technical difficulties associated with the methods they had to use.

Water

The values for water content in the earlier editions were obtained by drying; foods rich in sugars, such as fruits and vegetables, were dried at 50°C and other foods at around 100°C. This method measures the loss of all volatile constituents and could not be used where alcohol, acetic acid or volatile oils were present. Accordingly values for total solids were given for the alcoholic beverages and omitted for condiments with volatile oils. In the analyses for the third edition samples rich in sugar either were dried in a vacuum oven at 60°C or else had water measured by distillation with benzene in a Dean and Stark type apparatus (AOAC 1955). The two procedures gave similar results.

Methods used for this edition

The water content was calculated in most foods from the loss in mass on freeze-drying, and the moisture content of the freeze-dried solids determined by drying in an oven at 100°C until constant in mass. Samples rich in sugar and some beverages were dried in a vacuum oven at 70°C. Other samples that were not freeze-dried, such as Marmite and Bovril, were mixed with sand and ethanol, pre-dried on a water bath and heated to constant mass in an oven at 100°C.

Total nitrogen

In all the previous editions total nitrogen was measured by a micro-Kjeldahl method. There were some differences in the 'catalyst' mixtures used, which tended to follow the fashion of the day. Copper sulphate was used by McCance and Shipp (1933) for the meats and fish, and by McCance, Widdowson and Shackleton (1936) for the fruits, vegetables and nuts. In the analyses for the third edition the Chibnall, Rees and Williams (1943) catalyst was used (potassium sulphate/copper sulphate/sodium selenite).

The preparation of the sample for digestion differed according to the food. The measurements on the fruits, vegetables and nuts were made on small subsamples weighed out directly; for the other foods a large sample (10–20 g) was refluxed in dilute sulphuric acid to produce a homogeneous suspension from which aliquots were taken for digestion.

Methods used for this edition

Nitrogen was determined by a Kjeldahl procedure similar to the methods given in British Standards for meat and meat products (BS 4401:1969), milk (BS 1741:1963) and cheese (BS 770:1963).

Fat

Two types of procedure were used for the first and second editions. One was the conventional Soxhlet extraction of a dried sample of the food with petroleum spirit, and the other was a procedure devised by von Lieberman and Szekely (1898). In this the food was saponified in strong alkali and the fatty acids subsequently released by acidification. The fatty acids were extracted into petroleum spirit and the total fatty acids measured by titration. McCance and Widdowson (1940) found that the Soxhlet method under-estimated fat in many foods and preferred the alkaline extraction method because they felt that it measured triglycerides more nearly, in addition to giving a more complete extraction.

In the analyses for the third edition an alkaline extraction method was found to be suitable for many animal products, but with other foods an acid hydrolysis before extraction gave the most complete extraction of fat. The method used was a modification of the AOAC (1955) procedure for cereals followed by saponification of the total lipid and measurement of the fatty acids. In both these approaches it was found technically easier to measure the total fatty acids gravimetrically.

Methods used for this edition

Several types of procedure were used for the determination of fat. For meat and meat products the Werner Schmid method, which involved acid hydrolysis followed by extraction of the fat with diethyl ether, was found to be suitable (BS 4401, Pt 4, 1970 Method B). Another acid hydrolysis procedure, that of Weibull Stoldt, was found to be preferable to the Werner Schmid for certain products, such as some cakes, canned and dried soups and beverages (BS 4401, Pt 4, 1970 Method A). Free fat in the fat separated from meat and bacon was determined by Soxhlet extraction with light petroleum (BS 4401, Pt 5 1970). For milk and some milk products a Rose–Gottlieb procedure, now adopted as an international standard method, was used (BS 1743:1968), and for cheese Method 16.230 of the Official Methods of Analysis of the Association of Official Analytical Chemists (12th edition, 1975) was employed.

AOAC methods were also found to be applicable to fish (Method 18.040) and biscuits and breakfast cereals (14.019). The rapid method of Southgate (1971), modified by the use of chloroform instead of petroleum spirit for the final fat extraction, was used for liver and gave values higher than those obtained with the Werner Schmid method, similarly modified. However, for other offal the two methods gave similar values.

Carbohydrate

Available carbohydrate

The values in the first and second editions were obtained using methods based on those described by Widdowson and McCance (1935). These involved extraction of free sugars with aqueous ethanol in a Soxhlet extractor and measurement of the free sugars. Two types of reducing sugar method were used and the results obtained were used, by solving simultaneous equations, to give the glucose and fructose concentrations in the extract. Sucrose was measured by inversion, with either dilute citric acid or an invertase preparation. It is interesting to note that this procedure gave results very similar to those obtained with quantitative paper chromatography or specific enzymatic methods in the hand of modern authors (Southgate, 1976). Starch was measured in most foods after hydrolysis with takadiastase, again using two reducing sugar methods, in order to estimate the glucose : maltose ratio in the hydrolysate. The starch in some refined cereal foods was measured as reducing sugars after dilute acid hydrolysis.

In the third edition a similar approach was adopted, a combination of reducing sugar methods, inversion and paper chromatography being used for the analysis of the sugars in an aqueous alcoholic extract. Starch was measured by the use of a takadiastase preparation which gave very nearly complete conversion of starch to glucose (Southgate, 1969a), with dilute acid hydrolysis for some cereal foods as in earlier editions.

Methods used for this edition

The principles used for the new analyses were the same as those in earlier editions. They are described in some detail here as they are awaiting publication (Dean, in preparation). A preliminary examination of the foodstuff by paper chromatography was made to obtain an indication of the sugars that were present and their approximate concentration. This information also enabled a suitable standard for quantitative measurements to be selected—this needed to be a sugar not present in the sample and well separated on the chromatogram from other sugars that were present.

Sugars

For quantitative work the sugars were extracted from the foodstuffs either by treatment with 80 per cent v/v ethanol in a Soxhlet extraction apparatus for 3 hours or by boiling with 80 per cent v/v ethanol for 20 minutes. A known amount of a suitable internal standard was then added to the extract and the residue was set aside for starch determination. The ethanol was removed from the extract by rotary evaporation under reduced pressure and the residue obtained was dissolved in distilled water. This solution was then deproteinised by adding equivalent amounts of cadmium sulphate and barium hydroxide solutions and after filtration was concentrated by rotary evaporation. The solution was then deionised by passage through a mixed bed ion-exchange column; the eluant was reduced in volume to 50 ml and 1 ml of boric acid (0.2 M) was added to 1 ml of this solution. An aliquot of this mixture, usually 100 µl, containing about 50 µg of each sugar was then injected on to the column of an autoanalyser. The sugars were eluted from the anion-exchange column with boric acid solutions of increasing molarity, pH and chloride ion concentration from an autograd. The eluate from the column was mixed with orcinol in 70% v/v sulphuric acid (1 g per litre) in a bubble-segmented stream, and the colour developed at 95°C and measured continuously at

420 nm. The areas of the peaks on the chromatogram representing the various sugars were then converted to microgram amounts of sugar from previously prepared calibration lines and finally expressed as percentages of the sugar in the sample.

Starch

The residue remaining after extraction of the free sugars was further examined for starch and dextrin content. Both were hydrolysed with the selective enzyme glucamylase to glucose. After filtration of the digest the liberated glucose was treated with glucose oxidase and estimated colorimetrically at 37°C in an autoanalyser. For some products, such as some biscuits and breakfast cereals, the procedure was modified by heating the residue (50–200 mg) for 3 minutes with 20 ml distilled water in a boiling water bath and then autoclaving it for 1 hour at 121°C before proceeding with the enzymatic hydrolysis and colour reaction. Acid hydrolysis, in which 100 mg of the residue was heated with 10 ml 0.75 M sulphuric acid for 3 hours, was used before glucose determination for certain products, such as some cakes, breakfast cereals and dried soups.

Unavailable carbohydrates/dietary fibre

The method used for the unavailable carbohydrates or dietary fibre in fruit, vegetables and nuts in the earlier editions was to measure the weight of the residue insoluble in the 80 per cent v/v ethanol used to extract the free sugars. The values for starch and protein were then deducted from this residue to give a value for *unavailable carbohydrates*.

Method used for this edition

Unavailable carbohydrates were measured by the method of Southgate (1969b) as later modified (Southgate 1976). This method includes a starch hydrolysis and accounts for substantially all the residue insoluble in 80 per cent v/v ethanol except protein and ash, and it therefore gives total values that are virtually identical with those obtained by the previous method.

Alcohol

In the previous editions alcohol was measured by the standard Inland Revenue distillation method.

Method used for this edition

A similar method was used for the new analyses.

Inorganic constituents

In the first and second editions the values for inorganic constituents were obtained by microchemical methods (either colorimetric or titrimetric) on either a solution of the ash in dilute acid or an acid digestion (wet oxidation) of the sample.

Samples of the food were dried in silica crucibles and heated to drive off fumes; they were then heated in a special incinerative apparatus (McCance and Shipp, 1933) or a muffle furnace. Sodium, potassium, calcium, magnesium, iron and copper were measured on the acid extract of the ash.

Sodium was precipitated as the zinc uranyl acetate and the uranium measured colorimetrically by reaction with potassium ferrocyanide; *potassium* was precipitated as the cobalti-nitrite and the precipitate titrated with permanganate; *calcium* was precipitated as the oxalate and titrated with permanganate; *magnesium* was precipitated as the ammonium phosphate; *iron* was measured colorimetrically by reaction with thioglycollic acid or with thiocyanate, and *copper* colorimetrically after reaction with sodium diethyl dithiocarbamate.

Phosphorus was measured after oxidation of the sample in a perchloric/ sulphuric acid mixture by a colorimetric method involving reduction of the phosphomolybdic acid complex.

Chloride was measured by titration using Volhard's method after oxidation of the sample with nitric acid in the presence of excess silver nitrate.

Sulphur was measured by the method of Masters and McCance (1939).

For the third edition some modifications were introduced but the same overall principles were retained.

The samples were ashed in silica crucibles in an electric muffle furnace at 450°C after a preliminary heating to drive off fumes. This usually produced a light grey or white ash, but if the ash was still dark it was treated with nitric acid and reheated. The ash was extracted with acid as described by McCance, Widdowson and Shackleton (1936).

Sodium was then measured gravimetrically as the zinc uranyl acetate (Widdowson and Southgate, 1959); *potassium* was precipitated as the cobalti-nitrite and the precipitate titrated with permanganate or ceric sulphate (some measurements were also made with a flame photometer); *calcium* was precipitated as the oxalate and titrated with either permanganate or ceric sulphate; *magnesium* was precipitated as the ammonium phosphate and the phosphate measured by the Fiske and Subbarow (1925) method. *Iron* was measured by two colorimetric procedures, with thioglycollic acid or *ortho*-phenanthroline; these methods gave virtually identical results. *Copper* was measured colorimetrically after reaction with sodium diethyl dithiocarbamate. *Phosphorus* was measured by the Fiske and Subbarow (1925) method. The last three constituents were measured both in the ash extract and on perchloric/sulphuric acid oxidations of the food. No differences between the two results were seen provided that all the pyrophosphate formed in dry ashing had been hydrolysed before measurement of phosphorus was attempted. *Chloride* was measured by a procedure virtually identical with that used for the first and second editions. *Sulphur* values were obtained by the method of Masters and McCance (1939).

Methods used for this edition

A suitable portion of dried material was weighed into a silica dish and most of the organic matter was destroyed at low temperature either on a hot plate or under an infrared lamp. The dish was then brought slowly to 550°C in a muffle furnace and the sample allowed to ash overnight. The residue was extracted by heating it with dilute hydrochloric acid and the solution was filtered into a graduated flask. The filter paper and residue were ashed overnight as before at 550°C, the extract of the residue was combined with the first extract and the solution was diluted to 100 ml. After the solution had been thoroughly mixed it was used either directly or after dilution for the determination of *calcium, copper, iron, magnesium* and *zinc* by atomic absorption spectrometry, and for *sodium* and *potassium* determination by emission spectrometry, with a Perkin Elmer instrument Model 403. A portion of the extract was also used for the colorimetric determination of *phosphorus* by reduction of the complex formed wtih ammonium molybdate with hydroquinone and sodium sulphite.

Chloride A known weight of dried material was moistened with a solution of sodium carbonate and heated first under an infrared lamp and then in a muffle furnace at a temperature not exceeding 500°C. The residue was extracted with water and diluted to about 100 ml. Chloride was then determined according to the Volhard procedure by treating this solution with an excess of 0.1 N silver nitrate solution, adding 2 ml of nitrobenzene and determining the residual silver nitrate by titration with 0.1 N potassium thiocyanate solution, ferric alum being used as indicator.

Sulphur The dried sample was heated with sodium carbonate and sodium peroxide in a nickel crucible to destroy organic matter and convert sulphur to sulphate. After it had cooled, the fused mass was extracted with water and dilute hydrochloric acid, barium chloride solution was added and sulphur was determined gravimetrically as barium sulphate. Confirmation of the results was obtained by X-ray fluorescence (Isherwood and King, 1976).

Vitamins

The values for vitamins in the previous editions were taken from the literature with very few exceptions. The B vitamins in the bread samples were measured by methods similar to those used in this edition.

Methods used for this edition

Vitamin A: retinol and β-carotene

A suitable weight of sample was saponified with ethanolic potassium hydroxide solution and the unsaponifiable matter containing retinol and β-carotene was extracted with diethyl ether. Retinol was then separated from β-carotene, sterols and vitamins D and E by partition chromatography between a mobile phase of 2,2,4-trimethylpentane and a stationary phase of methanol containing 10 per cent water, Sephadex LH20 being used as the inert support for the stationary phase (Bell, 1971). Measurement of the absorbance of the retinol fraction in isopropyl alcohol (propan-2-ol) at 325 nm and of the β-carotene fraction in cyclohexane at 455 nm now enabled the vitamin content to be calculated, allowance for any irrelevant absorption in the ultraviolet being made by the correction of Wilkie (1964). In some cases such as canned fish, where purification of the extract by column chromatography on Sephadex LH20 was inadequate, an additional column of calcium phosphate was also employed.

Vitamins of the B group other than thiamin

Nicotinic acid, riboflavin, vitamin B_6, vitamin B_{12}, folic acid, pantothenic acid and biotin were determined by microbiological assay. The general scheme involved liberation of the vitamin from the sample by acid or enzymatic hydrolysis, dilution with nutrient medium, sterilisation of the solutions, inoculation with a suitable culture of microorganisms as shown in the table, and incubation and estimation of the growth of the microorganisms by turbidimetry. In some instances a microbiological assay technique was used to confirm values for thiamin obtained by the standard fluorimetric procedure. A detailed description of all the techniques employed has been given by Bell (1974).

Vitamin	Microorganism	Culture number		
		ATCC	NCIB	NCYC
Nicotinic acid	*Lactobacillus plantarum*	8014	6376	—
Riboflavin	*Streptococcus zymogenes*	10100	7432	—
Vitamin B_6	*Saccharomyces carlsbergensis*	9080	—	74
Vitamin B_{12}	*Lactobacillus leichmannii*	7830	8118	—
Folic acid	*Lactobacillus casei*	7469	6375	—
Pantothenic acid	*Lactobacillus plantarum*	8014	6376	—
Biotin	*Lactobacillus plantarum*	8014	6376	—
Thiamin	*Lactobacillus viridescens*	12706	8965	—

Thiamin

Thiamin was determined by the fluorimetric method recommended by the Society of Public Analysts and Other Analytical Chemists: Analytical Methods Committee (1951).

Vitamin C

Preparation of extract The vitamin was extracted from the food by mixing an extracting solution consisting of 3 per cent acetic acid and 8 per cent metaphosphoric acids with a suitable amount of the sample in a Sunbeam blender for 2 minutes, centrifuging the mixture at 3000 revolutions per minute for 10 minutes and filtering the supernatant liquid through a filter paper.

Ascorbic acid A suitable aliquot of the prepared extract was titrated with a solution of 2,6-dichlorophenolindophenol that had been standardised against pure ascorbic acid. The procedure was similar to that described in the Official Methods of Analysis of the AOAC (1975).

Ascorbic plus dehydroascorbic acids Ascorbic acid was converted to dehydroascorbic acid by shaking an aliquot of the prepared extract with Norit charcoal. The solution was then filtered and an aliquot was reacted with *o*-phenylenediamine reagent to form a fluorescent compound which was measured on a fluorimeter at 455 nm using an exciting wavelength of 365 nm. Any fluorescence due to interfering substances was determined after complexing the dehydroascorbic acid with boric acid and an allowance was made for this interference. The procedure was similar to that described in the Official Methods of Analysis of the AOAC (1975).

Vitamin E

A suitable quantity of the sample was saponified with ethanolic potassium hydroxide solution in the presence of ascorbic acid and nitrogen. The unsaponifiable matter was extracted with diethyl ether and the extract was purified on a column of neutral alumina deactivated with 10 per cent water. Tocopherols were then determined colorimetrically by reaction with ferric chloride and 4,7-diphenyl-1,10-phenanthroline reagents. Gas–liquid chromatography on 2-metre glass columns of 3 per cent OV17 on Gas Chrom Q at 235° was used to confirm some of the results obtained by colorimetry and to provide information on the nature of the individual tocopherols present (Christie, Dean and Millburn, 1973).

Amino acids

The values in the previous editions were in the main derived from the literature. The values for potato that were included were obtained by a manual version of the procedure used for the new analyses.

Method used for this edition

A sample of the freeze-dried food was defatted in a Soxhlet apparatus with light petroleum (BP40–60°) for 2 hours and then thoroughly mixed. 200 mg of the defatted material was hydrolysed with 5 ml 6N hydrochloric acid in an evacuated sealed tube for 16 hours at 125°C. The liberated amino acids were separated by ion-exchange chromatography on a Technicon autoanalyser and determined by colorimetric reaction with ninhydrin. Cystine and methionine were determined on pre-oxidised samples by the method of Moore (1963); tryptophan was determined after alkaline hydrolysis by the method of Miller (1967). Results were corrected for hydrolytic losses, and after the nitrogen content of the defatted material had been determined values were calculated as mg of amino acid per g of nitrogen.

Fatty acids

No values for fatty acid composition were given in the previous editions. For this edition the fat was isolated by a chloroform–methanol–water extraction process (Bligh and Dyer, 1959). Solutions of the methyl esters of the fatty acids were prepared from the fat by conventional methods (IUPAC, 1976). The methyl esters of the fatty acids were separated and determined by gas–liquid chromatography (IUPAC, 1976).

Cholesterol

No values were given previously. For this edition a known weight of dried material was saponified with ethanolic potassium hydroxide solution for 1 hour and the unsaponifiable material was extracted with diethyl ether. The ether extract was washed with water and the traces of water were removed from the ether by rotary evaporation in the presence of ethanol. After removal of traces of ethanol by a further evaporation with light petroleum, the residue was dissolved in 10 ml of iso-octane (2,2,4,-trimethylpentane). An appropriate amount of an internal standard (5α-cholestan-3-one) was added to the extract and cholesterol was determined by gas–liquid chromatography on a 2-metre glass column, internal diameter 2 mm, containing 3 per cent OV17 on Gas Chrom Q at a temperature of 220°C.

References to appendix 1

AOAC (1955) *Official methods of analysis*, 8th edition. Association of Official Agricultural Chemists. Washington DC

AOAC (1975) *Official methods of analysis*, 12th edition. Association of Official Analytical Chemists. Washington, DC

Bell, J. G. (1971) Separation of oil-soluble vitamins by partition chromatography on Sephadex LH20. *Chem. and Ind.* 201–202

Bell, J. G. (1974) Microbiological assay of vitamins of the B group in foodstuffs *Lab. Pract.* **23**, 235–242, 252

Bligh, E. G., and Dyer, W. J. (1959) A rapid method of total lipid extraction and purification, *Canad. J. Biochem. Physiol.* **37**. 911–917

Chibnall, A. C., Rees, M. W., and Williams, E. F. (1943) The total nitrogen content of egg albumin and other proteins. *Biochem. J.* **37**, 354–359

Christie, A. A., Dean, A. C., and Millburn, B. A. (1973) The determination of vitamin E in food by colorimetry and gas–liquid chromatography *Analyst* **98**, 161–167

Dean, A. C. (in preparation)

Fiske, C. H., and Subbarow, Y. (1925) The colorimetric determination of phosphorus *J. biol. Chem.* **66**, 375–400

Isherwood, S. A., and King, R. T. (1976) Determination of calcium, potassium, chlorine, sulphur and phosphorus in meat and meat products by X-ray fluorescence spectroscopy *J. Sci. Food Agric.* **27**, 831–837

IUPAC (1977) Standard methods for the analysis of oils, fats and soaps 4th supplement to 5th edition. Method II D.19 Preparation of fatty acid methyl eters. Method II D.25 Gas liquid chromatography of fatty acid methyl esters

Masters, M., and McCance, R. A. (1939) The sulphur content of foods. *Biochem. J.* **33**, 1304–1312

McCance, R. A., and Shipp, H. L. (1933) *The chemistry and flesh foods and their losses on cooking*. Medical Research Council Special Report Series No. 187. HMSO, London

McCance, R. A., and Widdowson, E. M. (1940) *The chemical composition of foods*. Medical Research Council Special Report Series No. 235. HMSO, London

McCance, R. A., and Widdowson, E. M., and Shackleton, L. R. B. (1936) *The nutritive value of fruits, vegetables and nuts*. Medical Research Council Special Report Series No. 213. HMSO, London

Miller, E. L. (1967) Determination of the tryptophan content of feedingstuffs with particular reference to cereals. *J. Sci. Food Agric.* **18**, 381–386

Moore, S. M. (1963) On the determination of cystine as cysteic acid. *J. biol. Chem.* **238**, 235–237

Society of Public Analysts and Other Analytical Chemists : Analytical Methods Committee (1951) The chemical assay of aneurine in foodstuffs. *Analyst* **76**, 127–133

Southgate, D. A. T. (1969a) Determination of carbohydrates in foods. I Available carbohydrates. *J. Sci. Food Agric.* **20**, 326–330

Southgate, D. A. T. (1969b) Determination of carbohydrates in foods. II Unavailable carbohydrates. *J. Sci. Food Agric.* **20**, 331–335

Southgate, D. A. T. (1971) A procedure for the measurement of fats in foods. *J. Sci. Food Agric.* **22**, 590–591

Southgate, D. A. T. (1976) *Determination of food carbohydrates*. Applied Science Publishers, London

Von Lieberman, L., and Szekely, S. (1898) Eine neue Methode der Fettbestimmung in Futtermitteln, Fleisch, Koth, u.s.w. *Pflüg. Arch. ges. Physiol.* **72**, 360

Widdowson, E. M., and McCance, R. A. (1935) The available carbohydrates of fruits. Determination of glucose, fructose, sucrose, and starch. *Biochem. J.* **29**, 151–156

Widdowson, E. M., and Southgate, D. A. T. (1959) Haemorrhage and tissue electrolytes. *Biochem. J.* **72**, 200–204

Wilkie, J. B. (1964) Corrections for background in spectrophotometry using difference-in-absorbance values. Application to vitamin A. *Anal. Chem.* **36**, 896–900

Appendix 2: Note on the calculation of the energy value of foods and of diets*

by E. M. Widdowson

The energy value of a food is measured in Calories, which are physical units of heat. The number of Calories the body can derive from a food is, however, less than the number of Calories produced when the food is burned in a calorimeter because the calorie-producing nutrients, which are mainly protein, fat and carbohydrate, are not completely digested; the products of digestion, moreover, are not completely absorbed in the human gut, and the portion of the protein which is digested and absorbed is not completely oxidised to yield energy in the body.

The calorific value of a food is usually calculated from the amounts of protein, fat and carbohydrate it contains: these amounts are determined by chemical methods and the values are then multiplied by factors representing the number of Calories thought to be produced in the body by 1 gram of protein, fat or carbohydrate. The sum of these products gives the calorific value of the food. These calorie conversion factors do not represent the number of Calories which 1 gram of protein, fat, or carbohydrate would produce in a calorimeter. They are arrived at by applying to the values found by physical calorimetry various corrections allowing for losses occurring in digestion and absorption, and through incomplete oxidation. Since no two foods and no two people are ever exactly alike, and since these physiological corrections are based on averages the calorie conversion factors do not have the same accuracy as the values for Calories arrived at by physical calorimetry, or the values for protein, fat and carbohydrate found by chemical determination. Furthermore, different corrections are applied in different countries; and even within one country the method used may vary from one set of tables to another, and individual workers may use different methods from time to time. The problem is a complicated one, and there is no clear-cut answer to it.

The difference between the number of Calories which a diet would provide were the protein, fat and carbohydrate in it completely digested, and the number of Calories which it does in fact provide, is mainly due to the so-called 'unavailable carbohydrates' which are contained in plant foods. These are made up of hemicelluloses and fibre, and the digestive tract of man secretes no enzymes capable of digesting them, though micro-organisms in the gut may break down some of them and convert them to lower fatty acids, part of which may be absorbed and become a minor source of energy (McCance and Lawrence, 1929). In sheep and cattle, however, the large rumen provides space in which bacteria and protozoa can break down the hemicelluloses

* Reprinted from third edition

present in grasses and these contribute considerably to the nutrition of the animal. Although complex carbohydrates may, therefore, contribute a few Calories to man, their chief importance to the calorific value of a diet is a negative one. Fibre reduces the calorific value of a food or diet by hastening transport through the gut, and increasing the weight of the stools and the amount of nitrogen and fat in them. The more fibre a food or diet contains, the more nitrogen and fat will be excreted in the faeces and the less energy will therefore be derived from the protein and fat of the food or diet (McCance and Widdowson, 1947; McCance and Walsham 1948; McCance and Glaser, 1948).

History of calorie conversion factors

Most of the fundamental work on the calorific value of foods was carried out by Rubner and by Atwater and his colleagues more than 50 years ago. Rubner worked in Germany and Atwater in America during the last 20–25 years of the 19th century and the first part of the present one. In his early days, Atwater spent some time in Germany as Rubner's pupil, and it was undoubtedly this experience that inspired his later work. Rubner's most important papers for the present purpose were published in 1885 and 1901. He measured the heats of combustion of a number of different proteins, fats and carbohydrates in a bomb calorimeter, and also studied the heat of combustion of urine passed by a dog, a man, a boy and a baby. He realised that the heat of combustion of protein in the bomb calorimeter was greater than its calorific value to the body because the body oxidises protein only to urea, creatinine, uric acid and other nitrogenous end-products which are themselves capable of further oxidation. Rubner also analysed the faeces of the man who acted as his experimental subject and he found that the loss of energy in the nitrogenous substances in the urine and faeces were 16.3 and 6.9 per cent of the intake respectively, making a total loss of about 23 per cent. He deducted 23 per cent from the heats of combustion of animal and vegetable protein and arrived at a figure of 4.1 Calories per gram of mixed protein. Rubner made no allowances for losses in digestion and absorption of fat and carbohydrate, and his factors (9.3 Calories per gram of fat and 4.1 Calories per gram of carbohydrate) represent the average heats of combustion of a variety of fats and carbohydrates.

Atwater, working over 50 years ago, contributed more to our knowledge about the energy value of foods than any one else before or since his time. The heats of combustion of different proteins, fats and carbohydrates were measured in a bomb calorimeter (Atwater and Bryant, 1900). These authors also analysed the urine from forty-six persons and measured its heat of combustion. They found that for every gram of nitrogen in the urine there was unoxidized material sufficient to yield an average of 7.9 Calories. This is equivalent to 1.25 Calories per gram of protein in the food, if the person is in nitrogen equilibrium.

Atwater (1902) also made extensive studies of the 'availability' of nutrients, and he was careful to distinguish between what he called 'available' and 'digestible'. He regarded the faeces as being made up of two parts, the undigested and therefore unabsorbed food residues, and the 'metabolic products' of digestion, consisting of desquamated cells, bacteria and nitrogenous substances in the digestive juices. By 'digestible' nitrogen he meant the nitrogen in the food minus the nitrogen in the undigested, unabsorbed food residues, and this he could not measure. By 'available' nitrogen he meant the nitrogen in the food minus the nitrogen in the food residues together with the metabolic products of digestion, that is, the nitrogen in the food minus the nitrogen in the faeces.

Three men, aged 32, 29 and 22 years, served as subjects for Atwater's studies on 'availability'. Atwater made a total of fifty experiments on these men, each lasting for 3–8 days. The subjects ate what were described as mixed diets, which varied in the amount of fat and carbohydrate they contained, but none of the diets contained much roughage, i.e., unavailable carbohydrate. The foods were analysed for nitrogen and fat, and the faeces were analysed also.

Atwater and Bryant (1900) collected what they could find in the literature, including the results of their own work (Atwater and Benedict, 1897) on the 'availability' to man of single foods. From these data they prepared tentative coefficients for the 'availability' of the protein, fat and carbohydrate in the common classes of food, and they applied these coefficients to the mixed diets that their own subjects had eaten. They then compared the calculated 'availability' of the protein, fat and carbohydrate of the mixed diets with the 'availability' of these nutrients in the diets as found by experiment. They did the same with the results of sixty-one other experiments in which, apparently, ten men served as subjects, though no detailed description of these experiments was published. They found the 'coefficients of availability' of the protein, fat and carbohydrate in the mixed diets as determined by experiment to agree very well with the values as calculated by the proposed factors for availability of the protein, fat and carbohydrate in separate classes of foods.

For the calculation of the 'available energy' from mixed diets Atwater and Bryant (1900) suggested the use of the average factors 4.0, 8.9 and 4.0 for protein, fat and carbohydrate respectively. The figure 8.9 was later rounded off to 9.0 (Atwater, 1910). These factors, which Atwater had intended should be used only for calculating the Calories to be obtained from the protein, fat and carbohydrate in mixed diets, came to be widely used for calculating the available energy value of individual foods (Sherman, 1911, 1952; Chatfield and Adams, 1940; Platt, 1945).

In 1936, Morey published a paper in which she reviewed the work done by Rubner and Atwater at the turn of the century, and showed how the calorie conversion factors suggested by these two pioneers had been derived. It was Maynard (1944), however, who really opened up the whole subject again, and he was the first to draw attention to Atwater's original intention that the calorie conversion factors for protein, fat and carbohydrate should not be the same for all foods. Osmond (1948) was the first to adopt Maynard's suggestions as to the correct use of Atwater's factors in his tables of composition of Australian foods. Shortly after Maynard's paper was published, the Nutrition Division of the Food and Agriculture Organization of the United Nations appointed a Committee to discuss the question of calorie conversion factors, and the conclusions of the Committee were set out in a Report (1947), in which a table was given showing Atwater's suggested factors for calculating the physiological energy values of different classes of foods.

In 1955, the United States Department of Agriculture issued a handbook entitled *Energy Value of Foods* (Merrill and Watt, 1955) in which the fundamental work of Atwater was described in some detail, and the steps followed in his procedure for determining the energy value of foods were set out. The authors examined the results of work done since Atwater's time on the availability of the protein, fat and carbohydrate in individual foods, and prepared a more detailed table of factors for different classes of food. This table had formed the basis of the calculation of calorific values of foods in the current US Department of Agriculture's publication *Composition of Foods* (Watt and Merrill, 1950). At about the time that Merrill and Watt's (1955) handbook was published the British Nutrition Society held a symposium on the 'Assessment of the energy value of human and animal foods', when the differences and difficulties of the problem were discussed from a more general point of view (Blaxter and Graham, 1955; Widdowson, 1955; Hollingsworth, 1955).

Choice of calorie conversion factors for the 1st and 2nd editions of the present report

In the first edition of these tables an important departure from traditional practice was the method used for the determination of carbohydrate. In Atwater's own work, and in the work of those who followed him, the percentage of carbohydrate in foods was generally not determined directly but was calculated 'by difference', ie as the difference between 100 and the sum of the percentage of water, protein, fat and ash in the food. Thus it included not only sugars, dextrins and starch, which are known to be available to man, but also all the complex carbohydrates, most of which are not available as carbohydrate at all. When the first edition of this report was being prepared it was decided that the values found by direct determination of the available carbohydrates were likely to approximate more closely to the physiological values, and the method of calculating carbohydrate 'by difference' was abandoned. The glucose, fructose, sucrose, dextrins and starch were separately determined and their sum, expressed in terms of 'monosaccharides', was given as 'available carbohydrate'. Glucose and other monosaccharides have a heat of combustion of 3.75 Calories per gram, and this was the value assigned to the available carbohydrate fraction in the second edition of the present publication. The unavailable carbohydrate was considered to contribute no Calories to the diet.

The figure chosen for protein in the first and second editions was 4.1 Calories per gram, which was Rubner's factor for mixed meat and vegetable protein; this makes an allowance of about 7 per cent for nitrogen lost in the faeces, and the correction for unoxidized nitrogenous material in the urine is the same as Atwater's. Rubner's factor of 9.3 was chosen for fat; this is the average heat of combustion of animal and vegetable fats, and it makes little or no allowance for losses of fat in the faeces. The heat of combustion of ethyl alcohol is 7.07 Calories per gram, and a factor of 7.0 was used in the first and second edition of these tables.

Factors used in *Nutritive Values of Wartime Foods*

The second World War brought a need for tables giving the composition of the foods which were being produced and imported at that time, and the Council's Accessory Food Factors Committee undertook to compile such tables. These were published under the title *Nutritive Values of Wartime Foods* (Medical Research Council: Accessory Food Factors Committee, 1945). The figures for the protein, fat and carbohydrate in many of the foods were taken from the first edition of the present tables, the values for carbohydrate being based on direct chemical estimations of 'available carbohydrate', but expressed in terms of starch. To calculate the calorific value of the protein, fat and carbohydrate the factors 4, 9 and 4 Calories per gram respectively were used. In that publication the calorific value of the carbohydrate fraction of foods was definitely under-estimated since only the available fraction was considered, and a figure even lower than the physical, and probably also physiological, calorific value of starch (4.2 Calories per gram) was applied to it.

Choice of calorie conversion factors for the 3rd edition

Much thought has been given to the calorie conversion factors that should be used in the present edition of these tables. All the methods in current use are open to criticism. The use of different factors for protein and fat from various sources as worked out by Atwater and recommended by Maynard (1944), the FAO Committee (1947) and Merrill and Watt (1955) is undoubtedly a more correct approach than the use of the same factors for all foods, whether

4 and 9 or 4.1 and 9.3. On the other hand, the determination of the available carbohydrate fractions directly is acknowledged to be the better method, though there are few published tables in which this method has been used. The FAO Committee (1947) concluded that 'the correct chemical approach is by the extension of analytical work to include all substances covered by carbohydrates by difference'. Further studies of the digestibility of these substances are also required. Only when all the constituents of food have been determined and their physiological effects defined can their role in metabolism and their fuel value be accurately described'.

Work is in progress* along both lines suggested by the FAO Committee at the present time and, after much consideration, and with the advice of the Council's Diet and Energy Committee, the authors have decided that in order to avoid confusion the method of calculating the calorific values of foods shall remain unchanged in the present edition of these tables. The factors used, therefore, are 4.1 Calories per gram of protein, 9.3 Calories per gram of fat, 3.75 Calories per gram of available carbohydrate expressed as monosaccharides and 7.0 Calories per gram of alcohol. It is hoped that, when further evidence is available, a uniform method will be agreed upon, and used internationally.

Table 1 gives the calorific values of various foods as calculated by three different methods. It shows that in fact the agreement between the values arrived at by the different methods is in most instances quite close. The use of the factor 9.3 instead of 9 gives a slightly higher value for butter and other fats, but only in the case of fruit and vegetables where much of the carbohydrate is present in an 'unavailable' form do the figures really differ. Since these foods contribute a relatively small proportion of the calorific value of a whole diet it will not make much difference which factors are used to calculate the calorific value of mixed diets. In this respect it is of interest to note that calculations of the calorific value of National Food Supplies for the years 1947, 1955, 1956 and 1957 have been made by the various methods, and the results compared. The difference between the highest and lowest value was of the order of 2 per cent.

* Now published (Southgate and Durnin, 1970)

Table 1
Comparison of the calorific values of foods calculated by three methods
(Calories per 100 g)

	Third edition of the present tables: protein × 4.1 fat × 9.3 available carbohydrate (as monosaccharides) × 3.75	MRC War Memorandum No. 14 (1945): protein × 4.0 fat × 9.0 available carbohydrate (as starch) × 4.2	FAO (1947) Merrill and Watt (1955): specific factors for different foods
Cereals			
Bread, brown	242	245	251
Bread, white	243	242	242
Flour, Manitoba, wholemeal	339	336	327
Flour, Manitoba, white	352	350	353
Oatmeal	404	400	399
Rice, polished	361	359	368
Dairy-products			
Butter	793	768	748
Cheese, Cheddar	425	412	414
Cheese, Gorgonzola	393	380	382
Eggs	163	158	169
Milk, fresh, whole	66	65	68
Meat			
Beef, corned	231	224	231
Beef, frozen, raw	151	147	153
Beef, steak, raw	177	172	177
Liver, raw	143	139	144
Fruit			
Apples, English, eating	45	45	55
Apricots, dried	183	182	297
Bananas	77	76	103
Currants, black, raw	29	28	79
Currants, red, raw	21	21	60
Gooseberries, green, raw	17	17	35
Grapefruit	22	22	32
Oranges	35	35	49
Vegetables			
Beans, butter, raw	266	264	350
Beans, runner, raw	15	15	31
Cabbage, Savoy, raw	26	26	30
Carrots, old, raw	23	23	32
Peas, fresh, raw	64	63	81
Potatoes, old, raw	87	86	92
Nuts			
Peanuts	603	586	576
Walnuts	549	535	519

References to appendix 2

Atwater, W. O. (1902) On the digestibility and availability of food materials. *Conn. (Storrs) Agric. Exp. Sta. 14th Annu. Rep.*, 1901

Atwater, W. O. (1910) Principles of nutrition and nutritive value of food. *Fmrs' Bull. US Dep. Agric.* No. 142 (2nd review)

Atwater, W. O., and Benedict, F. G. (1897) Experiments on the digestion of food by man. *Conn. (Storrs) Agric. Exp. Sta. Bull.* No. 18

Atwater, W. O., and Bryant, A. P. (1900) The availability and fuel value of food materials. *Conn. (Storrs) Agric. Exp. Sta., 12th Annu. Rep.*, 1899

Blaxter, K. L., and Graham, N. McC. (1955) Methods of assessing the energy values of foods for ruminant animals. *Proc. Nutr. Soc.* **14**, 131–139

Chatfield, C., and Adams, G. (1940) Proximate composition of American food materials. *Circ. US Dep. Agric.* No. 549.

Food and Agriculture Organization of the United Nations. Committee on Calorie Conversion Factors and Food Composition Tables (1947) *Energy-yielding components of food and computation of calorie values.* United Nations, Food and Agriculture Organization, Washington DC

Hollingsworth, D. F. (1955) Some difficulties in estimating the energy values of human diets. *Proc. Nutr. Soc.* **14**, 154–160

McCance, R. A., and Glaser, E. M. (1948) The energy value of oatmeal and the digestibility and absorption of its proteins, fats and calcium. *Brit. J. Nutr.* **2**, 221–228

McCance, R. A., and Lawrence, R. D. (1929) *The carbohydrate content of foods.* Medical Research Council Special Report Series No. 135. HMSO, London

McCance, R. A., and Walsham, C. M. (1948) The digestibility and absorption of the calories, proteins, purines, fat and calcium in wholemeal wheaten bread. *Brit. J. Nutr.* **2**, 26–41

McCance, R. A., and Widdowson, E. M. (1947) The digestibility of English and Canadian wheats, with special reference to the digestibility of wheat protein by man. *J. Hyg. Camb.* **45**, 59–64

Maynard, L. A. (1944) The Atwater system of calculating the caloric value of diets. *J. Nutr.* **28**, 443–452

Medical Research Council: Accessory Food Factors Committee (1945) *Nutritive values of wartime foods.* Medical Research Council War Memorandum No. 14. HMSO, London

Merrill, A. L., and Watt, B. K. (1955) *Energy value of foods—basis and derivation.* US Department of Agriculture, Agriculture Handbook No. 74, Washington DC

Morey, N. B. (1936) An analysis and comparison of different methods of calculating the energy value of diets. *Nutr. Abstr. Rev.* **6**, 1–12

Osmond, A. (1946) *Tables of composition of Australian foods.* Special Report Series No. 2 of the National Health and Medical Research Council Nutrition Committee, Canberra

Platt, B. S. (1945) *Tables of representative values of foods commonly used in tropical countries.* Medical Research Council Special Report Series No. 253. HMSO, London

Rubner, M. (1885) Calorimetrische Untersuchungen. *Z. Biol.* **21**, 250

Rubner, M. (1901) Der Energiewert der Kost des Menschen. *Z. Biol.* **42**, 261

Sherman, H. C. (1911) *The chemistry of food and nutrition.* Macmillan Co., New York

Sherman, H. C. (1952) *The chemistry of food and nutrition,* 8th edition. Macmillan Co., New York

Southgate, D. A. T., and Durnin, J. V. G. A. (1970) Calorie conversion factors: an experimental reassessment of the factors used in calculation of the energy value of human diets. *Br. J. Nutr.* **24**, 517–535

Watt, B. J., and Merrill, A. L. (1950) *Composition of foods—raw, processed, prepared.* US Department of Agriculture, Agriculture Handbook No. 8, Washington DC

Widdowson, E. M. (1955) Assessment of the energy value of human foods. *Proc. Nutr. Soc.* **14**, 142–154

Appendix 3: Systematic names for fish and plant foods

Fish

Common name	Systematic name
White fish	
Cod	*Gadus morhua*
Haddock	*Melanogrammus aeglefinus*
Halibut	*Hippoglossus hippoglossus*
Lemon sole	*Microstomus kitt*
Plaice	*Pleuronectes platessa*
Saithe	*Pollachius virens*
Whiting	*Merlangius merlangus*
Fatty fish	
Eel	*Anguilla anguilla*
Herring, Bloater, Kipper	*Clupea harengus*
Mackerel	*Scomber scombrus*
Pilchards	*Sardinops sagax ocellata*
Salmon, Atlantic	*Salmo salar*
Salmon, red	*Oncorhynchus nerka*
Sardines	*Sardina pilchardus*
Sprats	*Sprattus sprattus*
Trout	*Salmo trutta*
Tuna, skipjack	*Euthynnus* sp; *Katsuwonus pelamis*
Whitebait	Young of *Clupea harengus* and *Sprattus sprattus*
Cartilaginous fish	
Dogfish	Probably *Squalus acanthias*
Skate	*Raja* sp
Crustacea	
Crab	*Cancer pagurus*
Lobster	*Homarus vulgaris*
Prawns	*Paleamon serratus*
Scampi	*Nephrops norvegicus*
Shrimps, brown	*Crangon crangon*
pink	*Pandalus montagui*
deep water	*Pandalus borealis*

	Common name	Systematic name
Fish *continued*	**Molluscs**	
	Cockles	*Cardium edule*
	Mussels	*Mytilus edulis*
	Oysters	*Ostrea edulis*
	Scallops	*Pecten maximus*
	Whelks	*Buccinum undatum*
	Winkles	*Littorina littorea*
Vegetables	Ackee	*Blighia sapida*
	Artichokes, globe	*Cynara scolymus*
	Artichokes, Jerusalem	*Helianthus tuberosus*
	Asparagus	*Asparagus officinalis* var *altilis*
	Aubergine	*Solanum melongena* var *ovigerum*
	Beans, French	*Phaseolus vulgaris*
	Beans, runner	*Phaseolus coccineus*
	Beans, broad	*Vicia faba*
	Beans, butter	*Phaseolus lunatus*
	Beans, haricot	*Phaseolus vulgaris*
	Beans, mung (green)	*Phaseolus aureus*
	Beans, red kidney	*Phaseolus vulgaris*
	Beansprouts	*Phaseolus aureus*
	Beetroot	*Beta vulgaris*
	Broccoli tops	*Brassica oleracea* var *botrytis*
	Brussels sprouts	*Brassica oleracea*
	Cabbage, red	*Brassica oleracea*
	Cabbage, Savoy	var of *Brassica oleracea*
	Cabbage, spring	*Brassica oleracea* var *capitata*
	Cabbage, white	var of *Brassica oleracea*
	Cabbage, winter	*Brassica oleracea* var *capitata*
	Carrots	*Daucus carota*
	Cauliflower	*Brassica oleracea* var *botrytis*
	Celeriac	*Apium graveolens* var *rapaceum*
	Celery	*Apium graveolens*
	Chicory	*Cichorium intybus*
	Cucumber	*Cucumis sativus*
	Endive	*Cichorium endivia*
	Horseradish	*Armoracia rusticana*
	Laverbread	*Porphyra umbilicalis*
	Leeks	*Allium ampelosprasum* var *porrum*
	Lentils	*Lens culinaris*
	Lettuce	*Lactuca sativa*
	Marrow	*Cucurbita pepo*
	Mushrooms	*Agaricus campestris*
	Mustard and cress	*Brassica* and *Lepidium* spp
	Okra	*Hibiscus esculentus*
	Onions	*Allium sepa*
	Parsley	*Petroselinum crispum*
	Parsnips	*Pastinaca sativa*
	Peas	*Pisum sativum*
	Peas, chick	*Cicer arietinum*
	Peas, red, pigeon	*Cajanus cajan*
	Peppers, green	*Capsicum annum*

	Common name	Systematic name
Vegetables *continued*		
	Plantain	*Musa paradisiaca*
	Potatoes	*Solanum tuberosum*
	Pumpkin	*Cucurbito pepo*
	Radishes	*Raphanus sativus*
	Salsify	*Tragopogon porrifolius*
	Seakale	*Crambe maritima*
	Spinach	*Spinacia oleracea*
	Spring greens	var *Brassica oleracea*
	Swedes	*Brassica napus* var *napobrassica*
	Sweet potatoes	*Ipomaea batatas*
	Sweetcorn	*Zea mays*
	Tomatoes	*Lycopersicon esculentum*
	Turnips	*Brassica rapa*
	Watercress	*Nasturtium officinale*
	Yam	*Dioscorea* sp
Fruit	Apple	*Malus pumila*
	Apricot	*Prunus armeniaca*
	Avocado pear	*Persea americana*
	Banana	*Musa* sp
	Bilberry	*Vaccinium myrtillus*
	Blackberry	*Rubus ulmifolius*
	Cherry	*Prunus avium*
	Cranberry	*Vaccinium oxycoccus*
	Currants, black	*Ribes nigrum*
	Currants, red	*Ribes rubrum*
	Currants, white	*Ribes sativum*
	Currants, dried	*Vitis vinifera*
	Damson	*Prunus domestica* subsp *insititia*
	Date	*Phoenix dactylifera*
	Fig	*Ficus carica*
	Gooseberry	*Ribes grossularia*
	Grape	*Vitis vinifera*
	Grapefruit	*Citrus paradisi*
	Greengage	*Prunus domestica* subsp *italica*
	Guava	*Psidiom guajava*
	Lemon	*Citrus limon*
	Loganberry	*Rubus loganobaccus*
	Lychee	*Litchi chinensis*
	Mango	*Mangifera indica*
	Medlar	*Mespilus germanica*
	Melon, Cantaloupe	*Cucumis melo*
	Melon, yellow	*Cucumis citrallus* and *C. melo*
	Melon, watermelon	*Citrullus lanatus*
	Mulberry	*Morus nigra*
	Nectarine	*Prunus persica* var *nectarina*
	Olive	*Olea europaea*
	Orange	*Citrus sinensis*
	Passion fruit	*Passiflora edulis*
	Paw paw	*Carica papaya*
	Peach	*Prunus persica*
	Pear	*Pyrus communis*
	Pineapple	*Ananas comosus*

	Common name	Systematic name
Fruit *continued*		
	Plum	*Prunus domestica* subsp *domestica*
	Pomegranate	*Punica granatum*
	Quince	*Cydonia vulgaris*
	Raisin	*Vitis vinifera*
	Raspberry	*Rubus idaeus*
	Rhubarb	*Rheum rhaponticum*
	Strawberry	*Fragaria* sp
	Sultana	*Vitis vinifera*
	Tangerine	*Citrus reticulata*
Nuts	Almonds	*Prunus amygdalus*
	Barcelona nuts	*Corylus maxima barcelonensis*
	Brazil nuts	*Bertholletia excelsa*
	Chestnuts	*Castanea vulgaris*
	Cob/hazel	*Corylus avellana* and *C. maxima*
	Coconut	*Cocos nucifera*
	Peanuts	*Arachis hypogoea*
	Walnuts	*Juglans regia*

References to appendix 3

Fish

Labelling of Food (Amendment) Regulations 1972. Statutory Instruments 1972 No. 1510 HMSO, London

McCance, R. A., and Shipp, H. L. (1933). *The chemistry of flesh foods and their losses on cooking*. Medical Research Council Special Report Series No. 187. HMSO, London

Vegetables, fruit and nuts

Bailey, L. H. (1949) *Manual of cultivated plants most commonly found in the continental United States and Canada*, Macmillan & Co. New York

Tutin, T. G., Heywood, V. H., Burgess, N. A., Valentine, D. H., Walters, S. M., and Webb, D. A., and Moore, D. M. (editors) *Flora Europaea* Vol. I: Lycopodiaceae to Plantaceae (1964); Vol. II: Rosaceae to Umbelliferae (1968). Cambridge University Press, Cambridge

Appendix 4: Recipes for cooked dishes

Revised by J. Thorn and A. A. Paul

Bread

39 Soda bread

500 g flour
1 level teaspoon salt
1 level teaspoon bicarbonate of soda
1 level teaspoon cream of tartar
290 ml milk

Sift the dry ingredients and quickly knead to a soft dough with the milk. Bake for 35 minutes at Mark 7, 220°C.

Biscuits

65 Biscuits, home made basic mixture

100 g margarine
100 g caster sugar
200 g flour
1 egg

Cream the fat and sugar. Mix in the egg, then the flour and knead the dough lightly until smooth. Roll out thinly, prick and shape. Bake 10-15 minutes at Mark 4, 180°C.

71 Shortbread

200 g flour
100 g butter
50 g caster sugar

Beat the butter and sugar to a cream. Mix in the flour and knead till smooth. Press into a flat tin to about 2 cm in thickness. Bake for about 45 minutes at Mark 3, 170°C.

Cakes

75 Fruit cake rich

200 g margarine
200 g brown sugar
4 eggs
20 g black treacle
20 ml brandy
250 g flour
$\frac{1}{4}$ level teaspoon salt
750 g mixed dried fruit
150 g mixed glacé fruit, chopped
1 level teaspoon mixed spice

Cream the fat and sugar. Beat in the eggs, treacle and brandy. Fold in the sifted flour and spices, and mix in the fruit. Turn into a 20 cm cake tin. Bake for 4 hours at Mark 2, 150°C.

76 Fruit cake rich, iced

1680 g fruit cake, rich
70 g apricot jam
410 g marzipan

Royal icing
300 g icing sugar
1 egg white
1 teaspoon lemon juice

Make the cake as in Recipe no 75. When cold spread with a thin layer of apricot jam and cover with marzipan. Make the royal icing by beating the egg whites and icing sugar; finally add the lemon juice.

Cakes
continued

78 Gingerbread

300 g flour
100 g margarine
100 g sugar
200 g treacle
2 eggs
2 level teaspoons ground ginger
$\frac{1}{2}$ level teaspoon bicarbonate of soda
75 ml milk

Melt the margarine, sugar and treacle in a pan, heating gently. Beat the egg well. Mix all the ingredients together and bake for about $1\frac{1}{4}$ hours at Mark 4, 180°C.

80 Rock cakes (basic recipe, rubbing-in method)

200 g flour
3 level teaspoons baking powder
100 g margarine
100 g sugar
1 egg
50 ml milk
100 g currants

Sift together the flour and baking powder, and rub in the fat, add the currants. Mix to a soft dropping consistency with the egg and milk. Drop the mixture in small portions on to a baking sheet. Bake for about 15 minutes at Mark 8, 230°C.

81 Sponge cake with fat (basic recipe, creaming method)

150 g flour
1 level teaspoon baking powder
150 g margarine
150 g caster sugar
3 eggs

Cream the fat and sugar until light and fluffy. Add the beaten egg a little at a time and beat well. Fold in the sifted flour and baking powder. Bake for about 20 minutes at Mark 5, 190°C.

82 Sponge cake without fat (basic recipe, whisking method)

4 eggs
100 g caster sugar
100 g flour

Whisk the eggs and sugar in a basin over hot water until stiff. Fold in the flour. Bake for about 25 minutes at Mark 5, 190°C, or, for Swiss rolls, 7 minutes at Mark 8, 230°C.

Buns and pastries

86 Eclairs

200 g choux pastry, cooked
150 g double cream

Icing
100 g icing sugar
50 g plain chocolate
30 ml water

Make the choux pastry (see below) into eclairs. Slit, fill with whipped cream and top with chocolate icing.

87 Jam tarts

200 g raw shortcrust pastry
200 g jam

Line about ten tart tins with thinly rolled pastry. Fill each tart with jam and bake in a hot oven, Mark 6, 200°C, for 10–15 minutes.

88 Mince pies

300 g raw shortcrust pastry
200 g mincemeat

Roll out the pastry and cut into rounds. Place half the rounds in tart tins. Fill with mincemeat and cover with remaining pastry. Bake for about 20 minutes at Mark 5, 190°C.

89 and 90 **Pastry, choux**

100 g flour
50 g margarine
150 ml water
2 eggs
$\frac{1}{4}$ level teaspoon salt

Boil the water, salt and margarine, add the flour and beat over heat to form a ball of smooth mixture. Cool and beat in the eggs. Pipe out as desired and bake for about 30 minutes at Mark 6, 200°C.

91 and 92 **Pastry, flaky**

200 g flour
75 g margarine
75 g lard
$\frac{1}{2}$ level teaspoon salt
10 ml lemon juice
80 ml water to bind

Make the pastry in the normal way, baking it at Mark 7, 220°C, or as directed.

93 and 94 **Pastry, shortcrust**

200 g flour
50 g margarine
50 g lard
$\frac{1}{2}$ level teaspoon salt
30 ml water to bind

Make the pastry in the normal way, baking it at Mark 6, 200°C, or as directed.

95 **Scones**

200 g flour
4 level teaspoons baking powder
$\frac{1}{4}$ level teaspoon salt
50 g margarine
10 g sugar
125 ml milk

Sift the flour, sugar and baking powder and rub in the fat. Mix in the milk. Roll out and cut into rounds. Bake in a hot oven, Mark 7, 220°C, for about ten minutes.

96 **Scotch pancakes (Drop scones)**

200 g flour
$\frac{1}{2}$ level teaspoon salt
$\frac{1}{2}$ level teaspoon bicarbonate of soda
1 level teaspoon cream of tartar
50 g margarine
25 g caster sugar
1 egg
200 ml milk
15 g margarine for griddle

Sift flour with salt and raising agent, rub in fat and mix in sugar. Add egg and milk to give a stiff batter. Cook by spoonfuls on hot greased griddle.

Puddings

97 **Apple crumble**

400 g cooking apples, weighed after preparation
100 g flour
$\frac{1}{2}$ level teaspoon cinnamon
50 g margarine
100 g sugar

Peel, core and slice the apples. Arrange in a dish and sprinkle with half the sugar. Rub the other ingredients together and pile on top. Bake for 40 minutes at Mark 5, 190°C.

98 **Bread and butter pudding**

75 g bread
20 g butter
500 ml milk
30 g sugar
2 eggs
30 g currants

Cut the bread very thinly and spread with butter. Beat the eggs with the sugar and add the milk. Place layers of bread and currants in a pie dish and pour the eggs and milk over the bread. Leave to soak for 30 minutes and then bake at Mark 4, 180°C, for 30–40 minutes.

99 Cheesecake

Base, for 18 cm tin
150 g digestive biscuit crumbs
75 g margarine

Top
350 g cream or curd cheese
2 eggs
100 g caster sugar
25 g cornflour
1 lemon (juice = 40 g) and finely grated rind
150 g double cream
½ teaspoon vanilla essence

Melt the margarine in a pan and combine with the biscuit crumbs. Press into the base of the tin. Combine the topping ingredients, beat well and pour into base. Bake for 45 minutes at Mark 4, 180°C, until only just firm in the centre.

100 Christmas pudding

100 g flour
300 g breadcrumbs, fresh
1 level teaspoon mixed spices
½ level teaspoon salt
125 g suet
150 g raisins
150 g sultanas
150 g currants
50 g chopped mixed peel
30 g ground almonds
150 g brown sugar
3 eggs
15 g treacle
150 ml stout

Sift the flour, spices and salt into a basin and mix in all dry ingredients. Whisk the eggs, treacle and stout and stir thoroughly into dry ingredients. Put into well greased basins, cover with greased paper and foil. Boil for 6 hours. Renew foil and store. Re-boil for about 2 hours when required.

101 **Custard, egg** baked or sauce

500 ml milk
2 eggs
30 g sugar
vanilla essence

Beat the eggs and sugar together. Add the milk and vanilla essence. Either stir over gentle heat until mixture thickens or bake in a dish standing in a pan of water at Mark 3, 170°C, for 40 minutes.

102 **Custard** made with powder

500 ml milk
25 g custard powder
25 g sugar

Blend the custard powder with a little of the milk. Add the sugar to the remainder of the milk and bring to the boil. Pour immediately over the paste, stirring all the time. Return to the pan, bring back to boiling point, stirring, then serve.

103 Custard tart

300 g raw shortcrust pastry
250 ml milk
1 egg
15 g sugar

Make the pastry and line a shallow tin. Make the custard and use as filling. Bake at Mark 6, 200°C, lowering to Mark 5, 190°C until the custard is set (about 40 minutes).

104 Dumpling

100 g flour
45 g suet
75 g water
1 level teaspoon baking powder
½ level teaspoon salt

Mix the dry ingredients together with the cold water to form a soft dough. Divide into balls, flour them and place in boiling water. Boil for 30 minutes.

106 **Fruit pie** with pastry top

200 g raw shortcrust pastry
450 g fruit (prepared)
80 g sugar
a little water if required

Place the prepared fruit, sugar and water in a pie dish. Cover with the pastry. Bake for 10 to 15 minutes at Mark 6, 200°C, to set the pastry, then about 20 minutes at Mark 4, 180°C, to cook the fruit.

110 **Jelly** made with water

130 g jelly cubes
440 ml water

Dissolve the jelly cubes in hot water. Add the rest of the cold water. Pour into a mould and allow to set.

111 **Jelly** made with milk

130 g jelly cubes
250 ml milk
200 ml water

Dissolve the jelly cubes in hot water. Cool, add milk slowly, stirring constantly. Leave to set in a mould.

112 **Lemon meringue pie**

200 g raw shortcrust pastry
2 lemons (juice = 80 g)
2 eggs
125 g caster sugar
25 g cornflour
15 g margarine
125 ml water

Boil the cornflour, water, grated rind and juice of lemons and 25 g of the sugar. Cool, stir in the egg yolks, and pour the mixture into the flan case. Make a meringue with the egg whites and the rest of the sugar; pile on top of the lemon mixture. Bake for 30 minutes at Mark 4, 180°C, until crisp and brown on top.

113 **Meringues**

4 egg whites
200 g caster sugar

Whisk the egg whites until stiff. Fold in the sugar. Pipe onto the baking sheet. Bake for 3 hours at Mark ½, 130°C.

114 **Milk puddings**

500 ml milk
50 g cereal (eg rice, sago, semolina, tapioca)
25 g sugar

Simmer until cooked or bake in a moderate oven Mark 4, 180°C, according to type of cereal.

116 **Pancakes**

100 g flour
250 ml milk
1 egg
50 g lard
50 g sugar

Sieve the flour into a basin. Break in the egg and add about 100 ml of the milk, stirring until smooth. Add the rest of the milk and beat to a smooth batter. Heat a little lard in a frying pan and pour in enough batter to cover the bottom. Cook both sides and turn onto sugared paper. Dredge lightly with sugar. Repeat until all the batter is used, to give about 10 pancakes.

117 Queen of puddings

250 ml milk
25 g butter
50 g breadcrumbs, fresh
100 g sugar
2 eggs, separated
rind of 1 lemon
50 g jam

Heat the milk and butter and pour over the breadcrumbs and 30 g of the sugar. Leave to soak for 30 minutes. Add the beaten egg yolks and grated lemon rind, and pour into a greased pie dish. Bake for about 20 minutes at Mark 4, 180°C. Spread the top with jam. Whisk the egg whites stiffly, then whisk in the rest of the sugar one teaspoonful at a time. Pile on top and bake at Mark 1, 140°C, until crisp and golden brown.

118 **Sponge pudding** steamed, basic mixture

100 g flour
1 level teaspoon baking powder
50 g margarine
50 g caster sugar
1 egg
30 ml milk

Cream the fat and sugar. Beat in the eggs a little at a time. Fold in the sifted flour and baking powder, adding milk to give a soft dropping consistency. Turn the mixture into a greased basin and steam for 1½ to 2 hours.

119 **Suet pudding** steamed, basic mixture

50 g flour
50 g breadcrumbs, fresh
50 g suet, shredded
30 g sugar
1 level teaspoon baking powder
¼ level teaspoon salt
80 ml milk

Mix the dry ingredients to a soft paste with the milk. Pour into a greased basin, cover with greased paper and steam for about 2½ hours.

120 Treacle tart

300 g raw shortcrust pastry
250 g golden syrup
50 g breadcrumbs, fresh

Line shallow tins with pastry, pour in the syrup and sprinkle with the bread-crumbs. Bake for 20–30 minutes at Mark 6, 200°C.

121 Trifle

75 g sponge cake
25 g jam
50 g fruit juice
75 g tinned fruit
25 ml sherry
250 g custard (made with powder)
25 g double cream
10 g nuts
10 g cherries
angelica

Slit the sponge cake, spread with jam and sandwich together. Cut into 4 cm cubes. Soak in the fruit juice and sherry. Mix with the fruit, cover with cold custard and decorate with the whipped cream, nuts and angelica.

122 Yorkshire pudding

100 g flour
1 level teaspoon salt
1 egg
250 ml milk
20 g dripping

Sieve flour and salt into a basin. Break in the egg and add about 100 ml of the milk, stirring until smooth. Add the rest of the milk and beat to a smooth batter. Pour into a tin containing very hot dripping. Bake for about 40 minutes at Mark 7, 220°C.

172 Omelette

2 eggs
10 ml water
10 g butter
$\frac{1}{2}$ level teaspoon salt
pepper

Beat the eggs with the salt and water. Heat the butter in an omelette pan. Pour in the mixture and stir until it begins to thicken evenly. While still creamy, fold the omelette and serve.

173 Scrambled eggs

2 eggs
15 g butter
20 ml milk
1 level teaspoon salt

Melt the butter in a small pan, stir in the beaten egg, milk and seasoning. Cook over gentle heat until the mixture thickens.

174 Cauliflower cheese

1 small cauliflower (700 g)
250 ml milk
100 ml cauliflower water
25 g margarine
25 g flour
100 g cheddar cheese, grated
$\frac{1}{2}$ level teaspoon salt
pepper

Prepare cauliflower and boil in water until just tender. Drain, saving 100 ml of the water, place cauliflower in a dish and keep warm. Make a white sauce from the margarine, flour, milk and cauliflower water. Add 75 g of the cheese and season. Pour over the cauliflower and sprinkle with the remaining cheese. Brown under the grill or in a hot oven, Mark 7, 220°C.

175 Cheese pudding

50 g breadcrumbs, fresh
250 ml milk
$\frac{1}{2}$ level teaspoon salt
cayenne pepper
75 g grated cheese
2 eggs

Heat the milk, pour over the breadcrumbs and allow to soak for about 30 minutes. Add the grated cheese, seasoning and egg yolks. Fold in the stiffly whipped whites and pour into a greased pie dish. Bake in a moderate oven, Mark 4, 180°C, for half an hour until well risen and golden brown.

176 Cheese soufflé

50 g margarine
50 g flour
250 ml milk
$\frac{1}{2}$ level teaspoon cayenne pepper
$\frac{1}{2}$ level teaspoon dry mustard
4 eggs
100 g cheese, grated

Melt the butter over gentle heat; stir in the flour and add the milk slowly. Cook for a minute or two. Cool slightly, beat in egg yolks, seasoning and cheese. Whisk egg whites stiffly and fold into mixture. Bake in a greased 17 cm soufflé dish at Mark 6, 200°C, for about 35 minutes.

177 Macaroni cheese

100 g macaroni
350 ml milk
25 g margarine
25 g flour
100 g cheese, grated
1 level teaspoon salt

Boil the macaroni and drain well. Make a white sauce from the margarine, flour and milk. Add 75 g of the cheese and season. Add the macaroni and put into a pie dish. Sprinkle with the remaining cheese and brown under the grill or in a hot oven Mark 7, 220°C.

Egg and cheese dishes *continued*

178 **Pizza** cheese and tomato

Dough
200 g flour
1 level teaspoon salt
1 level teaspoon sugar
15 g fresh yeast or 2 level teaspoons dried yeast
150 ml warm water
200 g tomatoes
150 g cheese
8 black olives (40 g)
20 g oil

Make the dough in the usual way, proving once. Knead and roll out to shape. Leave for 10 minutes. Arrange sliced or pulped tomatoes on top, then cheese and olives. Brush with oil. Bake for 30 minutes at Mark 8, 230°C.

179 Quiche Lorraine

200 g raw, shortcrust pastry
100 g bacon, streaky
100 g cheese
2 eggs
200 ml milk

Line a 20 cm flan ring with shortcrust pastry. Fill with the chopped bacon, fried and grated cheese. Beat the eggs in warmed milk and pour into the pastry case. Bake for 10 minutes at Mark 6, 200°C, then for 30 minutes at Mark 4, 180°C.

180 Scotch eggs

4 eggs
250 g raw pork sausage meat
25 g breadcrumbs, dried
20 g flour
15 g beaten egg

Hard boil the eggs, cool and shell. Dip in seasoned flour and cover with sausage meat. Brush with beaten egg and coat with crumbs. Deep fry for 8–10 minutes.

181 Welsh rarebit

2 slices buttered toast (50 g)
50 g grated cheese
¼ level teaspoon dry mustard
20 ml milk
¼ level teaspoon salt
cayenne pepper and pepper

Mix the cheese and seasoning with the milk. Spread on the toast and brown under the grill.

Meat dishes

422 and **423** **Sausage rolls** flaky and short

100 g raw flaky pastry
40 g pork sausage meat

or

100 g raw shortcrust pastry
50 g pork sausage meat

Make the pastry, roll out and cut into 10 cm squares. Place some sausage meat into the middle of each. Fold over and seal. Bake for 20–30 minutes at Mark 7, 220° C.

424 Steak and kidney pie

350 g raw flaky pastry
400 g raw stewing steak
200 g raw kidney
100 ml water
2 level teaspoons salt
15 g flour

Make the pastry. Prepare steak and kidney, cut into pieces and roll in seasoned flour. Place in a pie dish with water. Cover with pastry. Bake pie for 20 minutes at Mark 6, 200° C, then lower the heat to Mark 2, 150° C and cover with greaseproof paper. Cook for 2–2½ hours more.

426 Beef steak pudding

Suet crust
200 g flour
100 g suet
$1\frac{1}{2}$ level teaspoon baking powder
$\frac{1}{2}$ level teaspoon salt
130 ml water

500 g raw stewing steak
130 g onion (peeled and chopped)
50 g flour
25 ml water or stock
1 level teaspoon salt; pepper

Make the suet crust pastry, and line a pudding basin, leaving sufficient for a lid. Cut the meat into slices and roll in the seasoned flour. Put into the basin with the onion. Add a little water and cover with the remaining pastry. Steam for about 3 hours.

427 Beef stew

250 g raw stewing steak
75 g onion
75 g carrot
15 g dripping

300 ml water or stock
15 g flour
1 level teaspoon salt; pepper

Melt the dripping in a casserole and brown the pieces of meat. Remove the meat and brown the onion. Add the flour and cook the roux. Gradually blend in the water, add the meat, carrots and seasoning, bring to the boil and finish cooking at Mark 4, 180° C, for about 2 hours.

428 Bolognese sauce

25 g oil
75 g onion
75 g carrot
50 g celery
200 g minced beef

10 g tomato paste
200 g canned tomatoes
250 ml water or stock
1 level teaspoon salt; pepper, herbs

Brown the onion, carrot and celery in oil. Add the minced beef stirring thoroughly to brown. Add the tomatoes, stock and seasoning and simmer for 45 minutes with the lid on.

429 Curried meat

250 g cooked meat
200 g onion, peeled and chopped
50 g oil
75 g apple, peeled and chopped
50 g sultanas

15 g desiccated coconut
20 g flour
20 g curry powder
400 ml water
2 level teaspoons salt

Fry the onions in the oil. Add the apple, sultanas and coconut, then the flour and curry powder and fry for a minute or two. Add the water and bring to the boil. Simmer for 5 minutes. Add the cooked meat, cut into pieces and heat thoroughly.

430 Hot pot

250 g raw stewing steak
250 g potatoes
150 g onions

100 g carrots
125 ml stock
2 level teaspoons salt; pepper

Cut the steak into small pieces and arrange in layers with slices of carrot and onion. Add water and seasoning. Cover with a layer of sliced potatoes. Cover and bake at Mark 4, 180° C, for $2\frac{1}{2}$ hours, removing the lid for the last 30 minutes to brown the potatoes.

431 Irish stew

250 g neck of mutton (weighed with bone)
250 g potato
125 g onion

350 ml water
1 level teaspoon salt; pepper

Meat dishes *continued*

Cut up the meat, potato and onion and put into a saucepan. Add water and bring to the boil. Skim well and allow to simmer slowly for 1½ hours.

433 Moussaka

250 g minced beef
250 g aubergines or potatoes
150 g onions
30 g oil
100 ml water or stock
20 g tomato paste
1 level teaspoon salt

Sauce
150 ml milk
15 g flour
15 g oil
50 g cheese, grated
½ egg

Fry the sliced onions in the oil until soft and remove from pan. Fry the aubergines until transparent then brown the meat. Arrange layers of aubergines, meat and onions in a casserole. Add the tomato paste and the seasoned stock. Pour the cheese sauce over the top, and cooked for 1 hour at Mark 5, 190°C.

434 Shepherd's pie

350 g cooked minced beef
100 g onion boiled and chopped
150 ml water
500 g boiled potato
50 ml milk
20 g margarine
2 level teaspoons salt; pepper

Mix the beef and onion, moisten with water and add seasoning. Place in a pie dish. Mash the potato with the milk and margarine. Pile on top of the meat and bake in the oven for 25 minutes to brown. Mark 5, 190°C.

Fish dishes

548 Fish pie

200 g cooked white fish
400 g mashed potato

Sauce
150 ml milk
15 g margarine
15 g flour
½ level teaspoon salt

Flake the fish and mix with the white sauce. Pipe a potato border round a dish, pour in the fish mixture. Brown in the oven, Mark 6, 200°C, for 30 minutes.

549 Kedgeree

200 g smoked fillet, steamed
50 g rice
25 g margarine
2 eggs
½ level teaspoon salt; pepper

Boil the rice. Hard boil one egg. Melt the margarine and stir in the flaked fish, rice, seasoning and one beaten egg. Stir in chopped hard boiled egg and heat thoroughly.

Vegetable dishes

571 Beans, mung cooked, dahl

120 g dry beans
30 g butter
35 g onion
1 level teaspoon ginger
1 level teaspoon turmeric
½ level teaspoon garlic
2 level teaspoon salt
spices
chilli powder
840 ml water for cooking

Soak the beans in water for a few minutes; strain. Add the 840 ml water, salt and turmeric, bring to the boil and simmer for 40 minutes. Cook the chopped onions in the butter, with the garlic, ginger, spices and chilli powder. Add to the cooked dahl and simmer for 5 minutes.

Vegetable dishes *continued*

605 **Lentils** masur dahl, cooked

110 g dry lentils
20 g butter
30 g onion
1 level teaspoon ginger
½ level teaspoon garlic
1 level teaspoon salt
1 level teaspoon spices and seasonings
turmeric
680 ml water for cooking

Soak the lentils for 10–15 minutes; strain. Add the 680 ml water, bring to the boil, add salt and turmeric and simmer for 40 minutes. Cook the chopped onions in the butter, with the garlic, ginger, spices and seasonings. Add to the cooked dahl and simmer for 5 minutes.

631 **Peas, chick** cooked, dahl

110 g dry peas, whole
20 g butter
60 g onion
50 g tomatoes
2 level teaspoons ginger
½ level teaspoon garlic
1 level teaspoon salt
1 level teaspoon spices and seasonings
turmeric
500 ml water for cooking

Soak the peas overnight in cold water; strain. Cook the chopped onion in the butter, with the ginger, garlic and seasonings. Add the tomatoes. Mix, add the spices and finally the strained peas. Add the water and simmer for 1½ hours.

Preserves, sauces and soups

852 **Lemon curd**

300 g sugar (including some sugar lumps)
100 g butter
4 lemons (juice = 150 ml)
4 eggs

Wash lemons and rub a few sugar lumps over the rind to extract the flavour. Squeeze lemons. Melt butter, lemon juice and all the sugar in a double pan. Add the eggs one by one and cook slowly, stirring all the time, until the mixture coats the back of a spoon. Pour into jars and cover.

854 **Marzipan**

300 g ground almonds
150 g caster sugar
150 g icing sugar
1 egg
20 ml lemon juice

Mix almonds and sugar, add beaten egg and knead all ingredients until smooth.

920 **Bread sauce**

250 ml milk
50 g fresh breadcrumbs
5 g margarine
1 small onion
2 cloves
mace
½ level teaspoon salt

Put the milk and the onion, stuck with cloves, in a saucepan and bring to the boil. Add the breadcrumbs, and simmer for about 20 minutes over gentle heat. Remove the onion, stir in the margarine and season.

922 **Cheese sauce**

350 ml milk
25 g flour
25 g margarine
75 g cheese
½ level teaspoon salt
pepper, cayenne

Melt the fat in a pan. Add the flour and cook gently for a few minutes stirring all the time. Add the milk and cook until the mixture thickens, stirring continually. Add the grated cheese and seasoning. Reheat to soften the cheese and serve immediately.

923 Chutney, apple

500 g cooking apples
400 g onions
100 g raisins
400 ml vinegar
450 g sugar
1 level teaspoon salt
2 level teaspoons curry powder
½ level teaspoon mustard
½ level teaspoon pepper
½ level teaspoon ground ginger

Peel and core the apples and peel the onions and chop into small pieces. Mix all the ingredients except the sugar and boil gently till soft. Add the sugar and boil for a further 30 minutes. Pour into jars and tie down.

924 Chutney, tomato

1 kg tomatoes
125 g cooking apples
500 g onions
100 g sultanas
450 ml vinegar
500 g sugar
1 level teaspoon salt
½ level teaspoon mustard
¼ level teaspoon pepper
2 level teaspoons curry powder

Peel the tomatoes, chop the apples and onions into small pieces. Mix all the ingredients except the sugar and boil gently until soft. Add the sugar and boil for a further 30 minutes. Pour into jars and tie down.

925 French dressing

25 ml vinegar
75 g olive oil
½ level teaspoon salt
½ level teaspoon pepper

Shake the ingredients together in a screw-topped jar or bottle.

926 Mayonnaise

1 egg yolk
125 g oil
¼ level teaspoon salt
¼ level teaspoon made mustard
20 ml vinegar
pepper

Beat yolk and seasoning in a bowl. Whisk oil in very gradually to form a thick emulsion, adding the vinegar.

927 Onion sauce

White sauce
350 ml milk
25 g flour
25 g margarine
200 g cooked onion
1 level teaspoon salt
pepper

Make the white sauce and add the chopped onion and seasoning.

933 Tomato sauce

400 g tomatoes
25 g carrot
50 g onion
25 g bacon, streaky
15 g margarine
250 ml stock
25 g flour
½ level teaspoon salt
herbs (bouquet garni)

Fry the chopped vegetables gently with the margarine and bacon. Stir in the flour, blended with some of the stock, then the rest of the stock and the herbs. Simmer for 40 minutes, then sieve or liquidise if desired. Reheat, adjust seasoning and serve.

934 and 935 White sauce sweet or savoury

350 ml milk
25 g flour
25 g margarine
30 g sugar
or
½ level teaspoon salt

Melt the fat in a pan. Add the flour and cook for a few minutes, stirring constantly. Add the milk and salt or sugar and cook gently until the mixture thickens.

943 Lentil soup

100 g lentils
25 g carrot
50 g turnip
50 g onion
1 ham bone
25 g margarine
25 g flour
1 litre stock
125 ml milk
herbs (bouquet garni)
salt and pepper to taste

Melt the dripping and toss the lentils and sliced vegetables in it over a gentle heat. Add the stock, seasoning, herbs and ham bone and bring to the boil. Simmer for 2–2½ hours stirring at intervals. Remove the bone. Sieve or liquidise and return to the pan with the flour blended to a smooth cream with the milk. Simmer for 5 minutes, and adjust seasoning.

Appendix 5: Key to the use of amino acid and fatty acid numbers

This key shows the relationship between the numbers in section 1 and the numbers in sections 2 and 3. Primarily it is intended for use in the calculation of amino acid and fatty acid composition per 100g of food.

Description of the key

The number in the first column is the code number assigned in section 1; this is followed by the name of the item. A name given in bold italics *does not* appear in section 1 of the tables and is usually a grouped item appearing in one of the other sections.

The following columns give the appropriate code numbers in the amino acid section and fatty acid sections respectively.

In many cases amino acids or fatty acids have not been measured in a particular foodstuff; where it is reasonable to assume that little error would result from the use of values for a related food the code number of this food is given in parentheses.

The value zero indicates that no protein or fat is present in the food.

A dash means that it is not possible to give any guidance about which is the most appropriate value to use. Where fruits and vegetables are concerned a dash in the fatty acid column for all practical purposes can be regarded as zero. This generalisation applies for many foods and the symbol ∅ is used where little or no error would result from ignoring this food in amino acid or fatty acid calculations.

The symbol 'a' indicates that the value for fatty acids will depend on the fat used in cooking and that no general guidance can be given.

The symbol 'R' indicates that the values have been derived from calculations from a recipe. The appropriate items to use in calculating amino acids and fatty acids are given in part 2 of this key.

Calculation of amino acids and fatty acids per 100 g food

Amino acids per 100 g are obtained by multiplying the total nitrogen (g/100 g, given in section 1) by the amino acids composition (mg amino acid per g N) of the appropriate item in section 2 as indicated in the key.

For example:

Item 252 **Beef rump steak** grilled, lean and fat
Total N = 4.36 g/100 g
Corresponding item in amino acid section, 2237

	Amino acids			
	Ile	Leu	Lys	etc
Amino acids in 2237 (mg/gN)	320	500	570	
Amino acids in 252 (mg/100 g) = (4.36 × values in 2237)	1395	2180	2485	etc

The values for cooked dishes are based on the information given in part 2 of the key. The amino acids in the ingredients are multiplied by the amount of total nitrogen contributed by the ingredients to 100 g of the dish.

For example:

Item 116 **Pancakes**

Ingredients	Item in section 2	Total N contributed by ingredients
Flour	2011	0.43
Milk	2123	0.32
Egg	2165	0.25

The amino acids in the cooked dish are derived as follows:

	Amino acids			
	Ile	Leu	Lys	etc
From flour (0.43 × values in 2011)	103	189	52	
milk (0.32 × values in 2123)	112	205	163	
egg (0.25 × values in 2165)	88	130	98	
Amino acids in item 116 (mg/100 g) =	303	524	313	etc

Fatty acids per 100 g are calculated in a similar way using the compositions given in section 3, ***except*** that the factor giving the proportion of fatty acids in the total fat must also be used in the calculation (see p17).

For example:

Item 252 **Beef rump steak** grilled, lean and fat
Fat 12.1 g/100 g; factor 0.953
Corresponding item in fatty acid section 3240

	Fatty acids			
	14:0	15:0	16:0	etc
Fatty acids in item 3240 (g/100 g total fatty acids)	3.2	0.6	26.9	
Fatty acids in item 252 (g/100 g) = (12.1 × 0.953 × values in 3240/100)	0.37	0.07	3.10	

For cooked dishes the values in the fatty acid section are multiplied by values for the fatty acids contributed by the ingredients of the dish.

For example:

Item 116 **Pancakes**

Ingredient	Item in section 3	Fat contributed	Factor
Flour	3008	0.29	0.67
Milk	3123	2.20	0.945
Eggs	3165	1.39	0.830
Lard	3185	12.42	0.956

The fatty acids in the cooked dish are derived as follows:

	Fatty acids		
	14:0	16:0	18:0
From flour (0.29 × 0.67 × values in 3008/100)	Tr	0.03	Tr
milk (2.20 × 0.945 × values in 3123/100)	0.23	0.54	0.23
eggs (1.39 × 0.830 × values in 3165/100)	Tr	0.33	0.11
lard (12.42 × 0.956 × values in 3185/100)	0.19	3.18	1.85
Fatty acids in 116 (g/100g)	0.42	4.08	2.19

Appendix 5 (part 1): Key to the use of amino acid and fatty acid numbers

		Food names	Amino acids: tables section 2	Fatty acids: tables section 3
Cereals	1	**Arrowroot**	—	—
	2	**Barley** pearl, raw	2002	3002
	3	boiled	2002	3002
	4	**Bemax**	2009	3207
	5	**Bran** wheat	2005	3005
	6	**Cornflour**	2006	3198
	7	**Custard powder**	2006	3198
	8	***Flour general***	—	3008
	9	wholemeal (100%)	2009	3008
	10	brown (85%)	2010	3008
	11	white (72%) breadmaking	2011	3008
	12	household plain	2011	3008
	13	self raising	2011	3008
	14	patent (40%)	2011	3008
	15	**Macaroni** raw	2011	3008
	16	boiled	2011	3008
	17	**Oatmeal** raw	2017	3017
	18	**Porridge**	2017	3017
	19	**Rice** polished, raw	2019	3019
	20	boiled	2019	3019
	21	**Rye** flour (100%)	2021	3021
	22	**Sago** raw	—	—
	23	**Semolina** raw	2011	3008
	24	**Soya** flour, full fat	2024	3205
	25	low fat	2024	3205
	26	**Spaghetti** raw	2011	3008
	27	boiled	2011	3008
	28	canned in tomato sauce	2011	—
	29	**Tapioca**	—	—
	30	**Bread** wholemeal	2009	3030
	31	brown	2010	3030
	32	Hovis	2010	(3030)
	33	white	2011	3033
	34	white, fried	2011	a
	35	toasted	2011	3033
	36	dried crumbs	2011	3033
	37	currant	2011	3033
	38	malt	2009	—
	39	soda	R	R
	40	**Rolls** brown, crusty	2010	3030
	41	soft	2010	(3030)
	42	white crusty	2011	(3033)
	43	soft	2011	(3033)
	44	starch reduced	2011	(3008)
	45	**Chapatis** with fat	2010	a
	46	without fat	2010	3008
	47	**All-bran**	2005	3005
	48	**Cornflakes**	2006	3198
	49	**Grapenuts**	2002	—

		Food names	Amino acids: tables section 2	Fatty acids: tables section 3
Cereals *continued*	50	**Muesli**	(2017)	(3017)
	51	**Puffed wheat**	2009	3008
	52	**Ready Brek**	2017	3017
	53	**Rice Krispies**	2019	3019
	54	**Shredded Wheat**	2009	3008
	55	**Special K**	—	—
	56	**Sugar Puffs**	2009	3008
	57	**Weetabix**	2009	3008
		Biscuits		
	58	**Chocolate** full coated	2011	3058
	59	**Cream crackers**	2011	—
	60	**Crispbread** rye	2021	3060
	61	wheat, starch reduced	2011	3061
	62	**Digestive** plain	2010	—
	63	chocolate	—	3063
	64	**Ginger nuts**	(2011)	3064
	65	**Home made**	R	R
	66	**Matzo**	(2011)	3066
	67	**Oatcakes**	(2017)	3067
	68	**Sandwich**	(2011)	3068
	69	**Semi-sweet**	(2011)	3069
	70	**Short-sweet**	(2011)	3070
	71	**Shortbread**	R	R
	72	**Wafers** filled	(2011)	3072
	73	**Water biscuits**	(2011)	—
	74	**Fancy iced cakes**	(2011)	3074
	75	**Fruit cake** rich	R	R
	76	rich, iced	R	R
	77	plain	—	3077
	78	**Gingerbread**	R	R
	79	**Madeira cake**	(2011)	3079
	80	**Rock cakes**	R	R
	81	**Sponge cake** with fat	R	R
	82	**Sponge cake** without fat	R	R
	83	jam filled		3083
	84	**Currant buns**	2011)	—
	85	**Doughnuts**	(2011)	—
	86	**Eclairs**	R	R
	87	**Jam tarts**	R	R
	88	**Mince pies**	R	R
	89	**Pastry, choux** raw	R	R
	90	cooked	R	R
	91	**flaky** raw	2011	R
	92	cooked	2011	R
	93	**shortcrust** raw	2011	R
	94	cooked	2011	R
	95	**Scones**	R	R
	96	**Scotch pancakes**	R	R
	97	**Apple crumble**	R	R
	98	**Bread and butter pudding**	R	R
	99	**Cheesecake**	R	R
	100	**Christmas pudding**	R	R
	101	**Custard** egg	R	R
	102	made with powder	2123	3123
	103	**Custard tart**	R	R
	104	**Dumpling**	R	R

		Food names	Amino acids: tables section 2	Fatty acids: tables section 3
Cereals *continued*	105	**Fruit pie** individual	—	—
	106	pastry top	R	R
	107	**Ice cream** dairy	2123	3107
	108	non-dairy	2123	3108
	109	**Jelly** packet, cubes	2959	0
	110	made with water	2959	0
	111	made with milk	R	3123
	112	**Lemon meringue pie**	R	R
	113	**Meringues**	2166	0
	114	**Milk pudding**	R	R
	115	canned, rice	(R)	(R)
	117	**Queen of puddings**	R	R
	116	**Pancakes**	R	R
	118	**Sponge pudding** steamed	R	R
	119	**Suet pudding** steamed	R	R
	120	**Treacle tart**	R	R
	121	**Trifle**	R	R
	122	**Yorkshire pudding**	R	R
Milk	123	***Milk, cows'***	2123	3123
	124	fresh, whole, summer	2123	3123
	125	fresh, whole, winter	2123	3123
	126	***Milk, fresh, whole, Channel Islands***	2123	3123
	127	Channel Islands, summer	2123	3123
	128	Channel Islands, winter	2123	3123
	129	sterilised	2123	3123
	130	longlife (UHT treated)	2123	3123
	131	fresh, skimmed	2123	3123
	132	condensed, whole, sweetened	2123	3123
	133	condensed, skimmed, sweetened	2123	3123
	134	evaporated, whole, unsweetened	2123	3123
	135	dried, whole	2123	3123
	136	dried, skimmed	2123	3123
	137	**Milk, goats'**	—	3137
	138	**human** mature	2138	3138
	139	transitional	(2138)	(3138)
	140	**Butter** salted	2123	3123
	141	***Cream, single***	2123	3123
	142	single, summer	2123	3123
	143	winter	2123	3123
	144	***Cream, double***	2123	3123
	145	double, summer	2123	3123
	146	winter	2123	3123
	147	***Cream, whipping***	2123	3123
	148	whipping, summer	2123	3123
	149	winter	2123	3123
	150	sterilised, canned	2123	3123
	151	**Cheese** Camembert type	2123	3123
	152	Cheddar type	2123	3123
	153	Danish Blue type	2123	3123
	154	Edam type	2123	3123
	155	Parmesan type	2123	3123
	156	Stilton type	2123	3123
	157	cottage	2123	3123
	158	cream	2123	3123
	159	processed	2123	3123
	160	cheese spread	2123	3123

		Food names	Amino acids: tables section 2	Fatty acids: tables section 3
Milk *continued*	161	**Yogurt** low fat, natural	2161	3123
	162	flavoured	2161	3123
	163	fruit	2161	3123
	164	hazelnut	2161	3123
	165	**Eggs** whole, raw	2165	3165
	166	white, raw	2166	0
	167	yolk, raw	2167	3165
	168	dried	2165	3165
	169	boiled	2165	3165
	170	fried	2165	a
	171	poached	2165	3165
	172	omelette	2165	3165
	173	scrambled	2165	3165
	174	**Cauliflower cheese**	R	R
	175	**Cheese pudding**	R	R
	176	**Cheese soufflé**	R	R
	177	**Macaroni cheese**	R	R
	178	**Pizza** cheese and tomato	R	R
	179	**Quiche Lorraine**	R	R
	180	**Scotch egg**	R	R
	181	**Welsh rarebit**	R	R
Fats and Oils	182	**Cod liver oil**	0	
	183	**Compound cooking fat**	0	3183
	184	**Dripping, beef**	0	3184
	185	**Lard**	0	3185
	186	**Low fat spread**	0	3186
	187	**Margarine**	2123	a
		hard animal and vegetable oils	2123	3188
		vegetable oils only	2123	3189
		soft animal and vegetable oils	2123	3190
		vegetable oils only	2123	3191
		polyunsaturated vegetable oils only	2123	3192
	193	**Suet** block	—	3194
	194	shredded	—	3194
	195	**Vegetable oils**	0	a
	196	coconut	0	3196
	197	cottonseed	0	3197
	198	maize, corn	0	3198
	199	olive	0	3199
	200	palm	0	3200
	201	peanut, groundnut	0	3201
	202	rapeseed (high erucic acid)	0	3202
	203	rapeseed (low erucic acid)	0	3203
	204	safflowerseed	0	3204
	205	soyabean	0	3205
	206	sunflowerseed	0	3206
	207	wheatgerm	0	3008
Meat	208	**Bacon, dressed carcase** raw	2210	3209
	210	**lean** average raw	2210	3209
	211	**fat** average raw	2210	3209
	212	**fat** average cooked	2210	3209
	213	**collar joint** raw	2210	3209
	214	boiled lean and fat	2210	3209
	215	boiled lean only	2210	3209

		Food names	Amino acids: tables section 2	Fatty acids: tables section 3
Meat *continued*	216	**Gammon joint** raw	2210	3209
	217	boiled, lean and fat	2210	3209
	218	boiled, lean only	2210	3209
	219	**Gammon rashers** grilled, lean and fat	2210	3209
	220	lean only	2210	3209
	221	**Bacon rashers raw**, back	2210	3209
	222	middle	2210	3209
	223	streaky	2210	3209
	225	**fried,** average lean	2210	3209
	226	back, lean and fat	2210	3209
	227	middle, lean and fat	2210	3209
	228	streaky, lean and fat	2210	3209
	230	**grilled,** average lean	2210	3209
	231	back, lean and fat	2210	3209
	232	middle, lean and fat	2210	3209
	233	streaky, lean and fat	2210	3209
	235	**Beef, dressed carcase** raw	2237	3240
	237	**lean** average, raw	2237	3240
	240	**fat** average, raw	(2237)	3240
	241	**fat** average, cooked	2237	3240
	242	**brisket** raw, lean and fat	2237	3240
	243	boiled, lean and fat	2237	3240
	244	**forerib** raw, lean and fat	2237	3240
	245	roast, lean and fat	2237	3240
	246	lean only	2237	3240
	247	**mince** raw	2237	3240
	248	stewed	2237	3240
	249	**rump steak** raw, lean and fat	2237	3240
	250	fried, lean and fat	2237	3240
	251	lean only	2237	3240
	252	grilled, lean and fat	2237	3240
	253	lean only	2237	3240
	254	**silverside** salted, boiled, lean and fat	2237	3240
	255	lean only	2237	3240
	256	**sirloin** raw, lean and fat	2237	3240
	257	roast, lean and fat	2237	3240
	258	lean only	2237	3240
	259	**stewing steak** raw, lean and fat	2237	3240
	260	stewed, lean and fat	2237	3240
	261	**topside** raw, lean and fat	2237	3240
	262	roast, lean and fat	2237	3240
	263	lean only	2237	3240
	264	**Lamb, dressed carcase** raw	2266	3269
	266	**lean** average, raw	2266	3269
	269	**fat** average, raw	2266	3269
	270	cooked	2266	3269
	271	**breast** raw, lean and fat	2266	3269
	272	roast, lean and fat	2266	3269
	273	lean only	2266	3269
	274	**chops, loin** raw, lean and fat	2266	3269
	275	grilled, lean and fat	2266	3269
	276	lean and fat (with bone)	2266	3269
	277	lean only	2266	3269
	278	lean only (with fat and bone)	2266	3269

Meat *continued*

	Food names	Amino acids: tables section 2	Fatty acids: tables section 3
279	**Lamb cutlets** raw, lean and fat	2266	3269
280	grilled, lean and fat	2266	3269
281	lean and fat (with bone)	2266	3269
282	lean only	2266	3269
283	lean only (with fat and bone)	2266	3269
284	**leg** raw, lean and fat	2266	3269
285	roast, lean and fat	2266	3269
286	lean only	2266	3269
287	**scrag and neck** raw, lean and fat	2266	3269
288	stewed, lean and fat	2266	3269
289	lean only	2266	3269
290	lean only (with fat)	2266	3269
291	**shoulder** raw, lean and fat	2266	3269
292	roast, lean and fat	2266	3269
293	lean only	2266	3269
294	**Pork, dressed carcase** raw	2296	3299
296	**lean** average, raw	2296	3299
299	**fat** average, raw	(2296)	3299
300	cooked	(2296)	3299
301	**belly** rashers, raw	2296	3299
302	grilled, lean and fat	2296	3299
303	**chops, loin** raw	2296	3299
304	grilled, lean and fat	2296	3299
305	lean and fat (with bone)	2296	3299
306	lean only	2296	3299
307	lean only (with bone)	2296	3299
308	**leg** raw, lean and fat	2296	3299
309	roast, lean and fat	2296	3299
310	lean only	2296	3299
311	**Veal, cutlet** fried	2237	3240
312	**fillet** raw	2237	3240
313	roast	2237	3240
314	**Chicken** raw, meat only	2314	3314
315	meat and skin	(2314)	3314
316	light meat	2314	3314
317	dark meat	2314	3314
318	boiled, meat only	2314	3314
319	light meat	2313	3314
320	dark meat	2314	3314
321	roast, meat only	2314	3314
322	meat and skin	2314	3314
323	light meat	2314	3314
324	dark meat	2314	3314
325	wing quarter (with bone)	2314	3314
326	leg quarter (with bone)	2314	3314
327	**Duck** raw, meat only	2327	3328
328	meat with fat and skin	(2327)	3328
329	roast, meat only	2327	3328
330	meat with fat and skin	(2327)	3328
331	**Goose** roast	(2327)	—
332	**Grouse** roast	(2327)	3332
333	roast (with bone)	(2327)	3332
334	**Partridge** roast	(2327)	3334
335	roast (with bone)	(2327)	3334
336	**Pheasant** roast	(2327)	3336
337	roast (with bone)	(2327)	3336

Meat
continued

	Food names	Amino acids tables section 2	Fatty acids tables section 3
338	**Pigeon** roast	(2327)	—
339	roast (with bone)	(2327)	—
340	**Turkey** raw, meat only	2340	3340
341	meat and skin	2340	3340
342	light meat	2340	3340
343	dark meat	2340	3340
344	roast, meat only	2340	3340
345	meat and skin	2340	3340
346	light mat	2340	3340
347	dark meat	2340	3340
348	**Hare** stewed	2350	3350
349	stewed (with bone)	2350	3350
350	**Rabbit** raw	2350	3350
351	stewed	2350	3350
352	stewed (with bone)	2350	3350
353	**Venison** roast	(2237)	—
354	**Brain, calf** and **lamb** raw	2354	—
355	**calf** boiled	2354	—
356	**lamb** boiled	2354	3356
357	***Heart, lamb and ox***	2357	—
358	**Heart, lamb** raw	2357	3358
359	**sheep** roast	2357	3358
360	**ox** raw	2357	3360
361	stewed	2357	3360
362	**pig** raw	2357	—
361	***Kidney, lamb, ox and pig***	2363	—
364	**Kidney, lamb** raw	2363	3364
365	fried	2363	a
366	**ox** raw	2363	3366
367	stewed	2363	3366
368	**pig** raw	2363	3368
369	stewed	2363	3368
370	***Liver, all species***	2370	—
371	**Liver, calf** raw	2370	3371
372	fried	2370	a
373	**chicken** raw	2370	3373
374	fried	2370	a
375	**lamb** raw	2370	3375
376	fried	2370	a
377	**ox** raw	2370	3377
378	stewed	2370	3377
379	**pig** raw	2370	3379
380	stewed	2370	3379
381	**Oxtail** raw	2381	3240
382	stewed	2381	3240
383	stewed (with bone)	2381	3240
384	**Sweetbread, lamb** raw	2384	3384
385	fried	2384	a
386	***Tongue, lamb and ox***	2386	—
387	**Tongue, lamb** raw	2386	3399
388	**sheep** stewed	2386	3399
389	**ox** pickled, raw	2386	—
390	boiled	2386	—
391	**Tripe** dressed	2391	3391
392	stewed	2391	3391
393	**Beef, corned**	2393	3240

		Food names	Amino acids: tables section 2	Fatty acids: tables section 3
Meat *continued*	394	**Ham**	2394	3394
	395	**Ham and pork** chopped	2395	3395
	396	**Luncheon meat**	2396	3396
	397	**Stewed steak with gravy**	2397	3240
	398	**Tongue, lamb and ox,** canned	2398	—
	399	***Tongue, lamb***	—	3399
	400	**Veal, jellied**	2400	—
	401	**Black pudding** fried	2401	—
	402	**Faggots**	2402	—
	403	**Haggis** boiled	2403	—
	404	**Liver sausage**	2404	3404
	405	**Frankfurters**	2405	(3299)
	406	**Polony**	2406	—
	407	**Salami**	2407	(3299)
	408	**Sausages, beef** raw	2408	3408
	409	fried	2408	a
	410	grilled	2408	3408
	411	**Sausages, pork** raw	2411	3411
	412	fried	2411	a
	413	grilled	2411	3411
	414	**Saveloy**	2414	—
	415	**Beefburgers** raw	2415	(3240)
	416	fried	2415	(3240)
	417	**Brawn**	2417	—
	418	**Meat paste**	2418	(3299)
	419	**White pudding**	2419	—
	420	**Cornish pastie**	2420	—
	421	**Pork pie** individual	2421	—
	422	**Sausage roll** flaky pastry	R	R
	423	short pastry	R	R
	424	**Steak and kidney pie** with pastry top only	R	R
	425	individual	2425	3425
	426	**Beefsteak pudding**	R	R
	427	**Beef stew**	R	R
	428	**Bolognese sauce**	R	R
	429	**Curried meat**	R	R
	430	**Hot pot**	R	R
	431	**Irish stew**	R	R
	432	**Irish stew** (with bone)	R	R
	433	**Moussaka**	R	R
	434	**Shepherds' pie**	R	R
Fish	435	***White and fatty fish all species***	2435	—
	438	**Cod** raw, fresh fillets	2435	3438
	439	frozen steaks	2435	3438
	440	baked	2435	3438
	441	baked (with bones and skin)	2435	3438
	442	fried in batter	2435	a
	443	grilled	2435	3438
	444	poached	2435	3438
	445	poached (with bones and skin	2435	3438
	446	steamed	2435	3438
	447	steamed (with bones and skin)	2435	3438
	448	**Cod smoked** raw	2435	3438
	449	poached	2435	3438
	450	**dried** salt, boiled	2435	3438

		Food names	Amino acids: tables section 2	Fatty acids: tables section 3
Fish *continued*	451	**Haddock, fresh** raw	2435	3451
	452	fried	2435	a
	453	fried (with bone)	2435	a
	454	steamed	2435	3451
	455	steamed (with bones)	2435	3451
	456	**smoked** steamed	2435	3451
	457	steamed (with bones and skin)	2435	3451
	458	**Halibut** raw	2435	3458
	459	steamed	2435	3458
	460	steamed (with bones and skin)	2435	3458
	461	**Lemon sole** raw	2435	3461
	462	fried	2435	a
	463	fried (with bone)	2435	a
	464	steamed	2435	3461
	465	steamed (with bones and skin)	2435	3461
	466	**Plaice** raw	2435	3466
	467	fried in batter	2435	a
	468	fried in crumbs	2435	a
	469	steamed	2435	3466
	470	steamed (with bones and skin)	2435	3466
	471	**Saithe** raw	2435	3471
	472	steamed	2435	3471
	473	steamed (with bones and skin)	2435	3471
	474	**Whiting** raw	2435	3474
	475	fried	2435	a
	476	fried (with bones)	2435	a
	477	steamed	2435	3474
	478	steamed (with bones and skin)	2435	3474
	480	**Eel** raw	2435	—
	481	stewed	2435	—
	482	**Herring** raw	2435	3482
	483	fried	2435	a
	484	**Herring** fried (with bones)	2435	a
	485	grilled	2435	3482
	486	grilled (with bones)	2435	3482
	487	**Bloater** grilled	2435	3482
	488	grilled (with bones)	2435	3482
	489	**Kipper** baked	2435	3482
	490	baked (with bones)	2435	3482
	491	**Mackerel** raw	2435	3491
	492	fried	2435	a
	493	fried (with bones)	2435	a
	494	**Pilchards** canned in tomato sauce	2435	3494
	495	**Salmon** raw	2435	(3498)
	496	steamed	2435	(3498)
	497	steamed (with bones and skin)	2435	(3498)
	498	canned	2435	3498
	499	smoked	2435	(3498)
	500	**Sardines** canned in oil, fish only	2435	(3501)
	501	fish plus oil	2435	3501
	502	canned in tomato sauce	2435	3502
	503	***Sprats, raw***	—	3503
	504	**Sprats** fried	2435	a
	505	fried (with bones)	2435	a
	506	**Trout, brown** steamed	2435	—
	507	steamed (with bones)	2435	—

		Food names	Amino acids tables section 2	Fatty acids tables section 3
Fish *continued*	508	**Tuna** canned in oil	2435	3508
	509	**Whitebait** fried	2435	a
	510	***Dogfish, raw***	2435	3510
	511	**Dogfish** fried in batter	2435	—
	512	fried (with waste)	2435	a
	513	***Skate, raw***	2435	3513
	514	**Skate** fried in batter	2435	a
	515	fried (with waste)	2435	a
	516	***Crustacea, all kinds***	2516	—
	517	***Crab, raw***	—	3517
	518	**Crab** boiled	2516	3517
	519	boiled (with shell)	2516	3517
	520	canned	2516	3517
	521	**Lobster** boiled	2516	3517
	522	boiled (with shell)	2516	3517
	523	**Prawns** boiled	2516	3517
	524	boiled (with shell)	2516	3517
	525	**Scampi** fried	2516	a
	526	***Shrimps, raw***	—	3526
	527	**Shrimps** boiled	2516	3526
	528	boiled (with shell)	2516	3526
	529	canned	2516	3526
	530	***Molluscs all kinds***	2530	—
	531	**Cockles** boiled	2530	—
	532	**Mussels** raw	2530	3532
	533	boiled	2530	3532
	534	boiled (with shell)	2530	3532
	535	**Oysters** raw	2530	3535
	536	raw (with shell)	2530	3535
	537	***Scallops raw***		3537
	538	**Scallops** steamed	2530	3537
	539	**Whelks** boiled	2530	—
	540	boiled (with shell)	2530	—
	541	**Winkles** boiled	2530	—
	542	boiled (with shell)	2530	—
	543	**Fish cakes** frozen	—	—
	544	fried	—	—
	545	**Fish fingers** frozen	R	a
	546	fried	R	a
	547	**Fish paste**	—	—
	548	**Fish pie**	R	R
	549	**Kedgeree**	R	R
	550	**Roe, cod** hard, raw	—	3550
	551	fried	—	a
	552	**herring** soft, raw	—	—
	553	fried	—	a
Vegetables	554	**Ackee** canned	—	—
	555	**Artichokes, globe** boiled	—	∅
	556	boiled (as served)	—	∅
	557	**Jerusalem** boiled	—	∅
	558	**Asparagus** boiled	2558	∅
	559	boiled (as served)	2558	∅
	560	**Aubergine** raw	—	∅
	561	**Beans, French** boiled	2561	∅
	562	**runner** raw	2561	3562

		Food names	Amino acids: tables section 2	Fatty acids: tables section 3
Vegetables *continued*	563	boiled	2561	3562
	564	**broad** boiled	2564	—
	565	**butter** raw	2565	—
	566	boiled	2565	—
	567	**haricot** raw	2567	—
	568	boiled	2567	—
	569	**baked** canned in tomato sauce	2567	3569
	570	**mung** green gram, raw	—	—
	571	cooked, dahl	—	—
	572	**red kidney** raw	2572	—
	573	**Beansprouts** canned	—	∅
	574	**Beetroot** raw	2574	∅
	575	boiled	2574	∅
	576	**Broccoli tops** raw	2576	∅
	577	boiled	2576	∅
	578	**Brussels sprouts** raw	2578	∅
	579	boiled	2578	∅
	580	**Cabbage, red** raw	2585	∅
	581	**Savoy** raw	2585	∅
	582	boiled	2585	∅
	583	**spring** boiled	2585	∅
	584	**white** raw	2585	∅
	585	**winter** raw	2585	∅
	586	boiled	2585	∅
	587	**Carrots, old** raw	2587	∅
	588	boiled	2587	∅
	589	**young** boiled	2587	∅
	590	canned	2587	∅
	591	**Cauliflower** raw	2591	∅
	592	boiled	2591	∅
	593	**Celeriac** boiled	—	∅
	594	**Celery** raw	2594	∅
	595	boiled	2594	∅
	596	**Chicory** raw	—	∅
	597	**Cucumber** raw	2597	3597
	598	**Endive** raw	—	∅
	599	**Horseradish** raw	—	∅
	600	**Laverbread**	—	∅
	601	**Leeks** raw	—	∅
	602	boiled	—	∅
	603	**Lentils** raw	2603	—
	604	split, boiled	2603	—
	605	masur dahl, cooked	2603	—
	606	**Lettuce**	2606	
	607	**Marrow** raw	—	∅
	608	boiled	—	∅
	609	**Mushrooms** raw	2609	3609
	610	fried	2609	a
	611	**Mustard and cress** raw	—	∅
	612	**Okra** raw	—	∅
	613	**Onions** raw	2613	∅
	614	boiled	2613	∅
	615	fried	2613	∅
	616	**spring** raw	2613	∅
	617	**Parsley** raw	—	∅

		Food names	Amino acids: tables section 2	Fatty acids: tables section 3
Vegetables *continued*	618	**Parsnips** raw	—	∅
	619	boiled	—	∅
	620	**Peas** fresh, raw	2620	3621
	621	boiled	2620	3621
	622	frozen, raw	2620	3621
	623	boiled	2620	3621
	624	canned, garden	2620	3621
	625	processed	2620	3621
	626	dried, raw	2620	3621
	627	boiled	2620	3621
	628	split, dried raw	2620	3621
	629	boiled	2620	3621
	630	**chick** Bengal gram raw	2630	—
	631	cooked dahl	2630	—
	632	channa dahl	2630	—
	633	**red** pigeon, raw	2633	—
	634	**Peppers, green,** raw	—	3634
	635	boiled	—	3634
	636	**Plantain** green, raw	—	∅
	637	boiled	—	∅
	638	ripe, fried	—	a
	639	**Potatoes, old** raw	2639	3639
	640	boiled	2639	3639
	641	mashed	2639	a
	642	baked	2639	3639
	643	baked (with skin)	2639	3639
	644	roast	2639	a
	645	chips	2639	a
	646	frozen	2639	a
	647	frozen, fried	2639	a
	648	**new** boiled	2639	3639
	649	canned	2639	3639
	650	**instant** powder	2639	3639
	651	made up	2639	3639
	652	**Potato crisps**	2639	a
	653	**Pumpkin** raw	—	∅
	654	**Radishes** raw	—	∅
	655	**Salsify** boiled	—	∅
	656	**Seakale** boiled	—	∅
	657	**Spinach** boiled	2657	3657
	658	**Spring greens** boiled	2669	∅
	659	**Swedes** raw	—	∅
	660	boiled	—	∅
	661	**Sweetcorn, on-the-cob** raw	2006	3198
	662	boiled	2006	3198
	663	canned, kernels	2006	3198
	664	**Sweet potatoes** raw	2664	3664
	665	boiled	2664	3664
	666	**Tomatoes** raw	2666	∅
	667	fried	2666	a
	668	canned	2666	∅
	669	**Turnips** raw	2669	3669
	670	boiled	2669	3669
	671	**Turnip tops** boiled	2671	∅
	672	**Watercress** raw	—	∅

		Food names	Amino acids: tables section 2	Fatty acids: tables section 3
Vegetables *continued*	673	**Yam** raw	2673	∅
	674	boiled	2673	∅
Fruit	675	**Apples, eating**	2675	3675
	676	(with skin and core)	2675	3675
	677	**cooking** raw	2675	3675
	678	baked without sugar	2675	3675
	679	baked (with skin)	2675	3675
	680	stewed, without sugar	2675	3675
	681	with sugar	2675	3675
	682	**Apricots** fresh, raw	2682	∅
	683	raw (with stones)	2682	∅
	684	stewed without sugar	2682	∅
	685	stewed without sugar (with stones)	2682	∅
	686	with sugar	2682	∅
	687	with sugar (with stones)	2682	∅
	688	dried, raw	2682	∅
	689	stewed, without sugar	2682	∅
	690	with sugar	2682	∅
	691	canned	2682	∅
	692	**Avocado pears**	2692	3692
	693	**Bananas** raw	2693	3693
	694	raw (with skin)	2693	3693
	695	**Bilberries** raw	—	∅
	696	**Blackberries** raw	—	∅
	697	stewed, without sugar	—	∅
	698	with sugar	—	∅
	699	**Cherries, eating** raw	—	∅
	700	raw (with stones)	—	∅
	701	**cooking** raw	—	∅
	702	raw (with stones)	—	∅
	703	stewed without sugar	—	∅
	704	without sugar (with stones)	—	∅
	705	with sugar	—	∅
	706	with sugar (with stones)	—	∅
	707	**Cranberries** raw	—	∅
	708	**Currants, black** raw	—	∅
	709	stewed without sugar	—	∅
	710	with sugar	—	∅
	711	**red** raw	—	∅
	712	stewed without sugar	—	∅
	713	with sugar	—	∅
	714	**white** raw	—	∅
	715	stewed without sugar	—	∅
	716	with sugar	—	∅
	717	**dried**	2736	∅
	718	**Damsons** raw	—	∅
	719	raw (with stones)	—	∅
	720	stewed without sugar	—	∅
	721	without sugar (with stones)	—	∅
	722	with sugar	—	∅
	723	with sugar (with stones)	—	∅
	724	**Dates** dried	2724	∅
	725	dried (with stones)	2724	∅
	726	**Figs, green** raw	2726	∅

		Food names	Amino acids: tables section 2	Fatty acids: tables section 3
Fruit *continued*	727	**dried** raw	2726	∅
	728	stewed without sugar	2726	∅
	729	with sugar	2726	∅
	730	**Fruit pie filling** canned	—	∅
	731	**Fruit salad** canned	—	∅
	732	**Gooseberries, green** raw	—	∅
	733	stewed without sugar	—	∅
	734	with sugar	—	∅
	735	**ripe** raw	—	∅
	736	**Grapes, black** raw	2736	∅
	737	raw (whole grapes)	2736	∅
	738	**white** raw	2736	∅
	739	raw (whole grapes)	2736	∅
	740	**Grapefruit** raw	(2773)	∅
	741	raw (whole fruit)	(2773)	∅
	742	canned	(2773)	∅
	743	**Greengages** raw	—	∅
	744	raw (with stones)	—	∅
	745	stewed without sugar	—	∅
	746	without sugar (with stones)	—	∅
	747	with sugar	—	∅
	748	with sugar (with stones)	—	∅
	749	**Guavas** canned	(2773)	∅
	750	**Lemons** whole	(2773)	∅
	751	juice, fresh	—	∅
	752	**Loganberries** raw	—	∅
	753	stewed without sugar	—	∅
	754	with sugar	—	∅
	755	canned	—	∅
	756	**Lychees** raw	—	∅
	757	canned	—	∅
	758	**Mandarin oranges** canned	(2773)	∅
	759	**Mangoes** raw	—	∅
	760	canned	—	∅
	761	**Medlars** raw	—	∅
	762	**Melons, Canteloupe** raw	2762	∅
	763	raw (with skin)	2762	∅
	764	**yellow, Honeydew** raw	(2762)	∅
	765	raw (with skin)	(2762)	∅
	766	**watermelon** raw	2766	∅
	767	raw (with skin)	2766	∅
	768	**Mulberries** raw	—	∅
	769	**Nectarines** raw	(2779)	∅
	770	raw (with stones)	(2779)	∅
	771	**Olives** in brine	—	3199
	772	in brine (with stones)	—	3199
	773	**Oranges** raw	2773	∅
	774	raw (with peel and pith)	2773	∅
	775	**Orange juice** fresh	2773	∅
	776	**Passion fruit** raw	—	∅
	777	raw (with skin)	—	∅
	778	**Paw-paw** canned	—	∅
	779	**Peaches** fresh, raw	2779	∅
	780	raw (with stones)	2779	∅
	781	dried, raw	2779	∅

		Food names	Amino acids tables section 2	Fatty acids tables section 3
Fruit *continued*	782	stewed without sugar	2779	∅
	783	with sugar	2779	∅
	784	canned	2779	∅
	785	**Pears, eating**	2785	∅
	786	(with skin and core)	2785	∅
	787	**cooking** raw	2785	∅
	788	stewed without sugar	2785	∅
	789	with sugar	2785	∅
	790	canned	2785	∅
	791	**Pineapple** fresh	2791	∅
	792	canned	2791	∅
	793	**Plums, Victoria, dessert** raw	—	∅
	794	raw (with stones)	—	∅
	795	**cooking** raw	—	∅
	796	raw (with stones)	—	∅
	797	stewed without sugar	—	∅
	798	without sugar (with stones)	—	∅
	799	with sugar	—	∅
	800	with sugar (with stones)	—	∅
	801	**Pomegranate juice**	—	∅
	802	**Prunes** dried, raw	—	∅
	803	raw (with stones)	—	∅
	804	stewed without sugar	—	∅
	805	without sugar (with stones)	—	∅
	806	stewed with sugar	—	∅
	807	(with stones)	—	∅
	808	**Quinces** raw	—	∅
	809	**Raisins** dried	2736	∅
	810	**Raspberries** raw	—	∅
	811	stewed without sugar	—	∅
	812	with sugar	—	∅
	813	canned	—	∅
	814	**Rhubarb** raw	—	∅
	815	stewed without sugar	—	∅
	816	with sugar	—	∅
	817	**Strawberries** raw	2817	∅
	818	canned	2817	∅
	819	**Sultanas** dried	2736	∅
	820	**Tangerines** raw	(2773)	∅
	821	raw (with peel and pips)	(2773)	∅
Nuts	822	**Almonds**	2822	3822
	823	(nuts with shells)	2822	3822
	824	**Barcelona nuts**	(2830)	(3830)
	825	(nuts with shells)	(2830)	(3830)
	826	**Brazil nuts**	2826	3826
	827	(nuts with shells)	2826	3826
	828	**Chestnuts**	—	3828
	829	(nuts with shells)	—	3828
	830	**Cob or hazel nuts**	2830	3830
	831	(nuts with shells)	2830	3830
	832	**Coconut** fresh	2832	3832
	833	milk	—	∅
	834	desiccated	2832	3832

		Food names	Amino acids tables section 2	Fatty acids tables section 3
Nuts *continued*	835	**Peanuts** fresh	2835	3835
	836	(nuts with shells)	2835	3835
	837	roasted and salted	2835	3835
	838	**Peanut butter** smooth	2838	3838
	839	**Walnuts**	2839	3839
	840	(nuts with shells)	2839	3839
Sugars	841	**Glucose** liquid	0	0
	842	**Sugar,** Demerara	0	0
	843	white	0	0
	844	**Syrup,** golden	0	0
	845	**Treacle,** black	0	0
	846	**Cherries,** glacé	0	0
	847	**Honey** comb	—	—
	848	in jar	—	—
	849	**Jam** fruit with edible seeds	—	—
	850	stone fruit	—	—
	851	**Lemon curd** starch based	—	—
	852	home made	2165	R
	853	**Marmalade**	2773	0
	854	**Marzipan**	2822	2822
	855	**Mincemeat**	—	—
	856	**Boiled sweets**	0	0
	857	**Chocolate** milk	2857	3857
	858	plain	2858	3858
	859	fancy and filled	—	—
	860	Bounty Bar	—	—
	861	Mars Bar	—	—
	862	**Fruit gums**	—	∅
	863	**Liquorice allsorts**	—	—
	864	**Pastilles**	—	∅
	865	**Peppermints**	∅	∅
	866	**Toffees, mixed**	—	—
Beverages	867	**Bournvita**	—	—
	868	**Cocoa powder**	2868	3868
	869	**Coffee and chicory essence**	(2872)	∅
	870	**Coffee** ground, roasted	—	—
	871	infusion 5 min	(2872)	∅
	872	instant	2872	∅
	873	**Drinking chocolate**	2868	3873
	874	**Horlicks malted milk**	—	—
	875	**Ovaltine**	—	—
	876	**Tea, Indian**	—	∅
	877	infusion	∅	∅
	878	**Coca-cola**	∅	∅
	879	**Grapefruit juice** canned, unsweetened	(2773)	∅
	880	sweetened	(2773)	∅
	881	**Lemonade**	0	0
	882	**Lime juice cordial**	∅	0
	883	**Lucozade**	∅	0
	884	**Orange drink**	∅	0
	885	**Orange juice** canned, unsweetened	2773	∅
	886	sweetened	2773	∅
	887	**Pineapple juice** canned	2791	∅

		Food names	Amino acids tables section 2	Fatty acids tables section 3
Beverages *continued*	888	**Ribena** undiluted	—	∅
	889	**Rosehip syrup** undiluted	—	∅
	890	**Tomato juice** canned	2666	∅
	891	**Brown ale** bottled	(2895)	∅
	892	**Canned** bitter	(2895)	∅
	893	**Draught** bitter	(2895)	∅
	894	mild	(2895)	∅
	895	**Keg**	2895	∅
	896	**Lager** bottled	2896	∅
	897	**Pale ale** bottled	(2895)	∅
	898	**Stout** bottled	(2895)	∅
	899	**Stout** extra	(2895)	∅
	900	**Strong ale**	(2895)	∅
	901	**Cider** dry	∅	0
	902	sweet	∅	0
	903	vintage	∅	0
	904	**Wine, red**	∅	0
	905	**rosé**	∅	0
	906	**white** dry	∅	0
	907	medium	∅	0
	908	sweet	∅	0
	909	sparkling	∅	0
	910	**Port**	∅	0
	911	**Sherry** dry	∅	0
	912	medium	∅	0
	913	sweet	∅	0
	914	**Vermouth** dry	∅	0
	915	sweet	∅	0
	916	**Advocaat**	2165	3165
	917	**Cherry brandy**	∅	0
	918	**Curaçao**	∅	0
	919	**Spirits 70% proof**	0	0
Sauces	920	**Bread sauce**	R	R
	921	**Brown sauce** bottled	—	—
	922	**Cheese sauce**	R	R
	923	**Chutney,** apple	2675	∅
	924	tomato	2666	∅
	925	**French dressing**	—	3199
	926	**Mayonnaise**	2165	a
	927	**Onion sauce**	R	R
	928	**Piccalli**	—	—
	929	**Pickle** sweet	—	—
	930	**Salad cream**	—	—
	931	**Tomato ketchup**	2666	∅
	932	**Tomato pureé**	2666	∅
	933	**Tomato sauce**	R	R
	934	**White sauce** savoury	R	R
	935	sweet	R	R
Soups	937	**Bone and vegetable broth**	—	—
	938	**Soup chicken, cream of** canned	—	—
	939	condensed	—	—
	940	condensed, as served	—	—

		Food names	Amino acids tables section 2	Fatty acids tables section 3
Soups *continued*	941	**chicken noodle** dried	—	—
	942	dried, as served	—	—
	943	**lentil**	R	R
	944	**minestrone** dried	—	—
	945	dried, as served	—	—
	946	**mushroom, cream of** canned	—	—
	947	**oxtail** canned, ready to serve	—	—
	948	dried	—	—
	949	dried, as served	—	—
	950	**tomato, cream of** canned	—	—
	951	condensed	—	—
	952	condensed, as served	—	—
	953	dried	—	—
	954	dried, as served	—	—
	955	**vegetable** canned		—
Miscellaneous	956	**Baking powder**	2011	∅
	957	**Bovril**	2957	∅
	958	**Curry powder**	—	—
	959	**Gelatin**	2959	∅
	960	**Ginger, ground**	—	—
	961	**Marmite**	2961	∅
	962	**Oxo cubes**	—	—
	963	**Mustard powder**	—	—
	964	**Pepper**	—	—
	965	**Salt, block**	0	0
	966	**table**	0	0
	967	**Vinegar**	0	∅
	968	**Yeast** bakers, compressed	2968	—
	969	dried	2968	—

Appendix 5 (part 2): Key extension to calculation of amino acid and fatty acid composition of cooked dishes

	Tables section 2 code	N contributed per 100g	Tables section 3 code	Fat contributed per 100g
39 **Soda bread**	2011	1.17	3008	0.81
	2123	0.20	2123	1.49
65 **Biscuits** home made	2011	0.85	3008	0.6
	2165	0.24	3165	1.4
	2123	0.01	3187	20.0
71 **Shortbread**	2011	1.06	3008	0.7
	2123	0.02	3123	25.3
75 **Fruit cake, rich**	2011	0.26	3008	0.2
	2165	0.23	3165	1.3
			3187	9.5
	2736	0.13		
76 **Fruit cake, rich, iced**	2011	0.17	3008	0.1
	2165	0.21	3165	1.0
			3187	6.4
	2736	0.09		
	2822	0.24	3822	4.0
78 **Gingerbread**	2011	0.66	3008	0.4
	2165	0.25	3165	1.4
	2123	0.06	3123	0.4
			3187	10.4
80 **Rock cakes**	2011	0.64	3008	0.4
	2123	0.05	3123	0.3
	2165	0.18	3165	1.0
			3187	14.6
	2736	0.05		
81 **Sponge cake, with fat**	2011	0.49	3058	0.4
	2165	0.56	3165	3.1
	2123	0.01	3187	23.0
82 **Sponge cake, without fat**	2011	0.50	3008	0.3
	2165	1.14	3165	6.4
86 **Eclairs**	2011	0.25	3008	0.2
	2123	0.07	3123	13.6
	2165	0.28	3165	1.6
			3187	5.8
	2858	0.13	3858	2.8
87 **Jam tarts**	2011	0.56	3008	0.4
			3187	6.5
			3185	8.0
	2682	0.05		
88 **Mince pies**	2011	0.70	3008	0.5
			3187	8.2
			3185	10.0
	2736	0.04	3675	2.0

	Tables section 2 code	N contributed per 100 g	Tables section 3 code	Fat contributed per 100 g
89 **Pastry, choux** raw	2011	0.43	3008	0.3
	2165	0.49	3165	2.7
			3187	10.0
90 **Pastry, choux** cooked	2011	0.66	3008	0.5
	2165	0.76	3165	4.2
			3187	5.4
91 **Pastry, flaky** raw	2011	0.77	3008	0.5
			3187	13.6
			3185	16.5
92 **Pastry, flaky** cooked	2011	1.02	3008	0.7
			3187	17.9
			3185	21.4
93 **Pastry, shortcrust** raw	2011	1.03	3008	0.7
			3187	12.2
			3185	14.9
94 **Pastry, shortcrust** cooked	2011	1.20	3008	0.8
			3187	14.1
			3185	17.3
95 **Scones**	2011	1.10	3008	0.7
	2123	0.20	3123	1.5
			3187	12.4
96 **Scotch pancakes**	2011	0.71	3008	0.5
	2123	0.22	3123	1.6
	2165	0.20	3165	1.1
			3187	8.4
97 **Apple crumble**	2011	0.29	3008	0.2
			3187	6.7
	2675	0.03		
98 **Bread and butter pudding**	2011	0.15	3033	0.2
	2123	0.47	3123	5.7
	2165	0.36	3165	1.9
	2736	0.01		
99 **Cheesecake**	2011	0.26	(3069)	3.2
	2123	0.21	3123	24.2
	2165	0.22	3165	1.2
			3187	6.3
100 **Christmas pudding**	2011	0.46	3008	0.1
			3033	0.4
	2165	0.23	3165	1.3
			3194	8.5
	2736	0.11		
	2822	0.08	3822	1.3
101 **Custard, egg**	2123	0.52	3123	3.8
	2165	0.39	3165	2.2
102 **Custard powder**	2123	0.60	3123	4.4
103 **Custard tart**	2011	0.51	3008	0.4
	2123	0.26	3123	1.9
	2165	0.20	3165	1.1
			3187	6.1
			3185	7.4

		Tables section 2 code	N contributed per 100 g	Tables section 3 code	Fat contributed per 100 g
104	**Dumpling**	2011	0.51	3008	0.4
				3194	11.3
106	**Fruit pie, pastry**	2011	0.28	3008	0.2
	top			3187	3.3
		2675	0.06	3185	4.1
111	**Jelly, made with**	2959	0.25		
	milk	2123	0.22	3123	1.6
112	**Lemon meringue**	2011	0.38	3008	0.3
	pie	2165	0.37	3165	2.0
				3187	6.8
				3185	5.5
114	**Milk pudding**	see type	0.09		
		2123	0.57	3123	4.2
116	**Pancakes**	2011	0.43	3008	0.3
		2123	0.32	3123	2.2
		2165	0.25	3165	1.4
				3185	12.4
117	**Queen of puddings**	2011	0.14	3033	0.2
		2123	0.25	3123	5.7
		2165	0.38	3165	2.0
118	**Sponge pudding**	2011	0.60	3008	0.4
	steamed	2123	0.05	3123	0.3
		2165	0.34	3165	1.9
				3187	13.8
119	**Suet pudding**	2011	0.61	3008	0.2
	steamed			3033	0.4
		2123	0.16	3123	1.1
				3194	16.4
120	**Treacle tart**	2011	0.65	3008	0.4
				3033	0.1
				3187	6.1
				3185	7.4
121	**Trifle**	2011	0.07		
		2123	0.29	3123	4.2
		2165	0.16	3165	0.9
		2682	0.01		
		2822	0.04	3822	1.0
122	**Yorkshire pudding**	2011	0.48	3008	0.3
		2123	0.36	3123	2.7
		2165	0.28	3165	1.6
				3184	5.5
172	**Omelette**	2165	1.64	3165	9.7
		2123	0.01	3123	3.5
173	**Scrambled eggs**	2165	1.59	3165	8.9
		2123	0.08	3123	13.8
174	**Cauliflower cheese**	2011	0.05		
		2123	0.69	3123	5.4
				3187	2.6
		2591	0.16		

	Tables section 2 code	N contributed per 100 g	Tables section 3 code	Fat contributed per 100 g
175 **Cheese pudding**	2011	0.16	3008	0.2
	2123	1.01	3123	8.1
	2165	0.46	3165	2.5
176 **Cheese soufflé**	2011	0.15	3008	0.2
	2123	0.94	2123	7.7
	2165	0.69	3165	3.9
			3187	7.20
177 **Macaroni cheese**	2011	0.36	3008	0.3
	2123	0.83	3123	6.6
			3187	2.8
178 **Pizza, cheese and tomato**	2011	0.50	3008	0.3
	2123	0.94	3123	7.6
			3199	3.5
	2968	0.05		
	2666	0.04		
179 **Quiche Lorraine**	2011	0.41	3008	0.3
	2123	1.02	3123	8.1
	2165	0.39	3165	2.2
	2210	0.55	3209	6.7
			3187	4.8
			3185	5.9
180 **Scotch eggs**	2011	0.17	3008	0.1
	2165	0.85	3165	4.7
	2411	0.84	3411	16.1
181 **Welsh rarebit**	2011	0.6	3033	0.9
	2123	1.9	3123	22.7
422 **Sausage roll, flaky pastry**	2011	0.64	3008	0.4
			3187	11.3
			3185	13.8
	2411	0.57	3411	10.7
423 **Sausage roll, short pastry**	2011	0.74	3008	0.5
			3187	8.8
			3185	10.8
	2411	0.62	3411	11.7
424 **Steak and kidney pie, pastry top**	2011	0.35	3008	0.2
			3187	5.6
			3185	6.8
	2237	1.53	3240	5.0
	2363	0.59	3366	0.6
426 **Beef steak pudding**	2011	0.37	3008	0.2
	2237	1.38	3240	4.5
			3194	7.4
	2613	0.02		
427 **Beef stew**	2011	0.05		
			3184	2.7
	2237	1.47	3240	4.8
	2613	0.03		

	Tables section 2 code	N contributed per 100 g	Tables section 3 code	Fat contributed per 100 g
428 **Bolognese sauce**			3198	4.7
	2237	1.14	3240	6.2
	2587	0.02		
	2595	0.01		
	2666	0.08		
429 **Curried meat**	2011	0.30	3008	0.3
			3198	5.6
	2237	1.40	3240	3.1
	2613	0.03		
	2736	0.02		
	2832	0.02	2832	1.1
430 **Hot pot**	2237	1.29	3240	4.2
	2587	0.02		
	2639	0.14		
	2613	0.04		
431 **Irish stew**	2266	0.65	3269	7.3
	2639	0.15		
	2613	0.03		
432 **Irish stew**	2266	0.59	3269	6.7
(with bone)	2639	0.14		
	2613	0.04		
433 **Moussaka**	2011	0.03		
	2123	0.34	3123	2.72
	2237	0.91	3240	4.9
	2613	0.03		
	2639	0.10		
	2666	0.02		
	2165	0.06	3165	0.3
			3198	5.5
434 **Shepherd's pie**	2123	0.02	3123	0.2
			3187	1.4
	2237	1.11	3240	4.5
	2613	0.01		
	2639	0.10		
545 **Fish fingers** frozen	2435	1.24	a	a
	2011	0.78		
546 fried	2435	1.66	a	a
	2011	0.50		
548 **Fish pie**	2011	0.03		
	2123	0.11	3123	1.5
			3187	3.9
	2435	0.85	3438	0.2
	2638	0.14		
549 **Kedgeree**	2019	0.04	3019	0.1
	2165	0.42	3165	2.3
			3187	4.3
	2435	1.57	3451	0.4
852 **Lemon curd**			3123	10.7
homemade	2165	0.53	3165	2.8

	Tables section 2 code	N contributed per 100 g	Tables section 3 code	Fat contributed per 100 g
920 **Bread sauce**	2011	0.25	3033	0.3
	2123	0.45	3123	3.3
			3187	1.4
922 **Cheese sauce**	2011	0.10	3008	0.1
	2123	1.21	3123	9.5
			3187	5.0
927 **Onion sauce**	2011	0.08	3008	0.1
	2123	0.35	3123	2.5
			3187	3.8
	2613	0.03		
933 **Tomato sauce**	2011	0.10		
			3187	1.3
	2210	0.14	3209	1.1
934 **White sauce, savoury**	2011	0.11		
	2123	1.21	3123	9.5
			3187	5.0
935 **White sauce, sweet**	2011	0.68		
	2123	0.55	3123	4.1
			3187	6.2
943 **Lentil soup**	2011	0.06		
	2123	0.09	3123	0.7
			3187	2.8
	2587	0.03		
	2603	0.53		

Appendix 6: Key to references and sources of data

The following key shows the sources of the data for virtually all the new entries in the tables. It identifies all the new analyses carried out at the Laboratory of the Government Chemist and the Dunn Nutritional Laboratory, and gives references to values derived from the literature. Where a food or nutrient does not appear in this key it means either that the values are used unchanged from the third edition or that no information is available.

The following symbols are used in the key.

Bold type : new analytical data

* from the Laboratory of the Government Chemist

+ from the Dunn Nutritional Laboratory

Prox	Proximate constituents	Cl	Chloride
DF	Dietary fibre	Vit A	Retinol and carotene
N	Total nitrogen	Vit D	Vitamin D
CHO	Carbohydrate	Thi	Thiamin
Na	Sodium	Rib	Riboflavin
K	Potassium	Nic	Nicotinic acid
Ca	Calcium	Vit C	Vitamin C
Mg	Magnesium	Vit E	Vitamin E
P	Phosphorus	Vit B_6	Vitamin B_6
Fe	Iron	Vit B_{12}	Vitamin B_{12}
Cu	Copper	Folic	Folic acid
Zn	Zinc	Pant	Pantothenic acid
S	Sulphur		

The numbers given after the nutrient refer to the number of the reference. The literature in mainly confined to the period 1960–1976, though a few earlier publications are included where relevant. Emphasis has been given to papers reporting original work, although comprehensive review papers and food tables of other countries are included where appropriate. The coverage is of necessity selective, and while a much larger number of references were consulted during the course of the revision, only those that proved to be of direct use are included in this key.

Section 1: Proximate and inorganic constituents and vitamins

Food no

Cereals and cereal products

2 **DF+** Rib 307 Vit E 279 Vit B_6 229 Folic 131 Pant 229
3 Vitamins *calculated from raw*
4 Vitamins *manufacturer's data on carton*
5 **Prox+** **DF+** Inorganics 29 Thi, Rib, Nic, Vit B_6, Pant, Biotin 97 Vit E 279 Folic 49
6 Zn 216 Folic 126
7 **Na+** Cl *Na* × *1.5*
9 **Prox* DF+ K* Ca* Mg* P* Fe* Cu* Zn* Thi* Rib* Nic*** Vit E 279 Vit B_6 155 **Folic*** Pant and Biotin 155
10 **Prox* DF+ K* Ca* Mg* P* Fe* Zn* Thi* Rib* Nic*** Vit B_6 155 **Folic*** Pant 155
11 **Prox* DF+ K* Ca* Mg* P* Fe* Zn* Thi* Rib* Nic*** Vit E 279 Vit B_6 155 **Folic*** Pant and Biotin 155
12 **Prox* DF+ K* Ca* Mg* P* Fe* Zn* Thi* Rib* Nic*** Vit E 279 Vit B_6 155 **Folic*** Pant and Biotin 155
13 **Prox* DF+ Na* K* Ca* Mg* P* Fe* Cu* Zn* Thi* Rib* Nic* Folic***
15 **Water+ N+** Rib 307 Vit B_6 229 Folic 131
16 Water, N *calculated from raw* Thi, Rib, Nic 307
17 **DF+** Vit E 128, 278, 279 Folic 137
18 DF *calculated* Folic 137
19 **DF+** Zn 216 Vit E 128, 278 Folic 131
20 DF *calculated* Zn 216 Folic 137
21 Na 307 Cu 238 Vit E 279 Folic 131
24 **Sugars+ Starch+ DF+** Na 307 Rib 307 Vit B_6 229 Pant 229
25 Sugars, starch, DF *calculated from raw* Na 307 Rib 307 Vit B_6 229 Pant 229
26 **Water+ N+** Rib 307 Vit B_6 229 Folic 131
27 *Calculated from raw*
30 **Water* DF+ Na* K* Fe* Thi* Rib* Nic* Vit B_6*** The remaining nutrients *calculated* (*except* sugars and S *from 3rd edition*)
31 **DF+** The remaining nutrients *calculated* (*except* sugars and S *from 3rd edition*)
32 **DF+** Remaining nutrients *manufacturer's and British Baking Industries Research Association analytical data*
33 Water 163 **DF+** Fat 163 Na 163 K 163 Ca 163 Fe 163 Thi 163 Rib 163 Nic 163 Vit B_6 163 Folic (free) 163 Remaining nutrients *calculated* (*except* sugars, N and S *from 3rd edition*)
34 Fat 145
40–43 Prox 172 Ca, Fe, S 172 Rest of inorganics *calculated* Thi, Rib, Nic, Vit B_6 172
44 Na *manufacturer's data*
45 **Prox* DF+ Inorganics* Thi* Rib* Nic* Folic***
46 **Water+ Fat+** remainder *calculated*
47 **All nutrients*** *except* **DF+**
48 **All nutrients*** *except* **DF+**
49 **Prox* DF+ Inorganics* Thi* Rib* Nic* Folic* Vit E*** Vits A, D and B_{12} *manufacturers' data*

Cereals and cereal products *continued*	*Food no*	
	50	**All nutrients*** *except* **DF+**
	51	**All nutrients*** *except* **DF+**
	52	**All nutrients*** *except* **DF+**
	53	**All nutrients*** *except* **DF+**
	54	**All nutrients*** *except* **DF+**, protein *manufacturers' data*, starch *calculated*
	55	**All nutrients*** *except* **DF+**
	56	**All nutrients*** *except* **DF+**
	57	**All nutrients*** *except* **DF+** sugars *manufacturers' data*
	58	**All nutrients*** *except* **DF+**
	60	**All nutrients*** *except* **DF+** Folic, Pant and Biotin *calculated*
	63	**All nutrients*** *except* **DF+**
	64	**All nutrients*** *except* **DF+**
	66	**All nutrients*** *except* **DF+**
	67	**All nutrients*** *except* **DF+** Folic 137 Pant and Biotin *calculated*
	68	**All nutrients*** *except* **DF+**
	69	**All nutrients*** *except* **DF+** Folic 137
	70	**All nutrients*** *except* **DF+** Folic 137
	72	**All nutrients*** *except* **DF+**
	74	**All nutrients*** *except* **DF+**
	77	**All nutrients*** *except* **DF+**
	79	**All nutrients*** *except* **DF+**
	83	**All nutrients*** *except* **DF+**
	105	**All nutrients*** *except* **DF+**
	107	**All nutrients***
	108	**All nutrients***
	115	**All nutrients***
Milk and milk products	124–125	Water 164 Lactose 164 Protein 164 Fat 164, 207 **Na*** and 73 **K*** and 73, 87 **Ca*** and 164, 196, 249 **Mg*** and 196 **P*** and 249 **Fe*** and 196 **Cu*** and 218, 238, 269 **Zn*** and 18, 216, 269 Cl 249 Vit A 291 Vit D 291 **Rib*** and 52, 249 **Vit C** (boiled Milk)* and 258 Vit E 208, 291 Vit B_6 (boiled milk) 152 Vit B_{12} (boiled milk) 152, 223 Folic 47, 94, 152, 199, 219, 223
	127–128	Water 248 Lactose 248 Protein 248 Fat 207 Vit A 291 Vit D 291 Rib 249, 292
	129	Thi 266 Rib 164 Nic 164 Vit C 266 Vit B_6 266 Vit B_{12} 266 Folic 266
	130	Vit A 48, 93 Thi 48, 93, 111, 266 Rib 93 Nic 93 Vit C 48, 93, 266, 294 Vit E 93 Vit B_6 48, 93, 111, 266 Vit B_{12} 48, 93, 111, 266 Folic 48, 93, 266 Pant 93 Biotin 93
	132	**Prox* Na* K* Ca* Mg* P* Fe* Cu* Zn* Vit A* Thi* Rib* Nic* Vit C* Vit E* Vit B_6* Folic***
	134	**Prox* Na* K* Ca* Mg* P* Fe* Cu* Zn* Vit A* Thi* Rib* Nic* Vit C* Vit E* Vit B_6* Folic***
	135	**Prox* Na* K* Ca* Mg* P* Cu* Zn* Cl* Vit A* Thi* Rib* Nic* Vit E* Vit B_6*** Folic 92, 94
	136	**Prox* Na* K* Ca* Mg* P* Fe* Cu* Zn* Thi* Rib* Nic* Vit C* Vit E* Vit B_6* Vit B_{12}* Folic***
	137	Prox 164, 233 Na, K, Ca, Mg, P 233 Fe 18 Cu 18, 238 Zn 18 Vit A 164 Vit D 164 Thi 121, 164 Rib 121, 164 Nic 121, 164 Vit C 121, 164, 223 Vit B_6 229 Vit B_{12} 223 Folic 94, 223 Pant 121, 164 Biotin 164
	138	**All*** (*except* S and Vit D)
	140	**Water+ Fat+ Na+** Zn 269 Vit A 291 Vit D 125, 291 Vit E 291

Milk and milk products *continued*

Food no	
150	**Prox* Na* Cl***
151	Ca 164 Zn 281 Rib 121 Nic 121, 272 Vit B_6 151, 272 Vit B_{12} 7, 50, 186 Folic 272 Pant 272 Biotin 272
152	Mg 87 P 87 Zn 108 Thi 121, 164 Rib 121, 164 Nic 121, 164, 272 Vit B_6 272 Vit B_{12} 7, 50, 186, 187 Folic 131, 272 Pant 272 Biotin 272
153	Thi 121 Rib 121 Nic 121, 272 Vit B_6 121, 151, 272 Vit B_{12} 7, 50, 121, 186, 187 Folic 272 Pant 272 Biotin 272
154	Nic 272 Vit B_6 272 Vit B_{12} 121 Folic 272 Pant 272 Biotin 272
155	Thi 121 Rib 121 Nic 272 Vit B_6 272
157	**Prox* Inorganics* Thi* Rib* Nic* Vit B_6* Folic***
158	**Prox* Inorganics***
159	**Prox* Inorganics* Vit A* Thi* Rib* Nic* Folic***
161–164	**Prox* Inorganics* Thi* Rib* Nic* Vit C* Vit E* Vit B_6*** Vit B_{12} 7, 186, 187 **Folic***

Eggs

165	Prox 293 Na 293 K 293 Ca 293 Fe 293 Cu 161, 238 Zn 161 Vit A 293 Thi 293 Rib 293 Nic 293 Vit E 293 Vit B_6 246 Vit B_{12} 293 Folic 53, 293 Pant 293
166	Cu 269 Zn 270 Rib *calculated from whole egg* Vit B_6 246 Vit B_{12} 186 Folic 53, 131
167	Cu 238 Zn 270 Vit A, Rib, Vit B_{12} Pant *calculated from whole egg* Vit B_6 246 Folic 53, 131
168	Zn 281 Vit A, Vit B_{12} *calculated from whole* Vit B_6 246 Pant 229
169	Folic 53

Fats and oils

182	Vit A *B.P. specification* Vit D *B.P. specification* Vit E 42, 179, 215
186–187	Water 301 Fat 301 Na 301 Vit A 301 Vit D 301 Vit E 301
194	**Water* Fat* Carbohydrate* Retinol* Carotene* Vit E***
196	Vit E 127, 278, 279
197	Vit E 127, 278, 279
198	Vit E 57, 75, 127, 278, 279
199	Vit E 57, 127, 278, 279
200	Vit E 278
201	Vit E 127, 179, 278
202	Vit E 127, 278
204	Vit E 25, 127, 278, 279
205	Vit E 25, 127, 278, 279
206	Vit E 278
207	Vit E 127, 278, 279

Meat and meat products

For all items marked *, analyses were made for water, total N, fat, Na, K, Ca, Mg, P, Fe, Cu, Zn, S, Cl, thiamin, riboflavin, nicotinic acid, vitamin E, vitamin B_6, vitamin B_{12}, folic acid, pantothenic acid and biotin. Where no analyses were made of a particular cut of meat, the values were calculated from a related cut or raw analyses, and they are given in parenthesis. Retinol, carotene and vitamin C were analysed in liver. The composition of carcase meat was obtained by applying the equations of Callow, 1948 (51) to recent dissection data on tissue proportions—Cuthbertson 1974a and b, 1975 (69, 70, 71). Where no information is given for poultry and offal items, they are the values retained from the 3rd edition.

213–223	*
226–228	*

Meat and meat products *continued*

Food no	
231–233	*
242–263	*
271–275	*
277	*
279–280	*
282	*
284–289	*
291–293	*
301–304	*
306	*
308–310	*
312	**Fe***
313	Fe *calculated from raw*
314	*
315	**Prox*** in skin **Inorganics*** in skin Thi, Rib and Nic in skin 260 Other B vitamins in skin *not accounted for.*
316–321	*
322	**Prox*** in skin Inorganics in skin *calculated*
323–324	*
325–326	Plate waste determined on 10 samples.
327	*
328	**Prox*** in skin Inorganics in skin *calculated*
329	*
330	**Prox*** in skin Inorganics in skin *calculated*
331	Cu 238 Vit B_6 188
332	Thi, Rib and Nic 255
336	Thi, Rib and Nic 255
338	Nic 154 Vit E 44
340	*
341	**Prox*** in skin Inorganics in skin *calculated*
342–344	*
345	**Prox*** in skin Inorganics in skin *calculated*
346–347	*
350	*
351	Vitamins *calculated from raw*
353	Thi 114
354	* Vit A 162 Vit C 162
355	* Vit C *calculated from raw*
356	* Vit C *calculated from raw*
358	* Vit A 162 Vit C 162
359	Vitamins *calculated from raw*
360	* Vit A 162 Vit C 162
361	* Vit C *calculated from raw*
362	Vit A 162 Vit C 162
364	* Vit A 162, 193 Vit C 162
365	* Vitamins A and C *calculated from raw*
366	* Vit A 162, 193 Vit C 162
367	* Vitamins A and C *calculated from raw*
368	* Vit A 162, 193 Vit C 162
369	* Vitamins A and C *calculated from raw*
371	* **CHO*** **Vit A*** **Vit C***
372	* **CHO*** **Vit A*** **Vit C***
373	* **CHO*** **Vit A*** Vit D 96 **Vit C***

Meat and meat products *continued*

Food no	
374	* CHO* Vit A *calculated from raw* Vit C*
375	* CHO* Vit A* Vit C*
376	* CHO* Vit A *calculated from raw* Vit C*
377	* CHO* Vit A* Vit C*
378	* CHO* Vit A *calculated from raw* Vit C*
379	* CHO* Vit A* Vit C*
380	* CHO* Vit A *calculated from raw* Vit C*
381	*
382	*
384	* Vit A 162 Vit C 162
385	* CHO* Vit C *calculated from raw*
387	* Vit C 162
388	Vitamins *calculated from raw*
389	* Vit C 162
390	Vitamins *calculated from raw*
391	* Vit C 162
392	* Vit C *calculated from raw*
393–421	*
425	*

Fish and fish products

Food no	
438	Prox* Inorganics* Thi* Rib* Nic* Vit E* Vit B_6* Vit B_{12}* Folic* Biotin 217
439	Prox* Inorganics* (*except* S) S *calculated* Thi* Rib* Nic* Vit B_6 Vit B_{12}* Folic*
440	Prox* Inorganics* (*except* S) S *calculated* Thi* Rib* Nic* Vit E* Vit B_6* Vit B_{12}* Folic*
443	Prox* Inorganics* (*except* S) S *calculated* Thi* Rib* Nic* Vit B_6* Vit B_{12}* Folic*
444	Prox* Inorganics* (*except* S) S *calculated* Thi* Rib* Nic* Vit E* Vit B_6* Vit B_{12}* Folic*
448	Prox* Inorganics* (*except* S) S *calculated* Thi* Rib* Nic* Vit B_6* Vit B_{12}* Folic*
449	Prox* Inorganics* (*except* S) S *calculated* Thi* Rib* Nic* Vit B_6* Vit B_{12}* Folic*
451	Ca, P, Fe 276 Cu 238 Zn 108 Thi, Rib, Nic 217 Folic 131 Biotin 217
458	Prox 273 Inorganics *calculated* Vit E 38 Folic 131 Pant 190 Biotin 217
461	Prox 273 Inorganics *calculated* Thi, Rib, Nic, Vit B_{12} 217 Folic 131
466	Prox* Inorganics* (*except* S) S *calculated* Thi* and 217 Rib* and 217 Nic* Vit B_6* Vit B_{12}* Folic* Pant 217
467	Prox* Inorganics* (*except* S) S *calculated* Thi* Rib* Nic*
468	Prox* Inorganics* (*except* S) S *calculated* Thi* Rib* Nic* Vit B_6* Vit B_{12}* Folic*
471	Water* 217 Remainder of Prox and Inorganics *calculated* Thi, Rib 217 Nic 34 Vit E 38 Vit B_6 217 Vit B_{12} 217 Pant 34 Biotin 217
480	Zn 281 Vit D 68 Nic, Vit B_6, Vit B_{12}, Pant 217
482	Prox* Inorganics* (*except* S) S *calculated* Thi* and 217 Rib* and 217 Nic* and 217 Vit E* Vit B_6* Vit B_{12}* Folic* Biotin 217
485	Prox* Inorganics* (*except* S) S *calculated* Thi* Rib* Nic* Vit E* Vit B_6* Vit B_{12}* Folic*
491	Prox 273 Inorganics (*except* Cu and Zn) *calculated* Cu 238 Zn 210 Rib, Vit B_{12}, Pant, Biotin 217
494	Prox* Inorganics* Retinol* Vit D* Thi* Rib* Nic* Vit E* Vit B_{12} 7
495	Water 66 Remainder of Prox and Inorganics (*except* Cu and Zn) *calculated* Cu and Zn 276 Thi, Vit B_6, Vit B_{12} 217 Folic 131 Pant and Biotin 217

Fish and fish products *continued*

Food no

498 **Prox* Inorganics*** (*except* S) S *calculated* **Thi* Rib* Nic* Vit E* Vit B_6* Vit B_{12}* Folic***

499 **Prox* Inorganics* Thi* Rib* Nic***

500 **Prox* Inorganics*** (*except* S) S *calculated* **Retinol* Thi* Rib* Nic* Vit E* Vit B_6* Vit B_{12}* Folic***

501 *Calculated on the proportions 0.83 fish; 0.17 olive oil*

502 **Prox* Inorganics*** (*except* S) S *calculated* **Retinol* Thi* Rib* Nic* Vit E* Vit B_6* Vit B_{12}* Folic***

508 **Prox* Inorganics*** (*except* S) Vit D 220 **Thi* Rib* Nic* Vit E*** Vit B_6 220 Vit B_{12} 7 Folic 131 Pant and Biotin 220

511 **Prox* Inorganics*** (*except* S) S *calculated* **Thi* Rib* Nic* Vit E***

514 **Prox* Inorganics*** (*except* S) S *calculated* **Thi* Rib* Nic* Vit E***

518 Cu and Zn 210 Folic 131 Biotin 217

520 **Prox* Inorganics*** (*except* Zn and S) Zn 210 **Thi* Rib* Nic***

521 Cu 238 Zn 216 Thi 217 Vit E 3 Vit B_{12} 217 Folic 131 Pant and Biotin 190

523 Cu and Zn 210

525 **Prox* Inorganics*** (*except* S) **Thi* Rib* Nic***

527 Zn 210 Vit B_{12} 217 Biotin 190

529 **Prox* Inorganics*** (*except* S) **Thi* Rib* Nic* Vit B_6* Vit B_{12}* Folic*** Pant 289

531 Cu and Zn 210

532 Cu and Zn 210 Vit E 3, 38

533 Fe and Cu *calculated*

535 Cu and Zn 210 Vit E 3 Folic 126

538 Folic 131 Pant and Biotin 190

539 Zn 210 Vit E 38

541 Cu and Zn 210

543 **Prox* Inorganics*** (*except* S) **Thi* Rib* Nic***

544 **Prox* Inorganics*** (*except* S) **Thi* Rib* Nic***

545 **Prox* Inorganics*** (*except* S) **Thi* Rib* Nic* Vit B_6* Vit B_{12}* Folic***

546 **Prox* Inorganics*** (*except* S) **Thi* Rib* Nic* Vit B_6* Vit B_{12}* Folic***

547 **Prox* Inorganics*** (*except* S) **Thi* Rib* Nic* Vit E***

550 Prox 217 Retinol 242 Vit E 166 Vit B_6 36 Vit B_{12} 217 Pant and Biotin 35

552 Prox 217 Vit C 192 Vit B_{12} and Pant *as cod milt* 34

Vegetables

The ranges which are given for carotene and vitamin C were derived in the main from data collected for the 3rd edition and no further information is given here.

554 **Prox* Inorganics*** (*except* S) **Thi* Rib* Nic* Vit C* Vit B_6* Folic***

555 Thi, Rib, Nic, Vit C, Vit B_6, Folic, Pant, Biotin 64

557 Zn 270 Vit E 31

558 Zn 254 Vit E 31 Folic 131

560 Carotene, Vit C, Thi, Rib, Nic 243 Vit B_6 229 Folic 131, 176

562 **Prox* DF$^+$ Inorganics* Vit E* Folic***

563 **Prox* DF$^+$ Inorganics* Vit E* Folic***

564 Fat 134 Carotene 78, 134 Vit E 31

565 Fat 55 Zn 216 Vit B_6 229 Folic 131 Pant 229

567 Fat 307 Zn 216 Vit B_6 and Pant 229

569 **Prox* DF$^+$ Inorganics* Thi* Rib* Nic* Vit C$^+$ Vit E* Vit B_6* Folic***

570 Water, Protein, Fat 55 CHO 306 Na and K 107 Ca 243 Mg and P 107 Fe 243 Cu, S, Cl 107 Carotene, Thi, Rib, Nic, Vit C 243 Folic 107

Vegetables
continued

Food no

572 Water, Protein, Fat 55 CHO *as haricot beans* Ca 55 P 107 Fe 55 Carotene 243 Thi, Rib, Nic 55 Vit C 243 Vit B_6 and Pant 229
573 **Prox* DF+ Inorganics*** Carotene 243 **Thi* Rib* Nic* Vit C*** and 243 **Vit B_6* Folic***
574 Zn 108, 305 Vit E 31 Folic 131
575 Folic 137
576 **Prox* DF+ Inorganics* Vit C* Vit E*** Vit B_6 244 **Folic***
577 **Prox* DF+ Inorganics* Vit C* Vit E*** Vit B_6 244 **Folic***
578 **Prox* DF+ Inorganics* Vit C* Vit E* Folic***
579 **Prox* DF+ Inorganics*** (*except* S) **Vit C* Vit E* Folic***
580 Carotene 307 Other vitamins *as winter cabbage ex* Vit B_6 244
581 Vit C 227
583 Carotene 253 Vit C 227 Folic 53
584 **Prox* DF+ Inorganics* Vit C* Vit E* Folic***
585–586 **Prox* DF+ Inorganics* Vit C* Vit E*** Vit B_6 244 **Folic*** Pant 229
587–588 Zn 216 Vit E 31 Vit B_6 229 Folic 53, 131
590 Zn 216 Folic 53
591–592 **Prox* DF+ Inorganics* Vit C* Vit E* Folic***
593 Thi, Rib, Nic, Vit C 134 Vit B_6 229
594–595 Zn 108 Vit E 31 Folic 131
596 Zn 254 Carotene, Thi, Rib, Nic, Vit C 134 Vit B_6 229 Folic 131
597 Zn 108, 270 Vit E 31 Folic 131, 137
598 Nic 134 Folic 53
599 Carotene, Thi, Rib, Nic 134 Vit B_6 229
600 **Prox* DF+ Inorganics* Thi* Rib* Nic* Vit C* Vit E* Folic***
601 Vit E 31
602 Vit E 31, 32 Folic 137
603 Fat 243 Zn 216 Carotene, Rib, Nic, Vit C 243 Vit B_6 229 Folic 176
604 **Water+ Fat+**
605 **Water+**
606 **Prox* DF+ Inorganics*** Vit E 31, 57 **Folic*** Pant 229
607 **Prox* Inorganics*** Carotene 224 **Folic***
609 Zn 270 Vit E 31 Folic 131
610 Folic 137
611 Vit C 119
612 Water, Protein, Fat 243 CHO 306 Na 107 K 107, 134 Ca 243 Mg and P 107, 134 Fe 243 Cu, S, Cl 107 Carotene, Thi, Rib, Nic, Vit C 243 Vit B_6 229 Folic 107 Pant 229
613 Zn 108, 270 Vit E 31 Folic 131 Pant 229
614 Folic 53 Pant 229
616 Folic 131
617 Zn 270 Vit E 31
618 Zn 254 Rib 134 Vit E 31 Folic 131
620 Zn 231 Vit E 31 Pant 229
621 Dried peas Vit C 264
622–625 **Prox* DF* Inorganics* Thi* Rib* Nic* Vit C* Vit E* Vit B_6* Folic***
626 Fat 307 Zn 216 Vit B_6 229 Folic 131 Pant 229
627 Fat 307
628–629 Fat 307
630 Water, Protein, Fat *Data from* 243 CHO *calculated* Na, K 107 Ca *data from* 243 P 107, 177, 178 Fe *Data from* 243 Cu, S, Cl, 107 Carotene 107 Thi, Rib, Nic 243 Vit C 107 Folic 107
631 **Water+ Fat+**

Vegetables *continued*

Food no	
632	Prox* DF$^+$ Inorganics* Thi* Rib* Nic* Folic*
633	Water, Protein, Fat 243 CHO *calculated* Na and K 107 Ca 243 Mg and P 107 Fe 243 Cu, S, Cl 107 Carotene, Thi, Rib, Nic, Vit C 243 Folic 107
634–635	Prox* DF$^+$ Inorganics* Carotene 243 Thi* Rib* Nic* Vit C* Vit E* Vit B_6* Folic* Pant 229
636	Water, Protein, Fat 243 CHO 157, 306 Ca 243 Mg and P 107 Fe 243 Cu and S 107 Carotene, Thi, Rib, Nic, Vit C 243 Vit B_6 229 Folic 107 Pant 229
637–638	Prox* DF$^+$ Inorganics* Thi* Rib* Nic* Vit C* Folic*
639	Fe* and 144 Zn 216, 231, 305 Vit E 31 Vit B_6 229 Folic 53, 131
640–642	Fe and Zn *calculated from raw* Vit B_6 229 Folic 53, 131
644–645	Fat$^+$ Fe and Zn *calculated from raw*
646–647	Prox* DF$^+$ Inorganics* Thi* Rib* Nic* Vit C* Vit B_6* Folic*
648	Fe and Zn *calculated from raw*
649	Prox* DF$^+$ Inorganics* Thi* Rib* Nic* Vit C* Vit E* Vit B_6* Folic*
650	Prox* DF$^+$ Inorganics* Thi* Rib* Nic* Vit C* Folic*
652	Prox* DF$^+$ Inorganics* Thi* Rib* Nic* Vit C* Vit E* Vit B_6* Folic*
653	Vit B_6 and Pant 229
654	Zn 254, 270 Vit E 31 Folic 131
657	Zn 108, 216, 270 Vit E 31 Vit B_6 229 Folic 131, 137
658	Vitamins *calculated*
659–660	Vit E 31 Folic 53
661–662	Prox* DF$^+$ Inorganics* Carotene 307 Thi* Rib* Nic* Vit C 307 Vit E* Vit B_6* Folic* Pant 229
663	Prox* DF$^+$ Inorganics* Carotene 307 Thi* Rib* Nic* Vit C 307 Folic*
664	Water 243 Remainder of prox and inorganics *calculated from boiled* Vit B_6 229 Folic 131
665	Vitamins *calculated from raw*
666	Zn 216, 231 Vit C* and 45, 298, 299 Vit E* Folic* Pant 229
668	Prox* DF$^+$ Inorganics* Thi* Rib* Nic* Vit C* Vit E* Vit B_6* Folic*
669	Vit E 31 Folic 131 Pant 229
671	Vit B_6 and Pant 229
672	Zn 254 Rib 134 Vit E 31 Vit B_6 229 Folic 137
673	Water, protein, fat 243 CHO 134, 306 Ca 134 Fe, Cu, Zn *calculated from boiled* Carotene, Thi, Rib, Nic, Vit C 243 Pant 229
674	Prox* DF$^+$ Inorganics* Thi* Rib* Nic* Vit C* Folic*

Fruit

The ranges which are given for carotene and vitamin C were derived in the main from data collected for the 3rd edition, and no further information is given here.

675	Cu Zn 240 Vit C 17, 20, 54, 84, 156, 159, 224, 247 Vit E 31
677	Vit C 54, 159, 247, 310
682	Zn 231 Nic 307 Vit B_6 229
688	Zn 254 Vit B_6 229 Folic 131
691	Zn 108 Vit C$^+$ Folic 137
692	Prox* Na 237 Fe 277 Vit C 243 Vit E* Vit B_6 229 Folic* Pant 229 Biotin 277
693	Zn 108, 216 Vit E 43 Vit B_6 229 Folic 53, 131 Pant 229
695	Prox 281 Inorganics 281 *except* Zn 108 Vitamins 281 *except* Folic 131
696	Vit E 31
699	Zn 108, 254, 281 Rib 247 Vit E 31 Folic 131 Pant 229
707	Thi, Rib 107 Vit B_6 229 Folic 131 Pant 229
708	Vit E 31

Fruit
continued

Food no	
711	Carotene 107 Vit E 31
717	Folic 131
718	Vit E 31
724	Zn 254 Vit B_6 229 Folic 131, 137
726	Zn 254 Vit B_6 229 Pant 229
727	Zn 254 Vit B_6 229 Folic 131 Pant 229
730	**Prox*** **Inorganics*** **Thi*** **Rib*** **Nic***
731	**Fe**$^+$
732	Zn 254 Vit E 31
734	**Vit C**$^+$ in canned
735	Zn 254 Vit E 31
736–738	Zn 281 Folic 31 Biotin 241
740	Zn 231 Vit B_6 229 Folic 53, 131, 284 Pant 229
742	**Prox*** **DF**$^+$ **Inorganics*** **Folic***
749	**Prox*** **DF**$^+$ **Inorganics*** Carotene, Thi, Rib, Nic *estimated from raw* 243 **Vit C*** Vit C in raw 243 Vit B_6 229
750	Vitamins 26
751	Vitamins 26 Folic 131
756	Prox 243 Na, K, Ca 307 Mg and P 107 Fe 243 S and Cl 107 Thi, Rib, Nic 243, Vit C 107, 243, 307
757	**Prox*** **DF**$^+$ **Inorganics*** **Vit C***
758	**Prox*** **DF**$^+$ **Inorganics*** **Vit C*** Vit B_6 26 **Folic***
759	Water and Protein 243 CHO 72 Na and K 307 Ca 243 Mg and P 307 Cu 315 Carotene, Thi, Rib, Nic, Vit C 243 Pant 229
760	**Prox*** **DF**$^+$ **Inorganics*** **Vit C***
761	Vit C 24
762–764	Zn 108 Carotene, Rib 307 Vit E 31, 43 Vit B_6 229 Folic 131
766	Prox 243 Na and K 134, 315 Ca 134, 243, 315 Mg and P 134, 315 Fe 134, 243, 315 Cu 315 Carotene and Thi 243 Rib and Nic 134, 243 Vit C 243 Folic 131 Pant 229
768	Thi 224 Vit C 24, 243
769	Folic 131
771	Carotene 307
773	Zn 108, 216 Vit B_6 229 Folic 49, 53, 131, 284
775	Zn 108, 216, 231 Vit B_6 229 Folic 49, 284
776	Carotene, Thi, Rib, Nic 243
778	**Prox*** **DF**$^+$ **Inorganics*** Thi, Rib, Nic *based on* 243 **Vit C***
779	Zn 216, 231 Folic 131
781	Rib and Nic 307 Vit B_6 229
784	**Fe**$^+$
785–787	Zn 86 Vit E 31 Folic 131 Pant 229
790	**Fe**$^+$
791	Zn 254, 281 Vit B_6 229 Folic 131 Pant 229
792	**K**$^+$ **Fe**$^+$ **Vit C**$^+$ Vit B_6 229 Folic 131 Pant 229
793	Zn 254, 281 Folic 131
801	Thi, Rib, Nic, Vit C 243
802	Vit B_6 229 Folic 131 Pant 229
808	Thi, Rib, Nic, Vit C 243
809	Zn 254 Rib 307 Vit B_6 229 Folic 131 Pant 229
810	Vit E 31 Vit B_6 and Pant 229
813	Prox and inorganics 78 Thi, Nic, Vit B_6, Pant 78
814	Rib, Nic 134 Vit E 31 Vit B_6 229 Folic 131
816	**Vit C**$^+$ in canned

	Food no
Fruit *continued*	
	817 Zn 254 Vit E 31 Vit B_6 229 Folic 131 Pant 229
	818 **Prox*** **DF**$^+$ **Inorganics*** **Vit C*** Vit B_6 229 **Folic*** Pant 229
	819 Vit E 30
	820 Zn 270 Vit B_6 229 Folic 131, 284 Pant 229
Nuts	822 Zn 269 Thi and Rib 307 Vit E 181, 279 Vit B_6 229 Folic 131 Pant 229
	826 Zn 269 Rib, Nic 307 Vit E 181 Vit B_6 and Pant 229
	828 Rib 307 Vit E 181 Vit B_6 229, 245 Pant 229
	830 Zn 270 Rib 307 Vit E 181 Vit B_6 229 Folic 131 Pant 229
	832 Zn 270 Vit E 181 Vit B_6 229 Folic 131 Pant 229
	833 Thi, Rib, Nic, Vit C 307 Vit B_6 229, 245 Pant 229
	834 Thi, Rib, Nic 307
	835 Zn 216, 270 Vit E 181, 279 Folic 131, 137
	837 **Na**$^+$ Cl *calculated* Vit B_6 229
	838 **Prox*** **DF**$^+$ **Inorganics*** **Thi* Rib* Nic* Vit E* Vit B_6* Folic***
	839 Zn 108, 270 Vit E 181, 279 Vit B_6 229 Folic 131 Pant 229
Sugars, preserves and confectionery	849 **Vit C*** in blackcurrant jam
	851 **Prox*** **DF**$^+$ **Inorganics***
	853 **DF**$^+$ Folic 137
	855 **Prox*** **DF**$^+$ **Inorganics***
	857 **Prox*** **Inorganics*** **Thi* Rib* Nic* Vit E***
	858 **Prox*** **Inorganics*** **Thi* Rib* Nic* Vit E***
Beverages	867 Water, Protein, Fat *manufacturers' data* **CHO**$^+$ Inorganics *manufacturers' data*
	868 **Prox*** **Inorganics*** **Thi* Rib* Nic* Vit E* Vit B_6* Folic***
	869 **Prox*** **Thi* Rib* Nic***
	872 **Water*** **Nitrogen*** **Inorganics*** **Thi* Rib* Nic*** and 225, 287 Pant and Biotin 287
	873 **Prox*** **Inorganics*** **Thi* Rib* Nic* Vit E* Vit B_6* Folic***
	874 Prox, Inorganics, Retinol, Vit D, Thi, Rib, Nic *manufacturers' data*
	875 Prox *manufacturers' data* **Na**$^+$ **K**$^+$ **Fe**$^+$ Vit D, Thi *manufacturers' data on label*
	876 Purine N 12 CHO 209, 282 Fat 209, 282 Zn 282 Carotene, Thi, Rib, Nic, Vit C, Pant 282
Soft drinks, fruit and vegetable juices	878 **Water*** **CHO**$^+$ *and manufacturers' data* **Inorganics*** *except* Fe *manufacturers' data*
	879 **Prox*** **Inorganics*** **Vit C* Folic***
	880 **Prox*** **Inorganics*** **Vit C***
	883 Prox, Inorganics and Vit C *manufacturers' data*
	884 Prox and Inorganics *manufacturers' data* **Vit C**$^+$
	885 **Prox*** **Inorganics*** **Vit C* Folic***
	886 **Prox*** **Inorganics*** **Vit C***
	888 Prox and inorganics *manufacturers' data* **Vit C**$^+$
	889 **Water*** **CHO*** **Na*** **K*** **Vit C*** **Folic***
	890 **Water*** **Protein*** **CHO**$^+$ **Inorganics*** **Vit C*** **Folic***
Alcholic beverages	891 **Thi* Rib* Nic* Vit B_6* Vit B_{12}* Folic***
	892 **Inorganics***
	893 **Thi* Rib* Nic* Vit B_6* Vit B_{12}* Folic***
	895 **Alcohol* Solids* Nitrogen* Inorganics* Thi* Rib* Nic* Vit B_6* Vit B_{12}* Folic***

	Food no	
Alcoholic beverages *continued*		
	896	Alcohol* Solids* Nitrogen* CHO+ Inorganics* Thi* Rib* Nic* Vit B_6* Vit B_{12}* Folic*
	897–900	Thi* Rib* Nic* Vit B_6* Vit B_{12}* Folic*
	901–902	Thi, Rib, Nic, Pant 109
	904	Thi* Rib* Nic* Vit B_6* Vit B_{12}* Folic*
	905–906	Prox* Inorganics* Thi* Rib* Nic* Vit B_6* Vit B_{12}* Folic*
	907–910	Thi* Rib* Nic* Vit B_6* Vit B_{12}* Folic*
	911	K+ Thi* Rib* Nic* Vit B_6* Vit B_{12}* Folic*
	912	Prox* Inorganics* Thi* Rib* Nic* Vit B_6* Vit B_{12}* Folic*
	913	K+ Thi* Rib* Nic* Vit B_6* Vit B_{12}* Folic*
	914–915	Prox* Inorganics* Thi* Rib* Nic* Vit B_6* Vit B_{12}* Folic*
	916–918	Prox*
Sauces and pickles	928	Prox* Inorganics* Thi* Rib* Nic*
	929	Prox* Inorganics* Thi* Rib* Nic*
	930	Prox *manufacturers' data*
	932	Water+ Na+
Soups	938	Prox* Inorganics* Thi* Rib* Nic* Vit B_6*
	939	Prox* Cl* Rib* Nic*
	940	*Calculated from soup as purchased*
	941	Prox* Inorganics* Thi* Rib* Nic*
	944	Prox* DF+ Inorganics* Thi* Rib* Nic*
	946	Prox* Inorganics* Thi* Rib* Nic* Vit B_6*
	947	Prox* Inorganics* Thi* Rib* Nic* Vit B_6*
	948	Prox* DF+ Inorganics* Thi* Rib* Nic* Vit B_6*
	950	Prox* Inorganics* Carotene* Thi* Rib* Nic* Vit B_6* Folic*
	951	Prox* Cl* Rib* Nic*
	952	*Calculated from soup as purchased*
	953	Prox* DF+ Inorganics* Thi* Rib* Nic* Folic*
	955	Prox* Inorganics* Carotene* Thi* Rib* Nic* Vit B_6* Folic*
Miscellaneous	957	Prox* Inorganics* Thi* Rib* Nic* Vit B_6* Vit B_{12}* Folic*
	959	Water 307 Nitrogen+
	961	Prox* Inorganics* Thi* Rib* Nic* Vit B_6* Vit B_{12}* Folic*
	968	Water, Fat 243 CHO+ DF+ Na, K 307 Ca 243 Mg, P 307 Fe 243 Thi, Rib, Nic 307 Vit B_6 and Pant 229
	969	Water, Fat 243 Nitrogen+ and 295 Na, K 307 Ca 243 Mg and P 307 Fe 243 Cu 184, 238 Zn 76 184, 270 Thi 307 Rib and Nic 243 Vit E 77 Vit B_6 229 Folic 49, 267, 286 Pant 229 Biotin 116

Section 2: Amino acids

As the new analytical values are marked with an asterisk in section 2, this key refers only to the literature sources.

	Food no	
Cereals	2002	91
	2005	91
	2006	91, 257
	2009	91, 173
	2010	91
	2011	91, 163
	2017	91, 262
	2019	91
	2021	91, 230
	2024	91
Milk	2123	91, 174, 311
	2166–2167	*Calculated from whole egg, using data of Lunven* 195
Meat		All values new analytical data
Fish	2435	16, 19, 37, 63, 83, 91
	2516	91
	2530	91
Vegetables	2558	78
	2561	91
	2564	91
	2565	91
	2567	91
	2572	91
	2574	91
	2576	91
	2578	91
	2585	91
	2587	91
	2591	91
	2594	91
	2597	91
	2603	91
	2606	91
	2609	78, 91, 205
	2613	91
	2620	91
	2630	91
	2633	91
	2639	91, 149
	2657	91
	2664	91
	2666	78, 91
	2669	91
	2671	91
	2673	91, 95

Section 3: Fatty acids

As the new analytical values are marked with an asterisk in section 3, this key refers only to the literature sources.

	Food no	
Cereals	3002	21, 189, 252
	3005	*as whole wheat*
	3008	46, 89, 128, 141, 221, 275
	3017	21, 128, 189
	3019	128
	3021	21, 189, 275
Milk and eggs	3123	104, 110, 115, 138, 139, 213, 235, 236, 258, 261, 285
	3137	104, 129, 233
	3138	*For range* 142, 165, 256, 283, 312
	3165	234, 250, 293
Fats and oils	3183	*Data from manufacturer* 301
	3184	*Taken as beef fat*
	3185	*Taken as pork fat*
	3186	*Data from manufacturer* 301
	3188–3192	*Data from manufacturer* 301
	3196–3198	39, 62, 211
	3199	39, 130
	3200	39, 61, 62, 211
	3201	41, 62, 98, 313
	3202–3203	39, 62, 211, 212
	3204–3206	39, 62
Meat	3240	14, 15, 60, 103, 123, 132, 133, 135, 232, 251, 288
	3269	67, 102, 103, 135, 232
	3299	13, 59, 60, 103, 132, 194, 232
	3332	135
	3334	135
	3336	135
	3340	58, 90, 135, 222
Fish	3438	1, 8, 146
	3451	23, 28
	3458	28, 226
	3461	171
	3466	171
	3471	171
	3474	171
	3482	4, 6, 80, 112, 180
	3491	5, 117
	3503	118
	3510	171
	3513	171
	3517	170
	3526	169

Section 4: Cholesterol

As the new analytical values are marked in table 4.1, this key refers only to the literature sources and values calculated from recipes.

	Food no	
Cereal products		All the items given (except for ice cream which was analysed) show the contributions from eggs and milk products which were calculated from the recipes.
Milk and milk products	4123	74, 88, 175
	4126	*Calculated from fat content*
	4129	*As fresh whole milk*
	4130	*As fresh whole milk*
	4131	175
	4132	*Calculated from fat content*
	4133	*Calculated from fat content*
	4134	*Calculated from fat content*
	4135	175
	4140	74, 175
	4141	*Calculated from fat content*
	4144	*Calculated from fat content*
	4147	*Calculated from fat content*
	4150–4151	*Calculated from fat content*
	4153–4156	*Calculated from fat content*
	4159–4160	*Calculated from fat content*
Eggs	4165	The value given is that of Tolan et al 1974 (293). This is a little lower than some other quoted values e.g., 81, 88, 153, 297, 309
	4167	*Calculated from whole egg*
	4168	*Calculated from whole egg*
	4169	*As raw egg*
	4171	*As raw egg*
Egg and cheese dishes	4172–4181	*Calculated from recipes*
Meat and meat products		Virtually all the values are new analytical data and the cooked dishes are calculated from the recipes.
Fish and fish products	4438	88, 153
	4440	*Calculated from raw cod*
	4443	*Calculated from raw cod*
	4444	*Calculated from raw cod*
	4446	*Calculated from raw cod*
	4451	88, 153
	4454	*Calculated from raw haddock*
	4458	88, 274
	4459	*Calculated from raw halibut*
	4461	*As other white fish*
	4464	*Calculated from raw lemon sole*

	Food no	
Fish and fish products ***continued***	4466	153
	4469	*Calculated from raw plaice*
	4471	*As other white fish*
	4472	*Calculated from raw saithe*
	4477	*Calculated from raw whiting* 274
	4482	88, 153
	4485	*Calculated from raw herring*
	4487	*As herring*
	4489	*As herring*
	4491	88, 274
	4498	153
	4500	88, 274
	4501	*Calculated from fish only*
	4517	88, 168, 271, 274, 290
	4520	*As fresh crab*
	4521	88, 168, 271, 274
	4523	271
	4525	271
	4526	88, 153, 168, 271, 274, 290
	4531	271
	4532	153
	4535	88, 271, 290
	4537	88, 271, 290
	4539	140, 271
	4541	140, 271
	4552	153
Products containing eggs		*Calculated from the recipes*

References to Sections 1–4

1 Ackman, R. G., and Burgher, R. D. (1964) Cod flesh: component fatty acids as determined by gas–liquid chromatography. *J. Fish. Res. Bd Can.* **21**, 367–371

2 Ackman, R. G., and Burgher, R. D. (1964) Cod roe: component fatty acids as determined by gas–liquid chromatography. *J. Fish. Res. Bd Can.* **21**, 469–476

3 Ackman, R. G., and Cormier, M. G. (1967) α-tocopherol in some Atlantic fish and shellfish with particular reference to live-holding without food. *J. Fish. Res. Bd Can.* **24**, 357–373

4 Ackman, R. G., and Eaton, C. A. (1966) Some commercial Atlantic herring oils; fatty acid composition, *J. Fish. Res. Bd Can.* **23**, 991–1006

5 Ackman, R. G., and Eaton, C. A. (1971) Mackerel lipids and fatty acids. *Can. Inst. Food Technol. J.* **4**, 169–174

6 Ackman, R. G., Eaton, C. A., and Hingley, J. (1975) Fillet fat and fatty acid details for Newfoundland winter herring. *Can. Inst. Food Sci. Technol. J.* **8**, 155–159

7 Adams, J. F., McEwan, F., and Wilson, A. (1973) The vitamin B_{12} content of meals and items of diet. *Brit. J. Nutr.* **29**, 65–72

8 Addison, R. F., Ackman, R. G., and Hingley, J. (1968) Distribution of fatty acids in cod flesh lipids. *J. Fish. Res. Bd Can.* **25**, 2083–2090

9 Alderman, G., and Stranks, M. H. (1967) The iodine content of bulk herd milk in summer in relation to estimated dietary iodine intake of cows. *J. Sci. Food Agric.* **18**, 151–153

10 Al-Aswad, M. B. (1971) The amino acids content of some Iraqi dates. *J. Food Sci.*, **36**, 1019–1020

11 Al-Rawi, N., Markakis, P., and Bauer, D. H. (1967) Amino acid composition of Iraqi dates. *J. Sci. Food Agric.* **18**, 1–2

12 Al-Samarrae, W., Ma, M. C. F., and Truswell, A. S. (1975) Mexthylxanthine consumption from coffee and tea. *Proc. Nutr. Soc.* **34**, 18A–19A

13 Anderson, B. A. (1976) Comprehensive evaluation of fatty acids in foods. VII. Pork products. *J. Amer. diet. Ass.* **69**, 44–49

14 Anderson, B. A., Kinsella, J. A., and Watt, B. K. (1975) Comprehensive evaluation of fatty acids in foods. II. Beef products. *J. Amer. diet. Ass.* **67**, 35–41

15 Anderson, D. B., Breidenstein, B. B., Kauffman, R. G., Cassens, R. G., and Bray, R. W. (1971) Effect of cooking on fatty acid composition of beef lipids. *J. Food Technol.* **6**, 141–152

16 Anon. (1958–1959) Amino acids in the edible parts of food fishes. *Ann. Rep. Fish. Res. Bd Can.* 132

17 Anon. (1968) Vitamin C in Tasmanian apples. *Food Technol. Aust.* **20**, 73

18 Archibald, J. G. (1958) Trace elements in milk: a review. Parts I and II. *Dairy Sci. Abstr.* **20**, 711–725 and 799–807

19 Arnesen, G. (1969) Total and free amino acids in fishmeals and vacuum-dried codfish organs, flesh, bones, skin and stomach contents. *J. Sci. Food Agric.* **20**, 218–220

20 Askew, H. O., and Kidson, E. B. (1947) Changes in vitamin C content and acidity of apples during cool storage. *N. Z. J. Sci. Technol.* [A] **28**, 344–351

21 Aylward, F., and Showler, A. J. (1962) Plant lipids. IV The glycerides and phosphatides in cereal grains. *J. Sci. Food Agric.* **13**, 492–499

22 Bannatyne, W. R., and Thomas, J. (1969) Fatty acid composition of New Zealand shellfish lipids. *N. Z. J. Sci.* **12**, 207–212

23 Beare, J. L. (1962) Fatty acid composition of food fats. *J. agric. Food Chem.* **10**, 120–123

24 Bicknell, F., and Prescott, F. (1942) *The vitamins in medicine*. Heinemann, London

25 Bieri, J. G., and Evarts, R. P. (1975) Vitamin E adequacy of vegetable oils. *J. Amer. diet. Ass.* **66**, 134–139

26 Birdsall, J. J., Derse, P. H., and Teply, L. J. (1961) Nutrients in California lemons and oranges. II Vitamin, mineral, and proximate composition. *J. Amer. diet. Ass.* **38**, 555–559

27 Boggess, T. S., Marion, J. E., Woodroof, J. G., and Dempsey, A. H. (1967) Changes in lipid composition of sweet potatoes as affected by controlled storage. *J. Food Sci.* **32**, 554–558

28 Bonnet, J. C., Sidwell, V. D., and Zook, E. G. (1974) Chemical and nutritive values of several fresh and canned finfish, crustaceans and mollusks. 2. Fatty acid composition. *Marine Fish Rev.* **36**, (2), 8–14

29 Booth, R. G., Carter, R. H., Jones, C. R., and Moran, T. (1946) The chemical composition of wheat and wheat products. In *The nation's food*, edited by A. L. Bacharach and T. Rendle. Society of Chemical Industry, London, pp. 162–182

30 Booth, V. H. (1963) Determination of tocopherols in plant tissues. *Analyst* **88**, 627–632

31 Booth, V. H., and Bradford, M. P. (1963a) Tocopherol contents of vegetables and fruits. *Brit. J. Nutr.* **17**, 575–581

32 Booth, V. H., and Bradford, M. P. (1963) The effect of cooking on α-tocopherol in vegetables. *Int. Z. Vitaminforsch.* **33**, 276–278

33 Bowen, H. J. M. (1959) The determination of chlorine, bromine and iodine in biological material by activation analysis. *Biochem. J.* **73**, 381–384

34 Braekkan, O. R. (1958) Vitaminer i norsk fisk. 3. Vitaminer i forskjellige organer fra de viktigste torskefisker (Gadidae) fanget langs Norsketysten. *Fiskdir. Skr. Ser. Teknol. Undersøkelser* **3** (6), 32 pp

35 Braekkan, O. R. (1958) Vitamins and the reproductive cycle of ovaries in cod (*Gadus morrhua*). *Fiskdir. Skr. Teknol. Undersøkelser* **3** (7), 19 pp

36 Braekkan, O. R., and Boge, G. (1962) Vitamin B_6 and the reproductive cycle of ovaries in cod (*Gadus morrhua*). *Nature, Lond.* **193**, 394–395

37 Braekkan, O. R., and Boge, G. (1962) A comparative study of amino acids in the muscle of different species of fish. *Fiskdir. Skr. Ser. Teknol. Undersøkelser* **4** (3), 19 pp

38 Braekkan, O. R., Lambertsen, G., and Myklestad, H. (1963) Alpha-tocopherol in some marine organisms and fish oils. *Fiskdir. Skr. Ser. Teknol. Undersøkelser* **4** (8), 11 pp

39 Brignoli, C. A., Kinsella, J. E., and Weihrauch, J. L. (1976) Comprehensive evaluation of fatty acids in foods. V. Unhydrogenated fats and oils. *J. Amer. diet. Ass.* **68**, 224–229

40 Broadhead, G. D., Pearson, I. B., and Wilson, G. M. (1965) Seasonal changes in iodine metabolism. 1. Iodine content of cows' milk. 2. Fluctuation in urinary iodine excretion. *Brit. med. J.* **i**, 343–348

41 Brown, D. F., Cater, C. M., Mattil, K. F., and Darroch, J. G. (1975) Effect of variety, growing location and their interaction on the fatty acid composition of peanuts. *J. Food Sci.* **40**, 1055–1060

42 Brown, F. (1953) Occurrence of vitamin E in cod and other fish-liver oils. *Nature, Lond.* **171**, 790–791

43 Bunnell, R. H., Keating, J. Quaresimo, A., and Parman, G. K. (1965) Alpha-tocopherol content of foods. *Amer. J. clin. Nutr.* **17**, 1–10

44 Bunyan, J., Edwin, E. E., Diplock, A. T., and Green, J. (1961) Ubiquinone and tocopherol in birds. *Nature, Lond.* **190**, 637

45 Burge, J., Mickelsen, O., Nicklow, C., and Marsh, G. L. (1975) Vitamin C in tomatoes: comparison of tomatoes developed for mechanical or hand harvesting. *Ecol. Food Nutr.* **4**, 27–31

46 Burkwall, M. P., and Glass, R. L. (1965) The fatty acids of wheat and its milled products. *Cereal Chem.* **42**, 236–246

47 Burton, H., Ford, J. E., Franklin, J. G., and Porter, J. W. G. (1967) Effects of repeated heat treatments on the levels of some vitamins of the B-complex in milk. *J. Dairy Res.* **34**, 193–197

48 Burton, H., Ford, J. E., Perkin, A. G., Porter, J. W. G., Scott, K. J., Thompson, S. Y., Toothill, J., and Edwards-Webb, J. D. (1970) Comparison of milks processed by the direct and indirect methods of ultra-high temperature sterilization. IV. The vitamin content of milks sterilized by different processes. *J. Dairy Res.* **37**, 529–533

49 Butterfield, S., and Calloway, D. H. (1972) Folacin in wheat and selected foods. *J. Amer. diet. Ass.* **60**, 310–314

50 Callieri, D. A. (1959) Studies on the vitamin B_{12}, cyanocobalamin-binding capacity, desoxyribosides, and methionine in some commercial milk products and cheese. *Acta chem. scand.* **13**, 737–749

51 Callow, E. H. (1948) Comparative studies of meat. 2. The changes in the carcass during growth and fattening, and their relation to the chemical composition of the fatty and muscular tissues. *J. agric. Sci., Camb.* **38**, 174–199

52 Causeret, J. (1959) Ebullition domestique et teneur en riboflavine du lait de vache. *Lait* **39**, 159–165

53 Chanarin, I. (1975) The folate content of foodstuffs and the availability of different folate analogues for absorption. *Getting the most out of food* No. 10. Van den Berghs and Jurgens Ltd, London, pp. 41–64

54 Chappell, G. (1940) The distribution of vitamin C in foods sold on the open market. *J. Hyg., Camb.* **40**, 699–732

55 Chatfield, C. (1949) *Food composition tables for international use.* FAO Nutrition Studies No 3. Food and Agriculture Organization, Rome

56 Chilean Iodine Educational Bureau. (1952) *Iodine content of foods:* annotated bibliography 1825–1951. Chilean Iodine Educational Bureau, London

57 Christie, A. A., Dean, A. C., and Millburn, B. A. (1973) The determination of vitamin E in food by colorimetry and gas–liquid chromatography. *Analyst* **98**, 161–167

58 Chung, R. A., Lien, Y. C., and Munday, R. A. (1967) Fatty acid composition of turkey meat as affected by dietary fat, cholesterol and diethylstilbestrol. *J. Food Sci.* **32**, 169–172

59 Chung, R. A., and Lin, C. C. (1965) The fatty acid content of pork cuts and variety meats as affected by different dietary lipids. *J. Food Sci.* **30**, 860–864

60 Chung, R. A., McKay, J. A., and Ramey, C. L. (1966) Fatty acid changes in beef, pork, and fish after deep-fat frying in different oils. *Food Technol.* **20**, 691–693

61 Clegg, A. J. (1973) Composition and related nutritional and organoleptic aspects of palm oil. *J. Amer. Oil Chem. Soc.* **50**, 321–323

62 Codex Alimentarius Commission (1976) *Report of the eighth session of the Codex Committee on Fats and Oils.* London, 24–28 November 1975. Alinorm 76/19. Food and Agriculture Organization and World Health Organization

63 Connell, J. J., and Howgate, P. F. (1959) The amino acid composition of some British food fishes. *J. Sci. Food Agric.* **10**, 241–244

64 Cook, B. B., and Sundaram, S. (1963) Nutrients in raw vs cooked globe artichokes. *J. Amer. diet. Ass.* **42**, 231–233

65 Coombs, T. L. (1972) The distribution of zinc in the oyster *Ostrea edulis* and its relation to enzymic activity and to other metals. *Marine Biol.* **12**, 170–178

66 Cowey, C. B., Daisley, K. W., and, Parry, G. (1962) Study of amino acids, free or as components of protein, and of some B-vitamins in the tissues of the Atlantic Salmon, *Salmo salar*, during spawning migration. *Comp. Biochem. Physiol.* **7**, 29–38

67 Cramer, D. A., Barton, R. A., Shorland, F. B., and Czochanska, Z. (1967) A comparison of the effects of white clover (*Trifolium repens*) and of perennial ryegrass (*Lolium perenne*) on fat composition and flavour of lamb. *J. agric. Sci., Camb.* **69**, 367–373

68 Cunningham, M. M. (1935) The vitamin D content of some New Zealand fish oils. With a note on the prophylactic method of biological assay. *N. Z. J. Sci. Technol.* **17**, 563–567

69 Cuthbertson, A. (1974a) Personal communication

70 Cuthbertson, A. (1974b) Personal communication

71 Cuthbertson, A. (1975) Personal communication

72 Dako, D. Y., Watson, J. D., and Amoakwa-Adu, M. (1974) Available carbohydrates in Ghanaian foodstuffs. 1. Distribution of sugars in fruits. *Plant Foods for Man* **1**, 121–125

73 Dawes, S. N. (1970) Sodium and potassium in cow's milk. II. Bulk milk, *N. Z. J. Sci.* **13**, 69–77

74 de Man, J. M. (1964) The free and ester cholesterol content of milk and dairy products. *Z. Ernährwiss.* **5**, 1–4

75 Dean, A. C. (1971) Separation of dimeric tocopherol products from vitamin E in food by dry-column chromatography. *Chemy Ind.* (24), 677–678

76 Dewar, W. A. (1967) The zinc and manganese contents of some British poultry foods. *J. Sci. Food Agric.* **18**, 68–71

77 Diplock, A. T., Green, J., Edwin, E. E., and Bunyan, J. (1961) Tocopherol, ubiquinones and ubichromenols in yeasts and mushrooms. *Nature, Lond.* **189**, 749–750

78 Doesburg, J. J., and Meijer, A. (1964) Analyse van Nederlandse Blikconserven. II. Groete- en vruchtenprodukten. *Voeding* **25**, 258–301

79 Dong, F. M., and Oace, S. M. (1975) Folate concentration and pattern in bovine milk. *J. agric. Food Chem.* **23**, 534–538

80 Drozdowski, B., and Ackman, R. G. (1969) Isopropyl alcohol extraction of oil and lipids in the production of fish protein concentrate from herring. *J. Amer. Oil Chem. Soc.* **46**, 371–376

81 Dua, P. N., Dilworth, B. C., Day, E. J., and Hill, J. E. (1967) Effect of dietary vitamin A and cholesterol on cholesterol and carotenoid content of plasma and egg yolk. *Poult. Sci.* **46** 530–531

82 Eastoe, J. E. (1955) The amino acid composition of mammalian collagen and gelatin. *Biochem. J.* **61**, 589–600

83 Ellinger, G. M., and Boyne, E. (1965) Amino acid composition of some fish products and casein. *Brit. J. Nutr.* **19**, 587–592

84 Eric, B., le Compte, J., and Reeve, R. F. (1970) Organoleptic assessment of irradiated Granny Smith apples from Western Australia. *Food Technol. Aust.* **22**, 298–300

85 Exler, J., Kinsella, J. E., and Watt, B. K. (1975) Lipids and fatty acids of important finfish: new data for nutrient tables. *J. Amer. Oil Chem. Soc.* **52**, 154–159

86 Faust, M., Shear, C. B., and Brooks, H. J. (1969) Mineral element gradients in pears. *J. Sci. Food Agric.* **20**, 257–258

87 Feeley, R. M., Criner, P. E., Murphy, E. W., and Toepfer, E. W. (1972) Major mineral elements in dairy products. *J. Amer. diet. Ass.* **61**, 505–510

88 Feeley, R. M., Criner, P. E., and Watt, B. K. (1972) Cholesterol content of foods. *J. Amer. diet. Ass.* **61**, 134–149

89 Fisher, N., Broughton, M. E., Peel, D. J., and Bennett, R. (1964) The lipids of wheat. II. Lipids of flours from single varieties of widely varying baking quality. *J. Sci. Food Agric.* **15**, 325–341

90 Fishwick, M. J. (1968) Changes in the lipids of turkey muscle during storage at chilling and freezing temperatures. *J. Sci. Food Agric.* **19**, 440–445

91 Food and Agriculture Organization (1970) *Amino-acid content of foods and biological data on proteins.* FAO Nutrition Studies No 24, Food and Agriculture Organization, Rome

92 Ford, J. E., Porter, J. W. G., Scott, K. J., Thompson, S. Y., le Marquand, J., and Truswell, A. S. (1974) Comparison of dried milk preparations for babies on sale in 7 European countries. II. Folic acid, vitamin B_6, thiamin, riboflavin and vitamin E. *Archs Dis. Childh.* **49**, 874–877

93 Ford, J. E., Porter, J. W. G., Thompson, S. Y., Toothill, J., and Edwards-Webb, J. (1969) Effects of ultra-high-temperature (UHT) processing and of subsequent storage on the vitamin content of milk. *J. Dairy Res.* **36**, 447–454

94 Ford, J. E., and Scott, K. J. (1968) The folic acid activity of some milk foods for babies. *J. Dairy Res.* **35**, 85–90

95 Francis, B. J., Halliday, D., and Robinson, J. M. (1975) Yams as a source of edible protein. *Trop. Sci.* **17**, 103–110

96 Fraser, D. R. (1976) Personal communication

97 Fraser, J. R. (1958) Flour survey 1950–1956. *J. Sci. Food Agric.* **9**, 125–136

98 Fristrom, G. A., Stewart, B. C., Weihrauch, J. L., and Posati, L. (1975) Comprehensive evaluation of fatty acids in foods. IV. Nuts, peanuts and soups. *J. Amer. diet. Ass.* **67**, 351–355

99 Galliard, T. (1968) Aspects of the lipid metabolism in higher plants. II. The identification and quantitative analysis of lipids from the pulp of pre- and post-climacteric apples. *Phytochemistry* **7**, 1915–1922

100 Galliard, T. (1973) Lipids of potato tubers. I. Lipid and fatty acid composition of tubers from different varieties of potato. *J. Sci. Food Agric.* **24**, 617–622

101 Gardner, D., and Riley, J. P. (1972) The component fatty acids of the lipids of some species of marine and freshwater molluscs. *J. mar. biol. Ass., UK* **52**, 827–838

102 Garton, G. A., and Duncan, W. R. H. (1969) Composition of adipose tissue triglycerides of neonatal and year-old lambs. *J. Sci. Food Agric.* **20**, 39–42

103 Giam, I., and Dugan, L. R. (1965) The fatty acid composition in free and bound lipids in freeze-dried meats. *J. Food Sci.* **30**, 262–265

104 Glass, R. L., Troolin, H. A., and Jenness, R. (1967) Comparative biochemical studies of milks. 4. Constituent fatty acids of milk fats. *Comp. Biochem. Biophys.* **22**, 415–425

105 Goldstein, J. L., and Wick, E. L. (1969) Lipids in ripening banana fruit. *J. Food Sci.* **34**, 482–484

106 Gontzea, I., and Sutzescu, P. (1968) *Natural antinutritive substances in foodstuffs and forages.* S. Karger, Basel.

107 Gopalan, C., Ramastri, B. V., and Balasubramanian, S. C. (1971) *Nutritive value of Indian foods.* Nat. Inst. Nutr., Indian Counc. Med. Res., Hyderabad

108 Gormican, A. (1970) Inorganic elements in foods used in hospital menus. *J. Amer. diet. Ass.* **56**, 397–403

109 Goverd, K. A., and Carr, J. G. (1974) The content of some B-group vitamins in single-variety apple juices and commercial ciders. *J. Sci. Food Agric.* **25**, 1185–1190

110 Gray, I. K. (1973) Seasonal variations in the composition and thermal properties of New Zealand milk fat. I. Fatty-acid composition. *J. Dairy Res.* **40**, 207–214

111 Gregory, M. E., and Burton, H. (1965) The effect of ultra-high temperature heat treatment on the content of thiamine, vitamin B_6 and vitamin B_{12} of milk. *J. Dairy Res.* **32**, 13–17

112 Gruger, E. H. (1967) Fatty acid composition. In *Fish oils: their chemistry, technology, stability, nutritional properties and uses,* edited by M. E. Stansby. Avi Publications, Westport, pp 3–30.

113 Gruger, E. H., Nelson, R. W., and Stansby, M. E. (1964) Fatty acid composition of oils from 21 species of marine fish, freshwater fish and shellfish. *J. Amer. Oil Chem. Soc.* **41**, 662–667

114 Guild, L., and Raines, R. (1972) Thiamin content and retention in venison. *J. Amer. diet. Ass.* **60**, 42–44

115 Hall, A. J. (1970) Seasonal and regional variations in the fatty acid composition of milk. *Diary Inds.* **35**, 20–24

116 Hardinge, M. G., and Crooks, H. (1961) Lesser known vitamins in foods. *J. Amer. diet. Ass.* **38**, 240–245

117 Hardy, R., and Keay, J. N. (1972) Seasonal variations in the chemical composition of Cornish mackerel *Scomber scombrus* L with detailed reference to the lipids. *J. Food Technol.* **7**, 125–137

118 Hardy, R., and Mackie, P. (1969) Seasonal variation in some of the lipid components of sprats (*Sprattus sprattus*). *J. Sci. Food Agric.* **20**, 193–198

119 Harkett, P. J. (1973) Personal communication

120 Harrison, M. T., McFarlane, S., Harden, R. McG., and Wayne, E. (1965) Nature and availability of iodine in fish. *Amer. J. clin. Nutr.* **17**, 73–77

121 Hartman, A. M., and Dryden, L. P. (1965) *Vitamins in milk and milk products.* American Dairy Science Association, Champaign, Illinois

122 Hartmann, B. G., and Hillig, F. (1934) Acid constituents of food products, with special reference to citric, malic and tartaric acids. *J. Ass. off. agric. Chem.* **27**, 522–531

123 Hecker, A. L., Cramer, D. A., and Hougham, D. F. (1975) Compositional and metabolic growth effects in the bovine. Muscle, subcutaneous and serum total fatty acids. *J. Food Sci.*, **40**, 144–149

124 Hegazi, S. M., and Salem, S. A. (1972) Amino acid pattern of the Egyptian apricot fruits (Hamawy). *J. Sci. Food Agric.* **23**, 497–499

125 Henry, K. M., Hosking, Z. D., Thompson, S. Y., Toothill, J., Edwards-Webb, J., and Smith, L. P. (1971) Factors affecting the concentration of vitamins in milk. III. Effect of season and solar radiation on the vitamin D potency of butter. *J. Dairy Res.* **38**, 209–216

126 Herbert, V. (1963) A palatable diet for producing experimental folate deficiency in man. *Amer. J. clin. Nutr.* **12**, 17–20

127 Herting, D. C., and Drury, E-J. E. (1963) Vitamin E content of vegetable oils and fats. *J. Nutr.* **81**, 335–342

128 Herting, D. C., and Drury, E-J. E. (1969) Alpha-tocopherol content of cereal grains and processed cereals. *J. agric. Food Chem.* **17**, 785–790

129 Hilditch, T. P., and Jasperson, H. (1944) The component acids of milk fats of the goat, ewe and mare. *Biochem. J.* **38**, 443–447

130 Hilditch, T. P., and Williams, P. N. (1964) *The chemical constitution of natural fats*, 4th edition. Chapman and Hall, London

131 Hoppner, K., Lampi, B., and Perrin, D. E. (1972) The free and total folate activity in foods available on the Canadian market. *Can. Inst. Food Sci. Technol. J.* **5**, 60–66

132 Hornstein, I., Crowe, P. F., and Heimberg, M. J. (1961). Fatty acid composition of meat tissue lipids. *J. Food Sci.* **26**, 581–586

133 Hornstein, I., Crowe, P. F., and Hiner, R. (1967) Composition of lipids in some beef muscles. *J. Food Sci.* **32**, 650–655

134 Howard, F. D., MacGillivray, J. H., and Yamaguchi, M. (1962) Nutrient composition of fresh California-grown vegetables. *Calif. agric. exp. Sta. Bull.* No. 788

135 Hubbard, A. W., and Pocklington, W. D. (1968) Distribution of fatty acids in lipids as an aid to the identification of animal tissues. 1. Bovine, porcine, ovine, and some avian species. *J. Sci. Food Agric.* **19**, 571–577

136 Hulme, A. C. (editor) (1971) *The biochemistry of fruits and their products*, vol. 2. Academic press, London and New York

137 Hurdle, A. D. F., Barton, D., and Searles, I. H. (1968) A method for measuring folate in food and its application to a hospital diet. *Amer. J. clin. Nutr.* **21**, 1202–1207

138 Hutton, K., Seeley, R. C., and Armstrong, D. G. (1969) The variation throughout a year in the fatty acid composition of milk fat from two dairy herds. *J. Dairy Res.* **36**, 103–113

139 Huyghebaert, A., and Hendrickx, H. (1970) The relation between the fatty acid composition and iodine value and refractive index of butterfat. *Milchwissenschaft* **25**, 506–510

140 Idler, D. R., and Wiseman, P. (1971) Sterols of molluscs. *Int. J. Biochem.* **2**, 516–528

141 Inkpen, J. A., and Quackenbush, F. W. (1969) Extractable and 'bound' fatty acids in wheat and wheat products. *Cereal Chem.* **46**, 580–587

142 Insull, W., and Ahrens, E. H. (1959) The fatty acids of human milk from mothers on diets taken *ad libitum*. *Biochem. J.* **72**, 27–33

143 Iverson, J. L., Firestone, D., and Horwitz, W. (1963) Fatty acid composition of oil from roasted and unroasted peanuts by gas–liquid chromatography. *J. Ass. off. agric. Chem.* **46**, 718–725

144 Jacobs, A., and Greenman, D. A. (1969) Availability of food iron. *Brit. med. J.* **i**, 673–676

145 Jamieson, M. M., Oxenham, J., and Robertson, J. (1961) Fat absorption by white bread during frying. *Proc. Nutr. Soc.* **20**, xxii–xxiii

146 Jangaard, P. M., Ackman, R. G., and Sipos, J. C. (1967) Seasonal changes in fatty acid composition of cod liver, flesh, roe and milt lipids. *J. Fish. Res. Bd Can.* **24**, 613–627

147 Jart, A. (1963) The fatty acid composition of filbert oil. *Acta chem. scand.* **11**, 1186–1187

148 Johansen, O., and Steinnes, E. (1976) Determination of iodine in plant material by a neutron activation method. *Analyst* **101**, 455–457

149 Kaldy, M. S., and Markakis, P. (1972) Amino acid composition in selected potato varieties. *J. Food Sci.* **37**, 375–377

150 Kaplan, E., Holmes, J. H., and Sapeika, N. (1974) Caffeine content of tea and coffee. *S. Afr. med. J.* **48**, 510–511

151 Karlin, R. (1961) Sur le taux de vitamine B_6 dans les fromages: variations au cours de la maturation. *Int. Z. Vitaminforsch.* **31**, 176–184

152 Karlin, R. (1969) Sur la teneur en folates des laits de grand mélange. Effets de divers traitements thermiques sur les taux de folates, B_{12} et B_6 de ces laits. *Int. Z. Vitaminforsch.* **39**, 359–371

153 Keller, G. H. M., and van de Bovenkamp, P. (1974) Cholesterolgehalte van voedingsmiddelen. *Voeding* **35**, 409–411

154 Kellog, W. L., Denton, C. A., and Bird, H. R. (1947) Nicotinic acid content of squab and pigeon tissues. *Poult. Sci.* **26**, 435–436

155 Kent-Jones, D. W. (1958) The case for fortified flour. *Proc. Nutr. Soc.* **17**, 38–43

156 Kenworthy, A. L., and Harris, N. (1963) Composition of McIntosh, Red Delicious and Golden Delicious apples as related to environment and season. *Quart. Bull. Mich. agric. exp. Sta.* **46**, 293–334

157 Ketiku, A. O. (1973) Chemical composition of unripe (green) and ripe plantain (*Musa paradisiaca*). *J. Sci. Food Agric.* **24**, 703–707

158 Keys, O. H. (1943) Vitamin C in applies and other materials. *N. Z. J. Sci. Technol.* [B] **24**, 146–148

159 Kieser, M. E., and Pollard, A. (1947) Vitamin C in English apples. *Nature, Lond.* **159**, 65

160 Kinsella, J. E. (1971) Composition of the lipids of cucumber and peppers. *J. Food Sci.* **36**, 865–866

161 Kirkpatrick, D. C., and Coffin, D. E. (1975) Trace metal content of chicken eggs. *J. Sci. Food Agric.* **26**, 99–103

162 Kizlaitis, L., Steinfeld, M. I., and Siedler, A. J. (1962) Nutrient content of variety meats. 1. Vitamin A, vitamin C, iron and proximate composition. *J. Food Sci.* **27**, 459–462

163 Knight, R. A., Christie, A. A., Orton, C. R., and Robertson, J. (1973) Studies on the composition of food. 4. Comparison of nutrient content of retail white bread made conventionally and by the Chorleywood Bread Process. *Brit. J. Nutr.* **30**, 181–188

164 Kon, S. K. (1972) *Milk and milk products in human nutrition.* FAO Nutrition Studies, No 27, 2nd edition. Food and Agriculture Organization, Rome

165 Krámer, M., Szöke, K., Lindner, K., and Tarján, R. (1965) The effect of different factors on the composition of human milk. 3. Effect of dietary fats on lipid composition of human milk. *Nutritio Dieta* **7**, 71–79

166 Kringstad, H., and Folkvord, S. (1949) The nutritive value of cod roe and cod liver. *J. Nutr.* **38**, 489–502

167 Kritchevsky, D. (1963) Sterols. In *Comprehensive biochemistry*, edited by M. Florkin and E. H. Stotz, vol. 10. Elsevier, Amsterdam, London and New York. pp 1–22

168 Kritchevsky, D., Tepper, S. A., Ditullo, N. W., and Holmes, W. L. (1967) The sterols of seafood. *J. Food Sci.* **32**, 64–66

169 Krzeczkowski, R. A. (1970) Fatty acids in raw and processed Alaska pink shrimp. *J. Amer. Oil Chem. Soc.* **47**, 451–452

170 Krzeczkowski, R. A., Tenney, R. D., and Kelley, C. (1971) Alaska king crab: fatty acid composition, carotenoid index and proximate analysis. *J. Food Sci.* **36**, 604–606

171 Laboratory of the Government Chemist (1964) Unpublished data

172 Laboratory of the Government Chemist (1967) Composition of bread rolls. In *Report of the Government Chemist 1966*. Ministry of Technology, HMSO, London. pp 43–44

173 Laboratory of the Government Chemist (1969) Unpublished data

174 Laboratory of the Government Chemist (1975) Unpublished data

175 Lacroix, D. E., Mattingly, W. A., Wong, N. P., and Alford, J. A. (1973) Cholesterol, fat and protein in dairy products. *J. Amer. diet. Ass.* **62**, 275–279

176 Lakshmiah, N., and Ramasastri, B. V. (1969) Folic acid content of some Indian foods of plant origin. *J. Nutr. Diet., India* **6**, 200–203

177 Lal, B. M., Prakash, V., and Verma, S. C. (1963) The distribution of nutrients in the seed parts of Bengal gram. *Experienta* **19**, 154–155

178 Lal, B. M., Rohewal, S. S., Verma, S. C., and Prakash, V. (1963) Chemical composition of some pure strains of Bengal gram (*Cicer arietinum* L.). *Ann. Biochem. exp. Med.* **23**, 543–548

179 Lambertsen, G., and Braekkan, O. R. (1959) The spectrophotometric determination of α-tocopherol. *Analyst* **84**, 706–711

180 Lambertsen, G., and Braekkan, O. R. (1965) The fatty acid composition of herring oils. *Fiskdir. Skr. Ser. Teknol. Undersøkelser* **4** (13), 14 pp

181 Lambertsen, G., Myklestad, H., and Braekkan, O. R. (1962) Tocopherols in nuts. *J. Sci. Food Agric.* **13**, 617–620

182 Lambertsen, G., Myklestad, H., and Braekkan, O. R. (1964) The determination of α- and γ-tocopherols in margarine. *J. Food Sci.* **29**, 164–167

183 Lange, W. (1950) Cholesterol, phytosterol and tocopherol content of food products and animal tissues. *J. Amer. Oil Chem. Soc.* **27**, 414–422

184 Larkin, D., Page, M., Bartlet, J. C., and Chapman, R. A. (1954) The lead, zinc and copper content of foods. *Foods Res.* **19**, 211–218

185 Lepage, M. (1967) Identification and composition of turnip root lipids. *Lipids* **2**, 244–250

186 Lichtenstein, H., Beloian, A., and Murphy, E. W. (1961) Vitamin B_{12}-microbiological assay methods and distribution in selected foods. *U.S. Dept. Agric. Home Econ. Res. Rep.* No 13. Washington DC

187 Lichtenstein, H., Beloian, A., and Reynolds, H. (1959) Comparative vitamin B_{12} assay of foods of animal origin by *Lactobacillus leichmanii* and *Ochromonas malhamensis*. *J. agric. Food Chem.* **7**, 771–774

188 Lieck, H., and Søndergaard, H. (1958) The content of vitamin B_6 in Danish foods. *Int. Z. Vitaminforsch.* **29**, 68–77

189 Lindberg, P., Bingefors, S., Lannek, N., and Tanhuanpää, E. (1964) The fatty acid composition of Swedish varieties of wheat, barley, oats and rye. *Acta agric. scand.* **14**, 3–11

190 Loughlin, M. E., and Teeri, A. E. (1960) Nutritive value of fish. II. Biotin, folic acid, pantothenic acid and free amino acids of various salt-water species. *Food Res.* **25**, 479–483

191 Love, R. M. (1970) *The chemical biology of fishes*. Academic Press, London and New York

192 Love, R. M., Lovern, J. A., and Jones, N. R. (1959) *The chemical composition of fish tissues*. DSIR Food Investigation Special Report No 69. HMSO, London

193 Lowe, J. S., Morton, R. A., and Vernon, J. (1957) Unsaponifiable constituents of kidney in various species. *Biochem. J.* **67**, 228–234

194 Luddy, F. E., Herb, S. F., Magidman, P., Spinelli, A. M., and Wasserman, A. E. (1970) Color and the lipid composition of pork muscles. *J. Amer. Oil Chem. Soc.* **47**, 65–68

195 Lunven, P., Le Clement de St Marcq, C., Carnovale, E., and Fratoni, A. (1973) Amino acid composition of hen's egg. *Brit. J. Nutr.* **30**, 189–194

196 Macy, I. G., Kelly, H. J., and Sloan, R. E. (1953) *The composition of milks*. National Academy of Science and National Research Council Publication No 254. Washington DC

197 Marion, W. W., Maxon, S. T., and Wangen, R. M. (1970) Lipid and fatty acid composition of turkey liver, skin and depot tissue. *J. Amer. Oil Chem. Soc.* **47**, 391–392

198 Mason, E. M., O'Donovan, E. M., and Kilbride, D. (1945) *An enquiry into the cause of goitre in County Tipperary*. An investigation of iodine content of foodstuff, soil and drinking water of that county compared with others of less goitrous counties in Ireland. Unpublished report to the Medical Research Council of Ireland, quoted by Chilean Iodine Educational Bureau 1952

199 Matoth, Y., Pinkas, A., and Sroka, Ch. (1965) Studies on folic acid in infancy. III. Folates in breast fed infants and their mothers. *Amer. J. clin. Nutr.* **16**, 356–359

200 Mattson, F. H., and Volpenhein, R. A. (1963) The specific distribution of unsaturated fatty acids in the triglycerides of plants. *J. Lipid Res.* **4**, 392–396

201 McCance, R. A., Sheldon, W., and Widdowson, E. M. (1934) Bone and vegetable broth. *Archs Dis. Childh.* **9**, 251–258

202 McCance, R. A., and Widdowson, E. M. (1935) Phytin in human nutrition. *Biochim. J.* **29**, 2694–2699

203 McCance, R. A., and Widdowson, E. M. (1942) Mineral metabolism of healthy adults on white and brown bread dietaries. *J. Physiol.* **101**, 44–85

204 McCarthy, M. A., Orr, M. L., and Watt, B. K. (1968) Phenylalanine and tyrosine in vegetables and fruits. *J. Amer. diet. Ass.* **52**, 130–134

205 McKellar, R. L., and Kohrman, R. E. (1975) Amino acid composition of the Morel mushroom. *J. agric. Food Chem.* **23**, 464–467

206 Merrill, A. L., and Watt, B. K. (1955) *Energy value of foods—basis and derivation. US* Department of Agriculture. Agriculture Handbook No 74, Washington DC

207 Milk Marketing Board Joint Milk Quality Committee (1973) *Milk compositional and hygienic quality control*. England and Wales. A progress report. Milk Marketing Board, Thames Ditton

208 Millar, K. R., and Sheppard, A. D. (1972) α-tocopherol and selenium levels in human and cows' milk. *N. Z. J. Sci.* **15**, 3–15

209 Millin, D. J., and Rustidge, D. W. (1967) Tea manufacture. *Process Biochem.* **2**, 9–13

210 Ministry of Agriculture, Fisheries and Food, Working Party on the Monitoring of Foodstuffs for Heavy Metals. Unpublished papers

211 Ministry of Agriculture, Fisheries and Food (1974) Personal communication

212 Ministry of Agriculture, Fisheries and Food (1975) Personal communication

213 Moore, J. H., and Williams, D. L. (1965) A note on the effect of a commercial drying process on the long chain fatty acids of milk. *J. Dairy Res.* **32**, 19–20

214 Moore, T. (1957) *Vitamin A*. Elsevier, London

215 Moore, T., Sharman, I. M., and Ward, R. J. (1959) Cod-liver oil as both source and antagonist of vitamin E. *Brit. J. Nutr.* **13**, 100–110

216 Murphy, E. W., Willis, B. W., and Watt, B. K. (1975) Provisional tables on the zinc content of foods. *J. Amer. diet. Ass.* **66**, 345–355

217 Murray, J., and Burt, J. R. (1969) The composition of fish. *Torry Advisory Note* No 38. Torry Research Station, Aberdeen

218 Murthy, G. K., Rhea, U. S., and Peeler, J. T. (1972) Copper, iron, manganese, strontium and zinc content of market milk. *J. Dairy Sci.* **55**, 1666–1674

219 Naiman, J. L., and Oski, F. A. (1964) The folic acid content of milk. Revised figures based on an improved assay method. *Pediatrics* **34**, 274–276

220 Neilands, J. B., Strong, F. M., and Elvehjem, C. A. (1947) The nutritive value of canned foods 25. Vitamin content of canned fish products. *J. Nutr.* **34**, 633–643

221 Nelson, J. H., Glass, R. L., and Geddes, W. F. (1963) The triglycerides and fatty acids of wheat. *Cereal Chem.* **40**, 343–351

222 Neudoerffer, T. S., and Lea, C. H. (1967) Effects of dietary polyunsaturated fatty acids on the composition of the individual lipids of turkey breast and leg muscle. *Brit. J. Nutr.* **21**, 691–714

223 Nicol, D. J., and Davis, R. E. (1967) The folate and vitamin B_{12} content of infant milk foods with particular reference to goats' milk. *Med. J. Aust.* **ii**, 212–214

224 Nobile, S., and Woodhill, J. M. (1973) A survey of the vitamin content of some 2,000 foods as they are consumed by selected groups of the Australian population. *Food Technol. Aust.* **25**, 80–100

225 Okungbowa, P., Ma, M. C. F., and Truswell, A. S. (1977) Niacin in instant coffee. *Proc. Nutr. Soc.* **36**, 26A

226 Olley, J., and Duncan, W. R. H. (1965) Lipids and protein denaturation in fish muscle. *J. Sci. Food Agric.* **16**, 99–104

227 Olliver, M. (1947) The cabbage as a source of ascorbic acid in the human diet. *Chemy Ind.* No 18, 235–240

228 Orr, J. B. (1931) *Report to the Nutrition Committee of the Medical Research Council on the correlation between iodine supply and the incidence of endemic goitre*. Medical Research Council Special Report Series No 154. HMSO, London

229 Orr, M. L. (1969) Pantothenic acid, vitamin B_6 and vitamin B_{12} in foods. US Department of Agriculture Home Economics Research Report No 36. Washington DC

230 Orr, M. L., and Watt, B. K. (1957) *Amino acid content of foods*. US Department of Agriculture Home Economics Research Report No 4. Washington DC

231 Osis, D., Kramer, L., Waitrowski, E., and Spencer, H. (1972) Dietary zinc intake in man. *Amer. J. clin. Nutr.* **25**, 582–588

232 Ostrander, J., and Dugan, L. R. (1962) Some differences in composition of covering fat, intermuscular fat, and intramuscular fat of meat animals. *J. Amer. Oil Chem. Soc.* **39**, 178–181

233 Parkash, S., and Jenness, R. (1968) The composition and characteristics of goats' milk: a review. *Dairy Sci. Abstr.* **30**, 67–87

234 Parkinson, T. L. (1966) The chemical composition of eggs. *J. Sci. Food Agric.* **17**, 101–111

235 Parodi, P. W. (1970) Fatty acid composition of Australian butter and milk fat. *Aust. J. Dairy Technol.* **25**, 200–205

236 Parodi, P. W. (1972) Observations on the variation in fatty acid composition of milkfat. *Aust. J. Dairy Technol.* **27**, 90–94

237 Pearson, D. (1975) Seasonal English market variations in the composition of South African and Israeli avocados. *J. Sci. Food Agric.* **26**, 207–213

238 Pennington, J. T., and Calloway, D. H. (1973) Copper content of foods. *J. Amer. diet. Ass.* **63**, 143–153

239 Pennock, J. F., Neiss, G., and Mahler, H. R. (1962) Biochemical studies on the developing avian embryo. 5. Ubiquinone and some other unsaponifiable lipids. *Biochem. J.* **85**, 530–537

240 Perring, M. A. (1968) Recent work at the Ditton Laboratory on the chemical composition and storage characteristics of apples in relation to orchard factors. *Rep. E. Malling Res. Sta. for 1967*, 191–198

241 Peynaud, E., and Lafourcade, S. (1958) Evolution des vitamines B dans le raisin. *Qualitas Pl. Mater. Veg.* **3**, 404–414

242 Plack, P. A., Kon, S. K., and Thompson, S. Y. (1959) Vitamin A_1 aldehyde in the eggs of the herring (*Clupae harengus* L.) and other marine teleosts. *Biochem. J.* **71**, 467–476

243 Platt, B. S. (1962) *Tables of representative values of foods commonly used in tropical countries.* Medical Research Council Special Report Series No 302. HMSO, London

244 Polansky, M. M. (1969) Vitamin B_6 components in fresh and dried vegetables. *J. Amer. diet. Ass.* **54**, 118–121

245 Polansky, M. M., and Murphy, E. W. (1966) Vitamin B_6 components in fruit and nuts. *J. Amer. diet. Ass.* **48**, 109–111

246 Polansky, M. M., and Toepfer, E. W. (1969) Vitamin B_6 components in some meats, fish, dairy products and commercial infant formulas. *J. agric. Food Chem.* **17**, 1394–1397

247 Pollard, A. (1950) Vitamins in fruit juices and related products. In *Recent advances in fruit juice production* edited by V. L. S. Charley. Commonwealth Bureau of Horticulture and Plant Crops Technical Communication No 21. pp 125–144

248 Porter, J. W. G. (1975) *Milk and dairy foods.* The value of foods series, general editors P. Fisher and A. E. Bender. Oxford University Press

249 Porter, J. W. G. (1976) Personal communication

250 Posati, L. P., Kinsella, J. E., and Watt, B. K. (1975) Comprehensive evaluation of fatty acids in foods. III. Eggs and egg products. *J. Amer. diet. Ass.* **67**, 111–115

251 Pothoven, M. A., Beitz, D. C., and Zimmerli, A. (1974) Fatty acid compositions of bovine adipose tissue and of *in vitro* lipogenesis. *J. Nutr.* **104**, 430–433

252 Price, P. B., and Parsons, J. G. (1974) Lipids of six cultivated barley (*Hordeum vulgare* L) varieties. *Lipids* **9**, 560–566

253 Pyke, M. (1942) The vitamin content of vegetables. *J. Soc. chem. Ind., Lond.* **61**, 149–151

254 Randoin, L., Le Gallic, P., Dupuis, Y., and Bernardin, A. (1961) *Tables de composition des aliments.* Institut Scientifique d'Hygiène Alimentaire. J. Lanore, Paris

255 Räsänen, L., Ahlstrom, A., and Kytovuori, P. (1972) Nutritional value of game birds. *Soumen Kemistilehti B.* **45**, 314–316

256 Read, W. W. C., and Sarrif, A. (1965) Human milk lipids, 1. changes in fatty acid composition of early colostrum. *Amer. J. clin. Nutr.* **17**, 177–179

257 Reiners, R. A., Morgan, R. E., and Shroder, J. D. (1970) Note on the amino acid composition of the protein in commercial corn starch. *Cereal Chem.* **47**, 205–206

258 Renner, E., and Baier, D. (1971) Einfluss von Temperatur und Sauerstoff auf den Gehalt an Ascorbinsäure und ungesättigten Fettsäuren in Milch. *Dt. Molk.-Ztg* **92**, 75–78

259 Renner, E., and Baier, D. (1971) Einfluss des Lichtes auf den Gehalt der Milch an Ascorbinsäure und ungesättigten Fettsäuren. *Dt. Molk-Ztg* **92**, 541–543

260 Rice, E. E., Strandine, E. J., Squires, E. M., and Lyddon, B. (1946) The distribution and comparative content of certain B-complex vitamins in chicken muscles. *Arch. Biochem.* **10**, 251–260

261 Richardson, T., and McGann, T. C. A. (1964) Fatty acids in Irish butterfat. *Irish J. agric. Res.* **3**, 151–157

262 Robbins, G. S., Pomeranz, Y., and Briggle, L. W. (1971) Amino acid composition of oat groats. *J. agric. Food Chem.* **19**, 536–539

263 Roberson, S., Marion, J. E., and Woodroof, J. G. (1966) Composition of commercial peanut butters. *J. Amer. diet. Ass.* **49**, 208–210

264 Robertson, J., and Sissons, D. J. (1966) The effects of maturity, processing, storage in the pod and cooking on the vitamin C content of fresh peas. *Nutrition, Lond.* **20**, 21–27

265 Rodgers, K., and Poole, D. B. (1958) The estimation of iodine in biological materials: a modification of the method of Ellis and Duncan. *Biochem. J.* **70**, 463–471

266 Rolls, B. A., and Porter, J. W. G. (1973) Some effects of processing and storage on the nutritive value of milk and milk products. *Proc. Nutr. Soc.* **32**, 9–15

267 Schertel, M. E., Boehne, J. W., and Libby, D. A. (1965) Folic acid derivatives in yeast. *J. biol. Chem.* **240**, 3154–3158

268 Schlettwein-Gsell, D., and Mommsen-Straub, S. (1972) Ubersicht Spurenelemente in Lebensmitteln. VII. Magnesium. *Int. Z. Vitaminforsch.* **42**, 324–352

269 Schroeder, H. A. (1971) Losses of vitamins and trace minerals resulting from processing and preservation of foods. *Amer. J. clin. Nutr.* **24**, 562–573

270 Schroeder, H. A., Nason, A. P., Tipton, I. H., and Balassa, J. J. (1967) Essential trace elements in man: zinc. Relation to environmental cadmium. *J. chron. Dis.* **20**, 179–210

271 Schulze, A., and Truswell, A. S. (1977) Sterols in British shellfish. *Proc. Nutr. Soc.* **36**, 25A

272 Shahani, K. M., Hathaway, I. L., and Kelly, P. L. (1962) B-complex vitamin content of cheese. II. Niacin, pantothenic acid, pyridoxine, biotin and folic acid. *J. Dairy Sci.* **45**, 833–841

273 Sidwell, V. D., Bonnet, J. C., and Zook, E. G. (1973) Chemical and nutritive values of several fresh and canned finfish, crustaceans and mollusks. 1. Proximate composition, calcium and phosphorus. *Marine Fish. Rev.* **35** (12), 16–19

274 Sidwell, V. D., Foncannon, P. R., Moore, N. S., and Bonnet, J. C. (1974) Composition of the edible portion of raw (fresh or frozen) crustaceans, finfish and mollusks. 1. Protein, fat, moisture, ash, carbohydrate, energy value, and cholesterol. *Marine Fish. Rev.* **36** (3), 21–35

275 Sinclair, A. J. (1974) Personal communication

276 Skramstad, K. H. (1969) Mineralstoffer i fisk. *Tidsskr. Hermetikkind* **55**, 14–20

277 Slater, G. G., Shankman, S., Shepherd, J. S., and Alfin-Slater, R. B. (1975) Seasonal variation in the composition of California Avocados. *J. agric. Food. Chem.* **23**, 468–474

278 Slover, H. T. (1971) Tocopherols in foods and fats. *Lipids* **6**, 291–296

279 Slover, H. T., Lehmann, J., and Valis, R. J. (1969) Vitamin E in foods: determination of tocols and tocotrienols. *J. Amer. Oil Chem. Soc.* **46**, 417–420

280 Smith, C. L., Kelleher, J., Losowsky, M. S., and Morrish, N. (1971) The content of vitamin E in British diets. *Brit. J. Nutr.* **26**, 89–96

281 Souci, S. W., Fauchman, W., and Kraut, H. (1962, 1964) *Die Zusammensetzung der Lebensmittel, Nährwert-Tabellen.* Wissenschaftliche Verlagsgesellschaft mbH., Stuttgart

282 Stagg, G. V., and Millin, D. J. (1975) The nutritional and therapeutic value of tea—a review. *J. Sci. Food Agric.* **26**, 1439–1459

283 Stevens, J. F. (1970) Faecal fatty acid patterns in the neonate. *J. med. Lab. Technol.* **27**, 327–331

284 Streiff, R. R. (1971) Folate levels in citrus and other juices. *Amer. J. clin. Nutr.* **24**, 1390–1392

285 Stull, J. W., and Brown, W. H. (1964) Fatty acid composition of milk. II. Some differences in common dairy breeds. *J. Dairy Sci.* **47**, 1412

286 Tamura, T., and Stokstad, E. L. R. (1973) The availability of food folate in man. *Brit. J. Haematol.* **25**, 513–532

287 Teply, L. J. (1958) Nutritional study of instant coffee powder. *Food Technol.* **12**, 485–486

288 Terrell, R. N., Lewis, R. W., Cassens, R. G., and Bray, R. W. (1967) Fatty acid compositions of bovine subcutaneous fat depots determined by gas–liquid chromatography. *J. Food Sci.* **32**, 516–520

289 Thomas, M. H., and Colloway, D. H. (1961) Nutritional value of dehydrated foods. *J. Amer. diet. Ass.* **39**, 105–116

290 Thompson, M. H. (1964) Cholesterol content of various species of shellfish. 1. Method of analysis and preliminary survey of variables. *US Fish. Wildlife Serv. Fish. Ind. Res.* **2**, 11–15

291 Thompson, S. Y., Henry, K. M., and Kon, S. K. (1964) Factors affecting the concentration of vitamins in milk. 1. Effect of breed, season and geographical location on fat-soluble vitamins. *J. Dairy Res.* **31**, 1–25

292 Thompson, S. Y., and Kon, S. K. (1964) Factors affecting the concentration of vitamins in milk. 2. Effect of breed, season and geographical location on riboflavin. *J. Dairy Res* **31**, 27–30

293 Tolan, A., Robertson, J., Orton, C. R., Head, M. J., Christie, A. A., and Millburn, B. A. (1974) Studies on the composition of food. 5. The chemical composition of eggs produced under battery, deep litter and free range conditions. *Brit. J. Nutr.* **31**, 185–200

294 Toothill, J., Thompson, S. Y., and Edwards-Webb, J. (1970) Observations on the use of 2,4-dinitrophenylhydrazine and 2,6-dichlorophenolindophenol for determination of vitamin C in raw and in heat-treated milk. *J. Dairy Res.* **37**, 29–45

295 Trevelyan, W. E. (1975) Determination of uric acid precursors in dried yeast and other forms of single cell protein. *J. Sci. Food Agric.* **26**, 1673–1680

296 Tsatsaronis, G. C., and Kehayoglou, A. H. (1971) Fatty acid composition of Capsicum oils by gas–liquid chromatography. *J. Amer. Oil Chem. Soc.* **48**, 365–367

297 Turk, D. E., and Barnett, B. D. (1971) Cholesterol content of market eggs. *Poult. Sci.* **50**, 1303–1306

298 Twomey, D. G., and Goodchild, J. (1970) Variations in the vitamin C content of imported tomatoes. *J. Sci. Food Agric.* **21**, 313

299 Twomey, D. G., and Ridge, B. D. (1970) Note on L-ascorbic acid content of English early tomatoes. *J. Sci. Food Agric.* **21**, 314

300 Underwood, E. J. (1971) *Trace elements in human and animal nutrition*, 3rd edition. Academic Press, London and New York

301 Van den Berghs and Jurgens Ltd (1976) Personal communication

302 Vought, R. L., and London, W. T. (1964) Dietary sources of iodine. *Amer. J. clin. Nutr.* **14**, 186–192

303 Walter, W. M., Hansen, A. P., and Purcell, A. E. (1971) Lipids of cured Centennial sweet potatoes. *J. Food Sci.* **36**, 795–797

304 Wangen, R. M., Marion, W. W., and Hotchkiss, D. K. (1972) Influence of age on the fatty acid composition of breast and thigh muscles of male turkeys. *Agric. Biol. Chem.* **36**, 2081–2086

305 Warren, H. V. (1972) Variations in the trace element contents of some vegetables. *J. Roy. Coll. gen. Practnrs* **22**, 56–60

306 Watson, J. D., Dako, D. Y., and Amoakwa-Adu, M. (1975) Available carbohydrates in Ghanaian foodstuffs. 2. Sugars and starch in staples and other foodstuffs. *Plant Foods for Man*, **1**, 169–176

307 Watt, B. K., and Merrill, A. L. (1963) *Composition of foods—raw, processed, prepared.* US Department of Agriculture, Agriculture Handbook No 8. Washington DC

308 Wayne, E. J., Koutras, D. A., and Alexander, W. D. (1964) *Clinical aspects of iodine metabolism.* Blackwells, Oxford

309 Weiss, J. F., Naber, E. C., and Johnson, R. M. (1964) Effect of dietary fat and other factors on egg yolk cholesterol. 1. The 'cholesterol' content of egg yolk as influenced by dietary unsaturated fat and the method of determination. *Arch. Biochem. Biophys.* **105**, 521–526

310 West, C., and Zilva, S. S. (1944) Synthesis of vitamin C in stored apples. *Biochem. J.* **38**, 105–108

311 Williams, A. P., Bishop, D. R., Cockburn, J. E., and Scott, K. J. (1976) Composition of ewe's milk. *J. Dairy Res.* **43**, 325–329

312 Woodruff, C. W., Bailey, M. C., Davis, J. T., Rogers, N., and Coniglio, J. G. (1964) Serum lipids in breast-fed infants and in infants fed evaporated milk. *Amer. J. clin. Nutr.* **14**, 83–90

313 Worthington, R. E., Hammons, R. O., and Allison, J. R. (1972) Varietal differences and seasonal effects on fatty acid composition and stability of oil from 82 peanut genotypes. *J. agric. Food Chem.* **20**, 727–730

314 Zanobini, A., Firenzouli, A. M., and Bianchi, A. (1974) Isolamento e dosaggio della vitamina D in avocado (*Persea gratissima*). *Boll. Soc. Ital. Biol. sper.* **50**, 887–891

315 Zook, E. G., and Lehmann, J. (1968) Mineral composition of fruits. II. Nitrogen, calcium, magnesium, phosphorus, potassium, aluminum, boron, copper, iron, manganese and sodium. *J. Amer. diet. Ass.* **52**, 225–231

Index of foods

*** Foods are indexed by food number *not* page number**

*** Foods are indexed by food number *not* page number**

*** Foods are indexed by food number *not* page number**

* **Foods are indexed by food number *not* page number**

*** Foods are indexed by food number *not* page number**

*** Foods are indexed by food number *not* page number**

*** Foods are indexed by food number *not* page number**

Printed in England for Her Majesty's Stationery Office by McCorquodale Printers Ltd. London
HM 7816 Dd 587267 K60 1/78 McC3339/3